Polymers in Information Storage Technology

Polymers in Information Storage Technology

Edited by
K. L. Mittal

IBM US Technical Education
Thornwood, New York

PLENUM PRESS • NEW YORK AND LONDON

Library of Congress Cataloging-in-Publication Data

Polymers in information storage technology / edited by K.L. Mittal.
        p.   cm.
     Proceedings of the American Chemical Society Symposium on Polymers
  in Information Storage Technology, held Sept. 25-30, 1988 in Los
  Angeles, Calif.
     Includes bibliographical references.
     ISBN 0-306-43390-7
     1. Optical storage devices--Materials--Congresses. 2. Polymers-
  -Congresses. 3. Magnetic recorders and recording--Congresses.
  I. Mittal, K. L., 1945-   . II. American Chemical Society Symposium
  on Polymers in Information Storage Technology (1988 : Los Angeles,
  Calif.)
  TA1635.P65   1990
  621.39'767--dc20   89-25539
                                                               CIP

Proceedings of the American Chemical Society Symposium on
Polymers in Information Storage Technology,
held September 25-30, 1988, in Los Angeles, California

© 1989 Plenum Press, New York
A Division of Plenum Publishing Corporation
233 Spring Street, New York, N.Y. 10013

Printed in the United States of America

D
621.39
POL

HS

PREFACE

This volume documents the proceedings of the Symposium on Polymers in Information Storage Technology held as a part of the American Chemical Society meeting in Los Angeles, September 25-30, 1988. It should be recorded here that this symposium was cosponsored by the Division of Polymeric Materials: Science and Engineering, and the Division of Polymer Chemistry.

Polymers are used for a variety of purposes in both optical and magnetic information storage technologies. For example, polymers find applications as substrate, for storing information directly, as protective coating, as lubricant, and as binder in magnetic media. In the last few years there has been a high tempo of research activity dealing with the many ramifications of polymers in the exciting arena of information storage. Concomitantly, we decided to organize this symposium and I believe this was the premier event on this topic. This symposium was conceived and organized with the following objectives in mind: (1) to bring together those actively involved (polymer chemists, polymer physicists, photochemists, surface and colloid chemists, tribologists and so on) in the various facets of this topic; (2) to provide a forum for discussion of latest R&D activity in this technology; (3) to provide an opportunity for cross-pollination of ideas; and (4) to identify and highlight areas, within the broad purview of this topic, which needed intensified or accelerated R&D efforts. The initial announcement of this symposium elicited an exciting response from the international community and the consensus was that such an event was very opportune. Also if the comments from the attendees are any barometer of the success of a symposium, then this symposium was a grand success and our stated objectives were amply fulfilled.

The final technical program comprised a total of 52 papers covering many ramifications of polymers in information storage and it was a veritable international event as the speakers hailed from many parts of the globe. Apropos, a number of luminaries in this field were specially invited to present subtopical overviews and these were augmented by original research contributions. It should also be recorded here that in addition to didactic presentations, there were brisk, lively and illuminating (not exothermic) discussions, both formally and informally, throughout the duration of the symposium.

As for this Volume, a total of 32 papers are included and these are divided into four parts as follows: Part I. Photochemical Aspects of Optical Recording; Part II. Physicochemical Considerations in Optical Recording; Part III. Polymer Physics: Relevance to Optical Recording; and Part IV. Bulk/Surface Chemical Considerations in Magnetic Recording. The topics covered include: organic materials for optical data storage; polymer-liquid crystal mixtures for optical data storage; nonlinear optical materials in neuro-optical network; polymeric data memories and polymeric substrates for information storage devices; re-writable dye-polymer optical storage media; materials for mastering and replication processes; magneto-optical recording media properties; optical disk substrates; magnetic disk coatings; high solids magnetic dispersions; polymer binders for magnetic recording media; magnetic and mechanical performance of particulate disk coatings; and lubricants. It should be recorded for posterity that all manuscripts were rigorously peer reviewed as peer review is _sine qua non_ to maintain high standard of publications, and most of these were considerably revised before inclusion in this book.

Even a cursory look at the Table of Contents will convince the reader that a bounty of information related to this field is made available in this book. Yours truly sincerely hopes that this volume would be of interest to both veterans (as a commentary on contemporary thinking) and novices (as a Baedeker to this field and as a fountain of new ideas).

Acknowledgements: First and foremost, I would like to record here that this Symposium was jointly organized by Dr. Armand Lewis (Kendall Co., Lexington, MA), Dr. Daniel Dawson (IBM Almaden Research Center, San Jose) and yours truly, and my sincere thanks and appreciation is extended to both of them. I am thankful to the appropriate management of IBM Corporation for permitting me to organize this symposium and to edit this volume. We are grateful to both Divisions of the American Chemical Society for sponsoring this event. The financial grant from the Petroleum Research Fund of the American Chemical Society is gratefully acknowledged, which was immensely helpful in providing support to a number of invited speakers. On a personal note, I would like to acknowledge the help, in more ways than one, of my wife, Usha, during the tenure of editing this volume. Special thanks are due to those behind the scenes (reviewers) who unflinchingly devoted their time in providing very valuable comments. Last, but most important, the enthusiasm, co-operation and contribution of the authors is deeply appreciated without which we would not have had the pleasure of adding this book to the literature.

K.L. Mittal
IBM U.S. Technical Education
500 Columbus Ave.
Thornwood, NY 10594

CONTENTS

PART I.   PHOTOCHEMICAL ASPECTS OF OPTICAL RECORDING

ORGANIC MATERIALS FOR OPTICAL DATA STORAGE

R. S. Jones and J. E. Kuder

Hoechst Celanese Corporation
86 Morris Avenue
Summit, New Jersey 07901

The different modes of optical data storage are defined, and
a review of write once optical storage is presented, including the basic recording mechanisms, the fundamental optical
design of the required apparatus, and the materials requirements necessary for commercially successful media. The advantages of organic materials as the active recording layer in
optical media are described. In the second part of the paper,
recent developments at Hoechst Celanese using specifically
designed and synthesized silicon naphthalocyanine materials
as data storage layers are described. Recording performance,
environmental stability, and comparison with other write once
media are presented.

INTRODUCTION

In the years following the Second World War, the industrialized
nations began a trend which continues today, away from the production of
goods to the production of information. The new technologies to handle
the flow and storage of information have required development of new
materials. The storage of information is dominated by magnetics, but this
dominating position is now being challenged by optical recording. The
challenge derives from the higher density and lower cost which optical
recording offers, as well as the ability to remove this high density
medium from its drive for distribution and the potential for greater
archivability. The most common embodiment, for reasons of random access,
is the optical disk, although effort has also gone into the development
of optical tape and optical cards.

Different types of optical disks have been developed to deal with
different types of information storage. The functional categories are:

1) Read only: already established on the scene for the
distribution of music, software and data;

2) Write once read many (WORM): aimed at archiving of
financial, medical, government, and personal records;

3) Erasable: which would permit the replacement of magnetic
hard disks.

Hoechst Celanese has an extensive research program on erasable and
write once organic media for diode laser drives[1] and for short wavelength
systems[2]. At the present time write once media have been developed to
the premanufacturing stage, and the characteristics of these types of
media will be described.

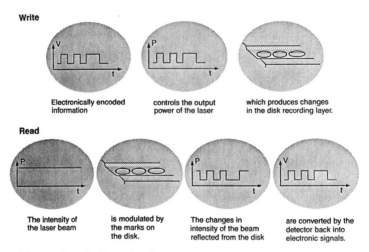

Write

Electronically encoded information — controls the output power of the laser — which produces changes in the disk recording layer.

Read

The intensity of the laser beam — is modulated by the marks on the disk. — The changes in intensity of the beam reflected from the disk — are converted by the detector back into electronic signals.

Figure 1.   Write, Read Processes of Optical Recording.

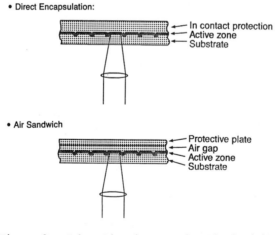

• Direct Encapsulation:

— In contact protection
— Active zone
— Substrate

• Air Sandwich

— Protective plate
— Air gap
— Active zone
— Substrate

Figure 2.   Schematic Diagram of Optical Disks.

In simplest terms, optical recording may be described as shown in
Figure 1. During writing, electronically encoded information controls
the output power of a laser. When the intensity of the beam focused onto
the recording layer is high, a mark is produced at that spot due to some
physical or chemical change in the recording layer. In reading, the
output power of the laser is constant and at a lower level than for writing. However, the beam is modulated when it encounters the marks written
on the disk. The changes in intensity of the beam reflected from the
disk are converted by a photodetector into electronic signals which correspond to the original information. From the interaction of the electronics and optics of the drive with the media unit it follows that
neither can be optimized independently of the other.

In its most frequently encountered form, the media unit has the
air sandwich construction. Here the recording layer is coated onto
a transparent substrate and is located within an air gap sandwich
structure (Figure 2).

As shown in Figure 3, during recording and reading, the laser beam is
focused through the substrate onto the recording layer where marking
occurs. Because polarized laser light is used to separate the incoming
from the reflected beam, considerable effort has gone into the development of molded substrates having birefringence which is sufficiently low
so as not to change the state of polarization of the beam. Besides providing mechanical support and protecting the recording layer from abuse,
the substrate also has molded into it grooves for tracking as well as
preformat features, which include sector, track and timing information.

The recording layer itself serves to capture the energy of the focused laser beam and consequently to undergo some physical or chemical
change that alters the intensity of the read beam which is reflected to
the detector. The materials first developed for this purpose were metals, in particular tellurium and its alloys, in which writing occurs by
the formation of micron-sized pits which diffract and scatter the read
beam. However, organic materials have a number of advantages over inorganic materials. They are less subject to degradation caused by air and
moisture. They have lower melting points and thermal conductivities and
thus potentially are capable of higher sensitivity and smaller recorded
marks. They can be coated from solution by spin coating techniques,
leading to lower fabrication costs as compared to vapor coating required
by thin metal films. Finally, the optical and thermomechanical properties of organic media can be tuned by design of molecular structure which
is the stock in trade of the synthetic chemist.

A notable difference between inorganic and organic media is that
while metals absorb radiation strongly and have high reflectivities at
all laser wavelengths, organic dyes are relatively narrow band absorbers
and have lower absorptivities. Thus, the dye must be selected to match
its absorption spectrum with the emission wavelength of the laser used.
The use of diode lasers emitting in the near infrared (roughly 780-840nm)
for optical recording has led to the design and synthesis of a number of
absorbers unknown to the classical dye chemists.

The use of a pure dye, undiluted by binder, allows for the most efficient capture of optical energy. This efficiency may be enhanced by a
factor of two or three if the dye coating is tuned to a quarterwave
thickness. Problems encountered with some dye-only media are their tendency to crystallize and their tendency to produce debris during writing,
both of which can lead to a decrease in signal to noise ratio upon reading. Debris formation can be controlled either by appropriate structure
modification or by the addition of a thin overcoat which does not impede
the marking process[3]. Both crystallization and debris formation can also
be reduced if the dye is coated in combination with a polymer to form a

Disk

Focusing
Objective

Quarter
Wave Plate

Polarizing
Beam Splitter   Photodetector

Collection
Lens

Anamorphic
Prism Pair

Collimating Lens

Diode Laser

Figure 3.   Schematic Diagram of an Optical Drive.

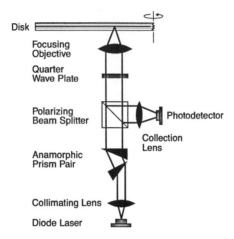

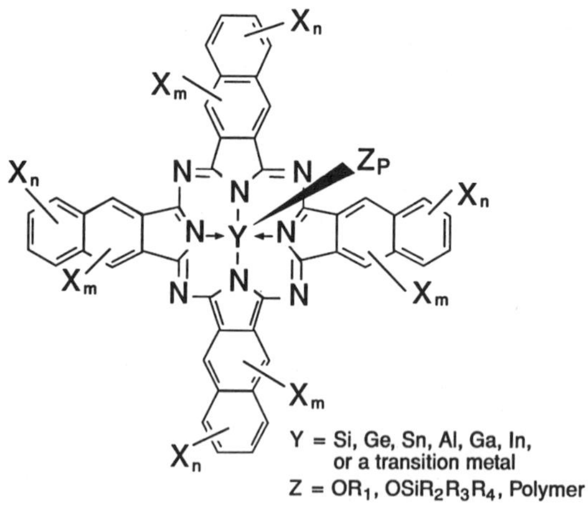

Y = Si, Ge, Sn, Al, Ga, In,
         or a transition metal
Z = OR_1, $OSiR_2R_3R_4$, Polymer

Figure 4.   Optical Media Dye Structure.

dye-binder recording medium. Not all binders are equally successful at preventing aggregation and crystallization of the dye molecules at the high concentration in which they are present in the polymer. The presence of covalent or ionic bonds connecting dye and binder can prevent their segregation in the recording layer.

As with tellurium based media, writing with organic media is generally accomplished by pit formation. A number of attempts have been made to model the pit forming process. The absorption of light by media whose thickness is on the order of a quarter of the wavelength of light is not adequately described by Beer's Law and must instead be treated by more sophisticated thin film optics. It has been suggested that the temperature at the center of the focused beam may reach as high as 1700 C[4], well above the decomposition temperature of organic materials. Pits are then formed when the gaseous products displace the surrounding molten material. An alternate model suggests that flow of molten material driven by surface tension gradients can adequately account for pit formation without invoking ablation of material[5]. In fact, recent studies which used Raman scattering as a temperature probe for laser irradiated organic media indicated a marking threshold only about 50C above the melting point of the absorbing film[6].

The effects of variations in thermomechanical properties on the pit formation process have also been considered. In one study[7], the same dye was used to form recording layers with a variety of polymers whose glass transition temperatures range from 18 to 190C. It was found that the threshold for marking was almost completely dominated by the optical density (and thus the efficiency of light absorption) of the film. Other work[8], however, indicates that the rheological properties of the medium may play a more significant role in the marking response. Here a constant dye-binder ratio (1:1) ensured the same optical response, while the melt viscosity of the binder covered ten decades of magnitude. Writing sensitivity was found to depend significantly on the melt viscosity of the binder.

In spite of the lack of total agreement as to the mechanism of pit formation, it seems certain that organic dye based media will play a significant role in the commercialization of WORM media. The second part of this paper deals with the performance characteristics of some representative organic WORM media developed in the Hoechst Celanese laboratories.

DESIGNED MATERIALS FOR OPTICAL MEDIA

Review of the technology of optical recording, and the requirements of data storage media in general, leads to the following list of requirements for optical media:

1. Optical uniformity, especially in terms of a low defect rate and low variability in the optical properties across the surface of the recording media.

2. High recorded signal level, generally expressed as carrier to noise ratio(CNR). CNR, measured in decibels(dB), is defined as 10 log $[P_S/P_N]$, where P_S is the power of a signal at a given frequency (the carrier) measured over a given frequency range (the bandwidth) and P_N is the measured power in the absence of a signal. The accuracy of the signal, compared with the input, can be measured using the harmonic to noise ratio (HNR), measured at twice the carrier frequency. A minimum in HNR indicates those recording conditions, using a 50% duty cycle write signal, which give a minimum of read signal distortion, relative to the data signal.

3. Stability under continuous playback conditions (read stability), to insure that recovery of data does not diminish the signal level or alter the data. Usually more than 1 million read passes without damage is considered acceptable.

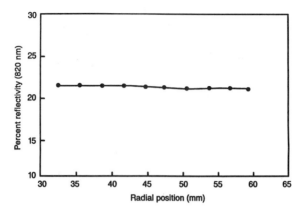

Figure 5.   Radial Uniformity of Grooved Disk Reflectivity.

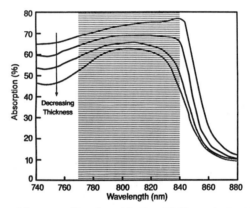

Figure 6.   Absorption vs. Wavelength for Different Coating Thicknesses.

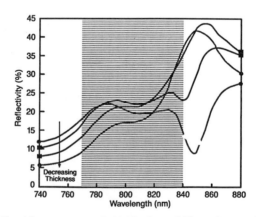

Figure 7.   Reflection vs. Wavelength for Different Coating Thicknesses.

4. Low incidence of errors in recorded data, expressed as bit error rate(BER). A raw bit error rate, before any error correction techniques are applied, must be 10^{-5} to 10^{-6}. An end of life BER after error correction is usually specified.

5. Environmental stability, including long shelf life, long archival data life, and tolerance to environmental extremes and fluctuations. Since optical media are new products, no real time data that meet the commonly quoted norm of ten years are available. Numerous accelerated aging schemes are used to estimate lifetimes under use conditions.

These requirements are demanding, nevertheless excellent recording performance has been obtained by using largely "off the shelf" materials. The use of previously available materials has its drawbacks, however, since finding all of the required performance criteria combined with commercially attractive manufacturing economics is extremely difficult. Alternatively, molecular architecture can be designed so that all properties, both those controlling performance and those necessary for ease and economy of manufacture, are found in the recording layer material.

The research program within Hoechst Celanese has an objective to develop organic materials designed for optical storage. With a baseline requirement of satisfying the performance demands for write once optical storage, the program also set forth three additional goals:

1. Development of materials with superior performance, both in terms of recording performance and environmental stability.

2. Identification of media technology adaptable to changes in hardware, especially to changes in laser wavelength.

3. Improvement in manufacturing economics, compared to earlier media.

This program has resulted in development of several classes of organic media, based on new dyes especially designed for this application. All of these materials provide attractive economics since they can be used as single layer recording media prepared by solvent coating procedures. Manufacturing processes which allow direct coating onto conventional polycarbonate substrates have been identified and developed. Materials sensitive to different wavelength regimes, from the UV/visible[2] to the near IR, have[9,10] been identified and developed. For the near IR naphthalocyanine dyes[9,10] which are suitable for all three types of media described above have been identified, and media development is now in the premanufacturing stage. Figure 4 shows a generalized structure of the dyes useful for these media. The X groups may be selected from a wide variety of substituents. Choice is typically dependent upon the effect on solubility, absorption, and thermal characteristics desired. The Y atom may be selected depending upon the absorption or chemical bonding characteristics needed, while Z can represent a polymer bound to the Y atom, or a lower molecular weight substituent. The Y group can influence absorption, thermal, or solubility characteristics. The values of the subscripts n, m, and p are selected according to the bonding requirements of the system and the desired properties. This molecular architecture offers great flexibility in design, since there are three independently adjustable variables to manipulate. For other wavelength regimes, the basic carbon skeleton may be changed.

PROPERTIES OF THE MEDIA

1. Optical Properties

Figure 5 shows the optical uniformity of a typical disk, measured as percent reflectivity. For a grooved substrate, variation in reflectivity, across the data zone, is less than $\pm 1\%$, with an average reflectivity of ~21%. Defect count is very low, as shown by bit error rate.

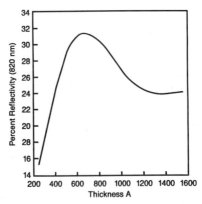

Figure 8. Reflectivity of HCC Medium as a Function of Film Thickness.

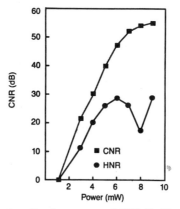

Figure 9. Representative Performance of HCC Media, at 2.5 MHz, 5.65 m/s, 830nm

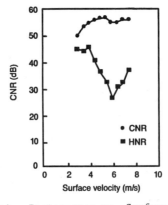

Figure 10. Representative Performance vs. Surface Velocity, at 1.25 MHz, 8.3 mW, 780 nm.

Laser manufacturing technology inherently leads to a certain variation in emitted wavelength for a given type of laser, although solid
state lasers are frequently described as "830 nm" or "780 nm". For this
reason optical uniformity over a wavelength band is necessary. Uniformity of absorption over a specified band, typically 815-840 nm for "830"
lasers, is rather easy to achieve(Figure 6), but reaching good uniformity
of reflectivity required development of specialized coating expertise.

Figure 7 shows that control of coating thickness is critical for reflectivity uniformity. A variation of $\pm1-2\%$ over 815 to 840 nm can be
routinely achieved. This requires precise control of the coating process
since reflectivity is very sensitive to small thickness changes (Figure
8).

2. Data Recording Characteristics

Figures 9 and 10 show recording characteristics at 830 and 780
nm, respectively. In Figure 9, the CNR plateaus at approximately 7
mW recording power at 52-55 dB, well above the present industry
standard of 45 dB. A clear threshold power of 1 mW

followed by a rapid rise of signal intensity, is obtained. Figure 10
shows an essentially flat CNR from 4 to 8 m/s. The lower CNR at low surface velocity indicates that recording energy density was too high for
optimum performance. Changing the structure of the recording layer material can have a large effect on the recording characteristics. Figure
11 shows that recording threshold power can be quadrupled by relatively
small structural changes in the recording material.

The recording phenomenon used in this medium is formation of a pit
in the smooth surface of the recording layer. Figure 12 is a scanning
tunneling microscope image[11] of a pit formed under a typical recording
condition. The pit is formed with a remarkably uniform rim and a symmetrical crater. These crater characteristics and direct measurement of pit
position and size precision suggest that these types of materials can
support more sophisticated methods of data encoding than is currently
practiced.

3. Read Stability

Figure 13 shows the read conditions which permit continuous
data recovery over a period of more than 10^6 read cycles for a type
of media. As would be predicted, the power that can be used to
read data without damage increases with increasing surface
velocity, i.e., with decreasing energy per unit area. Acceptable
read conditions, as well as recording threshold power, show a
strong dependence upon material structure. The recording material
with high threshold(Figure 11) also has a higher read
stability(Material B in Figure 13).

4. Bit Error Rate

Bit error rate is determined by comparing the recovered,
recorded data stream with the input stream. Sources of error are
numerous, and include defects, both detectable and undetectable, in
the unrecorded media, errors in generating the recording signal,
and poor recorded data quality, making detection difficult. Most
errors are related in some way to the quality of the recording
layer and the underlying substrate. Error correction schemes have
been developed which very efficiently correct for relatively short
error bursts. A raw, uncorrected BER of 10^{-5} or less, with almost
all of the errors of 1 to 5 bits in length is taken as the
criterion for end of life of the medium. These BER rates and
characteristics can only be achieved by producing substrates and
final media in the appropriate clean environment. Table 1 shows
the results of a randomly selected air sandwich disk made on the
premanufacturing line at Hoechst Celanese.

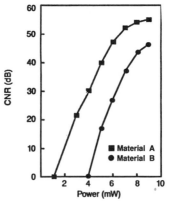

Figure 11. Comparative CNR vs. Dye Structure, at 830 mW.

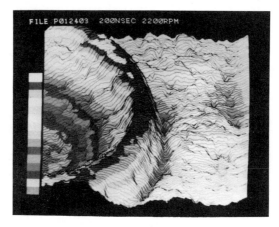

Figure 12. An STM image of a pit created by a 200 nsec laser pulse. The region scanned is ~1.0 μm x 1.0 μm. The total height variation in this image is 1020 A, divided into 15 different zones (68A/-zone).

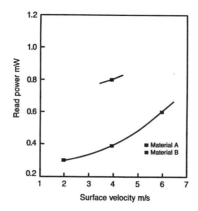

Figure 13. Read Stability of Optical Disks.

5. Environmental Stability

Ideally, since write once optical media are archival, the lifetime of recorded data should be measured in years or decades, while the shelf life of the unused media can be shorter. Because these devices are new, and are still rapidly evolving technologically, real shelf life and archival life values are unavailable. To estimate them, the industry has resorted to various high temperature, high humidity testing conditions, with the assumption that such conditions simulate accelerated aging conditions and not a change in the mode of failure. While this assumption can logically be argued against, the fact is that these tests are used on a routine basis, and until real time data are available the lifetimes of commercial optical media will be judged on the results of accelerated aging tests. Typical conditions include temperatures up to 80 C, and relative humidity up to 80-90 %. Tests usually include cycling of temperature and humidity in various ways.

Figure 14 shows data for ambient aging of a representative disk. Extensive laboratory trials have shown that disk CNR remains constant when reflectance and transmission values do. The effects of aging on BER levels must be measured directly, however.

Figure 15 shows comparative data for aging at 80 C, in low humidity, for four types of write once media: inorganic phase change, tellurium based pit forming media, cyanine dye, and Hoechst Celanese (HCC) media. Three types were essentially unchanged, while the cyanine dye disk was out of specification at the end of the test. Figure 16 shows a representative CNR curve for an HCC disk after aging for 306 days at 80 C. Comparison with Figure 9 shows that the two CNR curves are essentially identical. In Figure 17, representative CNR data as a function of aging time at 80 C and 80% RH are shown. BER has been shown to change less than one order of magnitude after 500 hours at 80 C/80% RH. From these results, and others at lower temperatures, it is estimated that Hoechst Celanese medium has a lifetime in excess of 10 years, and most probably several decades.

In addition to high temperature and humidity, accelerated aging can be performed with intense UV and visible light. Figure 18 shows the results of aging disks in a Weatherometer. There has been some concern that organic dye recording layers are unstable to UV-visible radiation. The Weatherometer results indicate that Hoechst Celanese materials have actinic stability comparable to inorganics.

CONCLUSION

Comparative data for various types of write once media show that all have similar recording response (Figure 19). Differences are found, however, in the environmental stability of media, the type of materials used in the recording layer(s), and the techniques and complexity of manufacture. In addition, the precision of recorded mark position and shape is exceptionally good for Hoechst Celanese materials. It is likely that these differences will be significant as the market for write once optical storage develops. The product which provides the performance and stability demanded for the application at the lowest possible cost will be favored in the long run. Dye based media, such as naphthalocyanine based ones described here, using solvent based coating technology to make a single layer product, are the most likely candidates.

Table 1. Bit Error Rate, Hoechst Celanese Media.

BER	2.9×10^{-6}
BITS WRITTEN	3.7×10^{7}
BIT ERROR COUNT	1.1×10^{2}
BURST LENGTH, BITS	
1-5	1.1×10^{2}
6-10	0
11-15	0
|	
>50	0

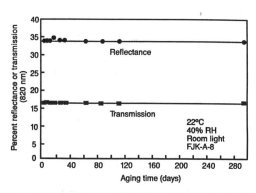

Figure 14. Optical Disk Stability, Ambient Conditions.

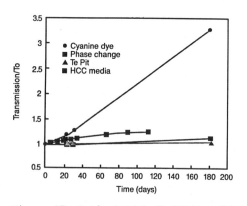

Figure 15. Optical Disk Stability, 80 C.

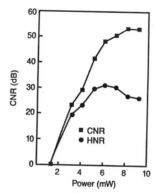

Figure 16. Performance After Aging 306 Days at 80 C.

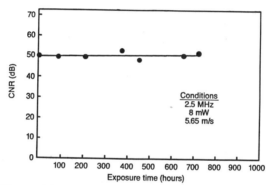

Figure 17. Optical Disk Stability, 80 C/80% RH.

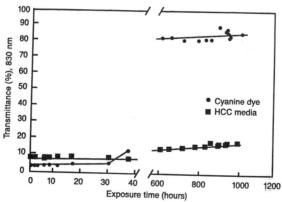

'Figure 18. Comparative Disk Stability, Weatherometer Aging.

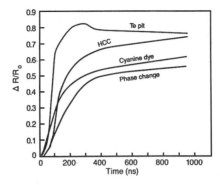

Figure 19. Comparative Performance of Various Media.

REFERENCES

1. H.A. Goldberg, R.S. Jones, P.S. Kalyanaraman, R.S.. Kohn, J.E.,
    Kuder, and D.E. Nikles, Dynamic laser marking experiments on an
    organic optical storage medium, Proc. SPIE, 695, 20(1986).

2. J.P. Shepherd, Organic optical storage media for short wavelength
    systems, Proc. SPIE, 899, 220(1988).

3. M. C. Gupta, Appl. Opt., 23, 3950(1984).

4. J. J. Wrobel, A. B. Marchant, and D. G. Howe, Appl. Phys. Lett.,
   44, 928(1982).

5. T. S. Chung, J. Appl. Phys., 60, 55(1986).

6. P. K. Chan and T. R. Hart, Appl. Optics, in press.

7. K. Y. Law and G. E. Johnson, J.Appl. Phys., 54, 4799(1983).

8. M.F. Molaire, The influence of melt viscosity on the writing
   sensitivity of organic dye-binder optical disk recording media,
   OSA Topical meeting on Optical Data Storage, Stateline, Nevada,
   March 11-13, 1987, Abstract ThA4-1, pp. 90-95

9. D. E. Nikles. R. S. Jones, and J. E. Kuder, U. S. Pat.
    4,605,607(1986).

10. M. E. Kenny, R. S. Jones, J. E. Kuder, D. E. Nikles, U. S. Pat.
    4,725,525(1988).

11. R. Reifenberger, private communication.

THE USE OF RADIATIONS FOR THE MODIFICATION OF POLYMERS

Adolphe Chapiro

C.N.R.S.
94320 Thiais, France

Gamma-rays or electron beams can be used efficiently
and economically for the production of new or modified
polymers. The chemical reactions underlying these
processes are described. Radiation curing is used on
a large scale for the high speed production of
improved coatings. Curing of magnetic formulations
looks promising. Radiation crosslinking is an
established technology in the wire and cable industry.
It imparts to the modified polymer improved resistance
to solvents and to high temperatures. A crosslinked
network also reduces the migration of "fillers" and
thereby stabilizes in time any message imprinted with
magnetic or colored pigments dispersed in a polymer.
Radiation grafting is a method for modifying more
profoundly the properties of a polymer. The method can
be used for numerous applications, such as introducing
polar groups into non-polar polymers, increasing or
reducing the wettability of a surface, imparting to a
polymeric surface better compatibility with a specific
coating and many others. The chemical modification may
be applied at will into the bulk of a material or
limited to a surface zone of any desired depth. The
justification for the use of a radiation treatment as
compared to other existing methods is discussed both
from technical and economical points of view. It is
shown that whereas U.V. radiation competes with
electrons for small production units, large scale and
high speed productions benefit highly from the lower
cost and greater efficiency of electron beams.

INTRODUCTION

"High energy" radiations, e.g. X- and gamma-rays or
accelerated electrons, are today widely used in the polymer industry
for the production of new or modified polymers exhibiting interesting
properties. Although the information on this specialized field is
in the open literature it is still poorly advertised. This is
particularly surprising owing to the fact that the corresponding
technology represents a multi-billion dollar market worldwide.
The purpose of this brief review is to describe some of the

techniques which are commonly used today in radiation processing, with special emphasis on those aspects which may be attractive to information storage technology.

It should be emphasized that for the treatment of thin films or thin layers of materials the most appropriate radiation sources are beams of "low" energy electrons produced by machines (electron accelerators) operating in the 300 to 500 keV range. The corresponding equipments are easily available today. They are reasonably priced and with the more powerful machines industrial treatments can be carried out within seconds or even within fractions of seconds.

The following processes will be considered here :
1 - radiation curing;
2 - radiation crosslinking;
3 - radiation grafting.

1 - RADIATION CURING

This operation involves the irradiation of specifically designed formulations deposited on a solid substrate : plastic film, paper, wood, metal, etc. The formulations contain one or several vinyl monomers or a blend of monomers with one or several "prepolymers". The latter increase the viscosity of the formulation to such an extent that the mixture can be deposited onto the substrate with a brush or a knife. They also significantly reduce the dose required for curing. Irradiation of such a system generates free radicals which polymerize the monomer converting the initial formulation, which has the consistency of a paste, into a hard coating.

The reaction is a free radical polymerization process involving the three usual steps of chain initiation, chain propagation and chain termination. However, the last step, which requires a bimolecular interaction of two polymeric radicals, is hindered by the high viscosity of the reaction medium and this results in longer kinetic chains and therefore in a higher efficiency of the overall process.

If the substrate is an organic material (plastic, paper, wood) free radicals are also generated from the surface of the irradiated substrate. These contribute to the polymerization process and thereby the coating is "grafted" to the substrate. The cured formulation then becomes chemically linked to the surface by covalent bonds, which results in outstanding adhesion. The composition of the formulation must be carefully established with consideration not only to its reactivity under irradiation but also to the properties of the final cured layer.

Numerous monomers and "reactive" prepolymers (i.e. containing reactive double bonds) are available from suppliers which makes it possible to "tailor" coatings with almost any desired properties. Pigments (magnetic powders) can obviously be added to the formulations up to the limit of the penetrating power of the electron beam, but for thin layers this is not a serious limitation.

Irradiation is preferably carried out under an inert atmosphere (nitrogen or combustion gases) or else after covering the formulation with a thin plastic or metal foil in order to prevent an oxidation process during the radiation treatment. Some formulations, however, cure as well when irradiated under normal atmospheric conditions.

The doses required for total cure vary with the composition. But in most cases, curing times can be reduced to very low values, the only limitation then being an excessive rise in temperature due to the exothermicity of the chemical reaction occurring in such a very short time. For slow or moderate curing times the polymerization proceeds at ordinary temperatures which may be of advantage for the properties of the final product.

2 - RADIATION CROSSLINKING

Crosslinking is the process which links polymeric chains to one another, thereby building a giant molecular network to which all macromolecules of the system are attached. Crosslinking can be achieved simultaneously with curing by adding to the formulation a polyfunctional monomer or a reactive prepolymer. Such conditions are met in most formulations presently used in the coating industry.

A radiation treatment specifically intended for crosslinking can also be applied to the substrate or to any polymeric material prior to or after coating. In such a process chemical bonds of the irradiated polymer are broken to generate polymeric free radicals, and these, after combination, form a tridimensional network. The efficiency of radiation crosslinking varies with the radiation sensitivity of the polymer. It is highest for halogen containing materials and lowest for aromatic and highly conjugated structures. The presence of branched structures in the polymer chain corresponds to weak points and induces chain scission under irradiation. Such "radiation degradation" generates polymers with lower molecular weights, and usually with reduced strength. A behaviour of this type is met in the following polymers : polyisobutylene, polymethacrylates, cellulosics and Teflon.

A crosslinked polymer is harder, has a much better resistance to solvents and to any mechanical aggression, such as wear; but crosslinking also reduces migration of pigments or of other "inert" fillers by reducing segmental motion in the polymeric matrix. This effect is of particular value for the stability in time of magnetic formulations into which a message is imprinted. The extent of migration under given conditions is monotonically reduced with the density of crosslinks and this can be easily controlled by properly selecting the radiation dose, the amount of added crosslinking agents and the like.

3 - RADIATION GRAFTING

This is a method for modifying more profoundly the properties of a polymer. The initial polymer ("trunk" polymer) is reacted with a monomer during or after a radiation treatment. Under such circumstances the monomer undergoes polymerization and the resulting chains form "branches" which are linked to the trunk polymer. The final system is a "molecular blend" of two polymers or a "polymeric alloy" which may exhibit combined properties of each of its components. For this reason graft copolymers represent a class of materials with almost unlimited potentialities. The radiation grafting method can be used for numerous applications only some of which are quoted here : one can introduce polar groups into non-polar polymers and thereby gradually modify their polarity; by a similar treatment one may at will increase or reduce the wettability of a polymeric surface, impart to the polymer a better compatibility with specific fillers, coatings, inks and many others. Here again the treatment can be applied continuously and with a high speed to a shaped polymer (film, fiber, etc.). The chemical modification may involve the bulk of the material or be limited purposely to a surface zone of any desired depth. The versatility of this technique makes it possible to solve a great variety of problems in "tailoring" and optimizing products for any given application.

4 - COMPETITION WITH OTHER METHODS

The radiation treatment can be considered as a convenient method for generating free radicals in any (organic) substance and thereby initiating free radical processes. As such, this method competes with other, more conventional methods for generating free radicals, such as initiation by chemical reagents (peroxides) or by U.V. light. With respect to chemical initiation radiation methods have the advantage of simplicity and cleanliness. They can be operated at any desired temperature, since radiation initiation is temperature independent. They are also attractive for economical reasons. Radiation, if used efficiently, is cheaper than the corresponding chemicals needed to produce the same effects and chemical initiation requires heat which is a most ineffective way of injecting energy into the system. With respect to U.V. light the differences are more subtle. Most systems will react almost indifferently to either high energy or U.V. radiation. The latter may require some additional "sensitizer" if the formulation does not absorb light of the proper length wave. The major difference, however, lies in production scale and in economics. For small size productions, U.V. light may be more attractive because investment in the radiation source is much smaller. However, for large or medium scale productions, requiring high speed outputs, the appropriate U.V. radiation sources are no longer available, whereas, existing electron accelerators generate beams of 30 KW or more. With such machines very large production speeds become possible and, in addition, economics are then in favor of the electron beam.

REFERENCES

A. Chapiro, "Radiation Chemistry of Polymeric Systems" John Wiley & Sons, New York, 1962.

A. Chapiro, Radiation-induced reactions, in "Encyclopedia of Polymer Science and Technology", John Wiley & Sons, New York, 1968, Vol. 11, pp.702-760.

S. Nablo, Radiation safety considerations in the use of Self-Shielded Electron Processors, Radiat. Phys. Chem. 22, n° 3-5, 369-377 (1983).

A. Charlesby, Some reflections on radiation research and technology, Radiat. Phys. Chem. 28, n° 5/6, 473-477 (1986).

OPTICAL DATA STORAGE USING PHASE SEPARATION OF POLYMER-LIQUID CRYSTAL

MIXTURES

William D. McIntyre and David S. Soane

Department of Chemical Engineering
University of California at Berkeley
Berkeley, CA 94720

A polymer-liquid crystal mixture has been studied for possible reversible optical storage applications. At ambient temperature, this system exhibits a phase-separated morphology. It can be heated past its lower critical solution temperature, briefly forming a single phase. The laser-heated sample is then allowed to cool back into a two-phase state at various cooling rates. Through control of the heating pulse shape, the mixture and its substrate can be given varying temperature profiles at the completion of heating, thus influencing the subsequent cooling rate. Faster cooling results in smaller final domain sizes. A sample spot with these smaller domains gives rise to higher light scattering intensity over a wide range of scattering angles.

The feasibility of this reversible-morphology principle for optical recording was demonstrated through an experiment in which a 4-8 μm coating was applied to clear ITO resistor-coated glass. Heat was put into the system by a pulse generator whose peak power and pulse duration could be controlled. Low intensity laser light scattering patterns were monitored in-situ. Light scattering patterns could be switched between two states by alternating the heating pulses, thus reversibly changing the sample morphology.

A one-dimensional model was developed to simulate heat transfer in an active medium coated on an aluminum reflector layer and heated by absorption of laser light. The model predicts that modulation of a writing laser power can render better control of the cooling rate than that achieved with the experimental setup.

To screen polymer-liquid crystal combinations as candidates for optical storage by the phase separation mechanism and to aid in choosing an appropriate solvent for spin coating, a thermodynamic model is desirable. The Flory lattice model for mixtures of solvent, rigid rod molecule, and polymer was modified with terms to account for enthalpic effects. Comparison of model predictions with experimental cloud points gave the temperature-dependent interaction parameters for the model.

PART I. EXPERIMENTAL STUDIES

Introduction

     Transition-metal/rare-earth alloys for magneto-optic systems and
metal alloys for phase-change systems are two commercially promising
approaches for reversible optical data storage. In magneto-optic systems,
a material with permanent magnetization is simultaneously heated with a
laser and exposed to an external magnetic field to affect the direction
of magnetization upon cooling. When a plane polarized laser is focussed
on the disc, differences of 0.2-0.4° in the direction of polarization of
the reflected light reveal the different directions of magnetization
stored in the medium (Kerr effect)[1].
     Phase-change systems rely on the modulation of the laser beam to
affect the cooling rate of an alloy composed typically of tellurium,
germanium, and tin which has been heated above the crystalline-amorphous
phase transition temperature[1]. Low cooling rates give a near-equilibrium
crystallinity, while higher cooling rates lead to an amorphous state which
can be kinetically stable. A typical write pulse is 1 μm^2 in area and has
a Gaussian intensity distribution, while an erase pulse uses an oval beam
about 10 μm^2 in area. The smaller beam size creates larger temperature
gradients, causing a higher cooling rate and a more amorphous final
state[1].
     The generally accepted requirements for performance of reversible
media in computer systems include the following[2,3]: one μm or less
diameter spot size with adequate resolution; ability to record and erase
information in one microsecond or less; over 10^6 write/erase cycles;
archival stability of ten years or more; and a bit error rate of less than
$1:10^6$. At the present time, few reversible media meet all of these
criteria. A particularly troublesome problem is that the transition-metal
and rare-earth elements used in the much-studied magneto-optic and phase-
change systems are subject to oxidation, reducing their durability.
Delamination between metal and plastic layers after repeated record/erase
cycles due to different thermal expansion characteristics has also been a
problem.
     In this work, a system consisting of amorphous polymers and small
liquid crystal (rigid rod) molecules has been investigated. To understand
the working principles of this system, we will first review the literature
on the phase behavior of mixtures. Phase separation can, in principle,
occur by one of two mechanisms: nucleation and growth or spinodal
decomposition[4]. The second derivative of the free energy-composition
curve determines which mechanism takes place. In the metastable region of
the phase diagram, nucleation and growth governs phase separation, while
in the unstable region, spinodal decomposition causes minute perturbations
to amplify with time. Nucleation and growth results in randomly spaced
phase domains of no fixed size. Spinodal decomposition theory predicts
periodic fluctuations of composition in a phase separated system, though
in actual systems, the periodicity can be destroyed by the process of
phase coarsening as the system matures.
     The spinodal decomposition process has been described for polymer
solutions with

$$\frac{\partial c}{\partial t} = \nabla \cdot \left\{ M \nabla \left[(\frac{\partial f}{\partial c}) - 2\kappa \nabla^2 c \right] \right\} \tag{1}$$

22

where: c = Volume fraction of polymer segments
        t = Time
        M = Diffusive mobility (ratio of the material
            flux to the chemical potential gradient)
        f = Molar free energy of mixing
        κ = Gradient energy coefficient

Eq. 1 is strictly valid for isotropic solutions but has been adopted for
anisotropic solutions of rigid rod solvent with polymer solute. Note that
during the early stages, the system behaves linearly and exhibits a well-
defined periodicity. The "resonant" periodicity with maximum growth is
described by

$$\lambda = 2 \ (2^{1/2}) \pi \left[- \frac{\left(\partial^2 f / \partial c^2 \right)_o}{2\kappa} \right]^{-1/2} \qquad (2)$$

where λ is the wavelength of the periodicity with maximum growth and the
second derivative is evaluated at the initial (homogeneous) composition,
denoted by the subscript o.

After one (or both) of the phase separation mechanisms takes place,
the final size of the equilibrium phase domains is further affected by the
thermodynamic driving force within the system to minimize interfacial free
energy by minimizing surface area where the two phases contact. Two modes
of phase coarsening often cited in the literature are phase coarsening by
diffusion and phase coarsening by hydrodynamic flow[5,6]. The degree to
which each of the mechanisms contributes to phase domain growth depends on
whether the minor phase domains are interconnected, and the viscosity and
mass diffusivity of the components.

For the Ostwald theory for interfacial tension-driven phase ripening
by diffusion, it has been shown that[5,7]

$$d^3 \propto Dt/T \qquad (3)$$

        where d = Minor phase droplet diameter
              D = Diffusivity
              T = Absolute Temperature
              t = Time

Since the diffusivity normally increases with temperature faster than
the first power[8], phase separation at higher temperatures results in
larger domain sizes if coarsening is allowed to occur. In phase ripening
by hydrodynamic flow, the increased viscosity of both phases at low
temperatures would be expected to lessen the phase domain growth. Thus, a
polymer blend which is cooled quickly past the phase separation
temperature can be expected to have smaller phase domains than the same
mixture when cooled slowly, regardless of the mechanisms of phase
separation and phase coarsening. Below the polymer glass transition
temperature, hydrodynamic flow will cease, while the rate of diffusion-
induced coarsening will also be limited by the low species diffusivity in
the glassy state. Polymer/liquid crystal mixtures are considered in this
study to behave similarly to polymer/polymer or polymer/solvent mixtures
in their available phase separation and coarsening mechanisms.

Blends of liquid crystal molecules and amorphous polymers which are
immiscible at room temperature become miscible when heated above the
critical temperature by an incident laser. The laser intensity, pulse
time, beam size, and beam shape can be controlled to influence the cooling

rate of the medium as it returns to the two phase region. Faster cooling rates have been shown to result in smaller size domains than slower ones, in accordance with theory. The domain size which is retained after the completion of cooling and vitrification of the system determines the pattern of light scattering from a low energy laser source.

The incentive for studying this particular type of organic mixture system derives from two potential advantages. First, polymer systems would not be as subject to oxidation and thermal expansion problems as the transition-metals and rare-earth elements used in most current media. Another advantage of organic media versus non-organic systems is that the optical discs could be spin coated, an inexpensive process in comparison to sputtering or vacuum deposition. The organic system under study can also be mixed with a dye to fine-tune its light absorption properties.

In the first part of this paper, an experimental apparatus is described which demonstrates reversible optical storage using a mixture of a rod-like liquid crystal, N-(4-ethoxybenzylidene-4'-n-butyl aniline) (EBBA) with poly(methyl methacrylate) (PMMA). Because of the importance of cooling rate on the effectiveness of the experiment, heat transfer in the experimental system is analyzed with a mathematical model in order to understand the transient behavior in the substrate/active layer structures.

In the second part, the development of a finite element numerical model of heat transfer in a stationary optical storage disc is reported. Predictions of this model are used with experimental results to indicate how a pulsed laser should be modulated to optimize this reversible phase-separation system. A computer algorithm is presented which predicts the phase behavior of solvent/rod/ polymer mixtures based on the lattice model developed by Flory[10], but modified with interaction terms to account for thermal effects. This computer algorithm can be used to screen chemical systems. It accommodates the liquid crystal molecule-polymer interaction parameter and rod/coil size effects in its basic framework. A combination of the finite element heat transfer program and the calculated phase diagram constitutes the starting point of system design and optimization.

Experimental

In the experiment, light scattering was performed on a thin film while it was subjected to thermal transients, simulating those which would result from a pulsed laser in actual optical storage. The two components of the mixture were dissolved in toluene (20 wt. % solids) and spin coated at 1750 rpm onto a clear, indium tin oxide (ITO) resistor coated glass. The resulting layer thickness was determined by weighing the substrate before and after spin coating. Successive spinning resulted in an additional coating of 2 μm increments at the spin coating conditions used. Final layer thickness ranged from 4 to 8 μm and was controlled by the number of applications. After spin coating, samples were annealed in a 120°C oven to reduce the large-scale anisotropy induced by the spin coating process and to evaporate residual toluene, which boils at 111°C.

The glass substrate design has been described by Sporer[11] and is sketched in Figure 1. The glass was coated with 265 Å of ITO by Donnelly Mirror Company (Holland, MI). The coating was etched using a 47% sulfuric acid, 47% water, 6% nitric acid solution at 70°C for 15 minutes except for a masked 2 mm wide strip. Two strips of thin gold spaced 2 mm apart were then evaporated at right angles to the ITO strips for use as electrodes. The resistance of each substrate was individually tested and recorded; the resistances ranged from 150 to 400 ohms. By spin coating the mixture directly onto the ITO and connecting the gold electrodes to a Hewlett-Packard 214A pulse generator, heat was transferred rapidly from the resistor into the polymer in a controlled manner. The square-shaped generator pulses ranged from 200 to 10000 ns and from 17 to 100 volts as determined by knob settings and confirmed on an oscilloscope.

To simulate a read laser and to monitor the progress of the phase changes imparted by the thermal excursion, a continuous 10 mW HeNe laser was projected normal to the glass, through the ITO, and into the sample, where it was scattered. The laser itself appeared not to affect the sample, as it could be projected through the sample for several hours with no change in the scattering pattern. This observation is consistent with reports that poly (methyl methacrylate) has negligible absorption at the

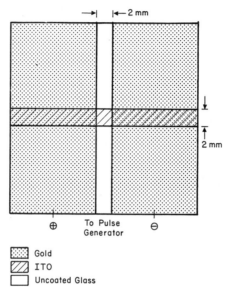

Figure 1. Schematic of the substrate used for fast heat transfer in a thin-film structure. The glass is coated with a 2 mm wide strip of indium-tin oxide (ITO), 265 angstroms thick. Gold strips, 2 mm apart, are then coated perpendicular to the ITO. The active layer is then spin coated on top of the substrate. The gold electrodes are connected to the pulse generator by alligator clips. The laser is directed through the center of the substrate, normal to the glass.

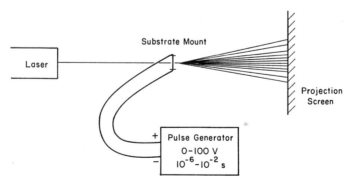

Figure 2. Schematic of the experimental setup. The phase separated substrate coating scatters the laser onto a projection screen where the results of thermal transients can be viewed. A photodetector mounted on a goniometer could replace the projection screen to quantify the scattering results.

wavelength of the laser[12]. The scattered light was projected onto a
screen, where it could be observed and the scattering angles measured.
Refraction of light at the sample/air interface caused the measured light
scattering angles to be larger than those actually generated by the
sample. Refraction was accounted for in determining the light scattering
angles through the use of Snell's law, using the refractive index[30] of
PMMA (n=1.49). The experimental set-up is depicted in Figure 2.

At the outset of experimentation, the polymer films were subjected to
widely different cooling rates to see how the light scattering patterns
from the cooled film would be affected. Two workable write and erase
pulses which would result in very different cooling rates were determined.
The first pulse was defined by setting the pulse generator at its maximum
duration and then increasing the voltage in one-volt increments until a
phase transition was noted through a change in the scattering pattern.
Next, the second pulse was defined by setting the voltage at the maximum
deliverable and increasing the duration in 100 ns increments until a
reversible change in the scattering pattern was observed.

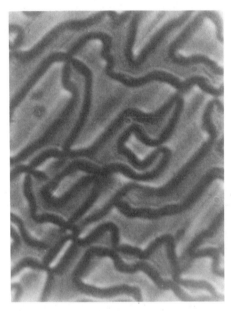

Figure 3. Photograph of film morphology which results when a 50% by
weight blend of high molecular weight PMMA and EBBA is placed in a 120°C
oven for ten minutes and cooled in ambient air. Domain width about 5 μm.

Results and Discussion

A preliminary experiment was made to demonstrate that cooling rate
affects vitrified domain size in a polymer/liquid crystal mixture.
Microscope slides were spin coated with a 50:50 mixture of EBBA and poly
(methyl methacrylate). The slides were placed in a 120°C oven for ten
minutes, then immediately cooled. One slide was cooled slowly, in ambient
air, while the other one was cooled in a bath of liquid nitrogen. Figure
3 shows a photograph of the slowly-cooled slide taken through an optical
microscope with cross polarized light. The size scale of refractive index
changes (the domain size) was 5 μm. Figure 4 shows the quickly-cooled
slide at the same magnification. The domain sizes are about 1 μm and
below. The smaller domain sizes of the quickly-cooled sample are
in accordance with theory.

The experimental apparatus was first employed using a 50:50 mixture by weight of EBBA and high molecular weight poly(methyl methacrylate) (M_v=620,000). Energy pulse characteristics were found which could be used to switch back and forth between large and small domain sizes, thus storing information in the polymer system which can be detected by light scattering. Successful pulses for three different film thicknesses are listed in Table I. Sample morphology was alternated reversibly between a larger scattering angle morphology and a smaller scattering angle morphology more than one hundred times with no signs of fatigue.

Photographs of "large" and "small" scattering patterns appear in Figures 5 and 6, respectively. The scattering patterns were intense up to a certain angle, with intensity dropping very quickly after this angle. The scattering pattern in Figure 5 extends outward to an angle of about 6.2°, while the one in Figure 6 extends out to 4.1° (after correcting for light refraction at the sample/air interface). In an optical storage system, the photodetector used to read information would be placed between these two angles. Figure 7 shows quantitatively how the scattering patterns in Figures 5 and 6 compare when measured with a photodetector mounted on a light scattering goniometer, confirming the visual observations.

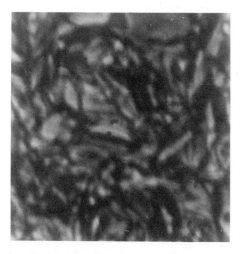

Figure 4. Photograph of the sample of Figure 3 except cooled in liquid nitrogen immediately after removal from the oven. Domains are about 1 μm across.

The samples which gave the scattering patterns in Figures 5 and 6 were observed under an optical microscope with cross polarized light. The phase boundaries were of irregular shape. The larger scattering pattern was generated by a smaller domain size, with domains of about 1.6 μm wide in the large scattering pattern sample and 2.0 μm wide in the small scattering pattern sample.

Electron microscopy was performed on samples to confirm that the observed scattering patterns were due to a refractive index difference of the phases within the bulk sample, rather than the result of a rough film surface. The micrographs showed that the sample surfaces were smooth on a submicron size scale.

Using literature values, Eq. 2 was solved to give the most rapidly growing wavelength (λ), i.e., resulting from spinodal decomposition. This was found to be 0.06 μm. Since this number is much smaller than the observed value of 1.6-2.0 μm, it appears that phase ripening is important in determining the phase domain sizes in the experimental system under study.

Figure 5. Photograph of scattering pattern which resulted from pulse 4 (50V, 0.001 s, 150 ohm). The intensity of scattered light is high for angles below 6.2°. The sample was 8 microns of 50% by weight blend of high molecular weight PMMA and EBBA.

Figure 6. Scattering pattern from the sample in Figure 5 after treatment by a pulse of 17V for 0.01 s. The intense part of the scattering pattern is at angles less than 4.1°.

The importance of phase ripening in determining phase domain size has been established in several studies of polymer blends[13,14,15]. As we shall see later, the experiments conducted here are done on active layers coated over glass, thus cooling rates are restricted. When the layers are deposited directly on a metal reflector which also serves as a heat sink, cooling rates can be greatly improved, creating the possibility of trapping morphology before coarsening.

A ring of maximum light intensity at a certain scattering angle was observed in samples upon removal from the oven, but not after energy pulses. Mixtures with components of two different refractive indices exhibit a ring of maximum intensity if they have a single characteristic dimension of spatial composition variation[13,16,17].

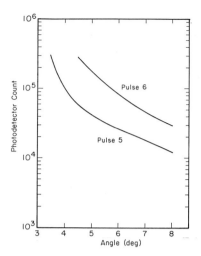

Figure 7. Plots of intensity versus scattering angle for the scattering patterns shown in Figures 5 and 6. The intensity was measured with a photodetector mounted on a goniometer.

This is only possible from spinodal decomposition, if phase coarsening is restrained[5,18].

The lack of a maximum intensity ring in samples which have been subjected to thermal transients indicates that the samples do not have a single periodicity of composition variations. A gradation of periodicities is consistent with the different thermal histories experienced at different depths of the film. This absence of a single periodicity in vitrified samples could also be due to phase ripening after the formation of a periodic structure in the initial stages of spinodal decomposition. Another remote possibility would be that the absence of a single characteristic periodicity is the result of phase separation by nucleation and growth mechanism.

Experimental Analysis using Heat Transfer Model

The mechanism of reversible morphology switching is governed by the thermal effects and kinetic processes occurring in the liquid crystal/amorphous polymer mixtures. In order to improve understanding of the influence of electric pulses on the samples, we performed numerical simulation on the heat transfer in such systems.

The experimental apparatus used in this study consisted of 0.05 cm thick glass coated with 265 Å of a resistive film and 4-8 μm of an organic mixture. Because the length and width of the glass and coating are on the scale of millimeters while the thicknesses are on the order of μm, the heat transfer problem can be approximated as one dimensional in the direction normal to the interfaces. To use an applicable analytical solution to the heat transfer equation in one dimension, the heating portion of the experiment was analyzed first. The profile at the end of the heating period was then used as the initial condition for the cooling part of the experiment, which was modeled using the finite element method. The system was treated as thermally isotropic in all heat transfer calculations. The thermal diffusivity and heat capacity of the organic mixture were taken as 0.00086 cm.

Carslaw and Jaeger[19] published the following analytical solution to the heat transfer equation in one dimension, where the initial temperature is uniform, one boundary is insulated, and the temperature history at the other boundary is known:

$$T = \frac{2}{\delta} \sum_{n=0}^{\infty} e^{-k_d(2n+1)^2\pi^2 t/4\delta^2} \cos \frac{(2n+1)\pi x}{2\delta} \Biggl\{$$

(4)

$$\frac{(2n+1)\pi k_d}{2\delta} (-1)^n \int_0^t e^{k_d(2n+1)^2\pi^2\lambda/4\delta^2} \phi(\lambda) \, d\lambda \Biggr\}$$

where T = Temperature rise above initial temperature
at position x and time t in the layer
x = Distance in the layer from the insulated x=0
boundary
k_d = Thermal diffusivity
t = Time
δ = Thickness of polymer or glass
λ = Dummy variable in time integration
$\phi(\lambda)$ = Temperature as a function of time at the
x=δ interface

If the polymer/air and the glass/air boundaries are considered insulated, the equation can be applied separately to the polymer and glass layers if $\phi(t)$ is known. $\phi(t)$ is the temperature history at the glass/ resistor and polymer/resistor interfaces, which should be approximately equal because the resistor is very thin. For this work, $\phi(t)$ was assumed to have the form of

$$\phi(t) = \alpha t^2 + \beta t \tag{5}$$

where α and β are empirical parameters.

To find the parameters, α and β, values were assumed and the temperature profiles in the glass and in the polymer layers were found at the completion of the energy pulse and halfway through the energy pulse. The energy uptake in the glass and the polymer layers was found by integration of the instantaneous temperature profiles with Simpson's rule. Using successively improved guesses, α and β were found such that the total energy in the experimental device matched the energy input to the resistor as calculated by Joule's law at the end of the pulse and halfway through the pulse. Total energy content was found at other fractions of the pulsewidth for the final values of α and β and compared to the electrical input. The energy values found by the two different methods differed by less than 20% at any given time.
To determine whether the assumption of insulated boundary conditions at the two air interfaces is valid, a heat transfer coefficient of 85 $J/m^2/hr/^\circ C$ was assumed. This value is the maximum given for the range of typical free convection heat transfer coefficients at solid/ air interfaces by Bird, et.al.[8] If the polymer were maintained at its highest temperature (69 C above ambient) for the entire length of the longest pulse (0.01 sec), with 4×10^{-6} m^2 interface area, the amount of heat lost at the interface would be, at the most, 6×10^{-8} joules. This is a relatively insignificant amount corresponding to a 1 $^\circ$C temperature drop in the outermost 0.01 μm of the polymer film.
The transient temperature profiles during cooling in the experiment were modeled using the Galerkin finite element method[20]. The model used quadratic basis functions and placed two elements (four subelements) in the polymer region and five elements (10 subelements) in the glass region. The heat transfer equation and the appropriate boundary conditions were

solved in dimensionless form. Characteristic variables were defined so as
to make the dimensionless temperature gradient match on both sides of the
polymer-glass interface. The residual equations for the finite element
program are developed. The integrals at each time step were evaluated
using ten-point Gaussian quadrature[21] and the Crank-Nicolson method[22].
The time step size was controlled by the method used by Bailey for
describing moving shock fronts[22], where the step size is doubled if all of
the temperatures in the previous time step change by less than 0.05%, but
the step size is halved if any of the temperatures change by more than
0.5%. The 15 X 15 matrix was solved by Gauss-Jordan elimination[21].
Details are documented elsewhere[23]. It should be noted that the one-
dimensional cooling problem is amenable to approximate analytic solution.
Since the polymer layer is much thinner than the glass substrate,
temperature profiles in the glass can be established by assuming an
effective heat transfer coefficient at the glass-polymer interface to
accommodate the transfer resistance contributed by the polymer layer. The
temperature history in the glass can, in turn, be used to calculate the
profiles in the polymer. This procedure was not implemented, since the
numerical code developed above to perform general two-dimensional
calculations was available for straightforward adaptation of one-
dimensional problems.

Results and Discussion

    Transient temperature profiles in the experiment for pulses 1, 2, 5,
and 6 of Table 1 were calculated. Figures 8 and 9 display the calculated
temperature profiles for pulses 1 and 2 in the glass and polymer at the
end of the heating period and at intermediate times as found from Eq. 4.
To ensure that the form of Eq. 5 for the interface temperature as a
function of time is reasonable, the temperature profiles in Figures 8 and
9 were integrated using Simpson's rule at 1/4 and 3/4 of the total pulse
time. The actual energy input by the pulse generator (which is proportional to time for a square pulse) and the energy uptake calculated from
the model-predicted temperature profiles in glass and polymer were found
to be in good agreement.
    The finite element cooling program used the final temperatures from
Figures 8 and 9 as initial conditions and resulted in the cooling curves
shown in Figures 10 and 11.
    Note from the curves that almost all of the temperature drop is in
the glass. The polymer temperature is nearly uniform during cooling,
except in the very early stages of cooling, where phase separation has not
started to occur. This is convenient for experimental analysis, because
the temperature at the center of the polymer can be taken as representative of the entire layer. It also eliminates the possibility that the
non-uniformity of vitrified domain sizes is a result of different thermal
histories across the thickness of the sample during cooling. Figure 12
shows the polymer center temperature versus the logarithm of cooling time
following the high energy and low energy pulses in the experiments.
    To demonstrate the effect of cooling rate at the phase separation
temperature on the vitrified light scattering pattern, the calculated
cooling rate ($^{\circ}$C/sec) is plotted versus center temperature in Figure 13.
The phase separation temperature for this mixture is 39°C (17°C above
ambient laboratory temperature), as determined by hot stage microscopy
using cross polarized light. Table II shows the cooling rate at 39°C for
each of the pulses. Cooling rates range from 200 to 4000 $^{\circ}$C/sec at this
temperature. Note from Figures 12 and 13 that the temperature gradient in
the glass at the end of the energy pulse has a strong influence on the
polymer cooling rate. If the temperature gradient in the glass is large,
then the polymer cools quickly.

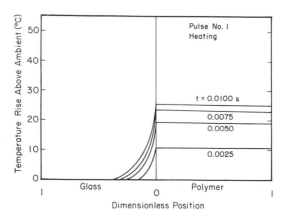

Figure 8. Temperature profiles at different times while heating the coated substrate with pulse 1 of Table I. Dimensionless position is plotted on a scale of 0 to 1 and is defined as x/l, where x is the distance from the ITO and l is the layer thickness (0.05 cm for glass, 0.0004 cm for polymer). The ITO layer was treated as uniform temperature and is represented by the centerline in the plot. The temperature indicated is degrees above ambient temperature. Ambient temperature was 22°C in the experiments.

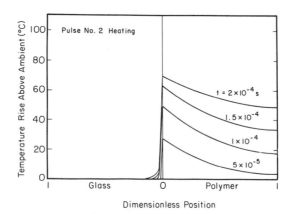

Figure 9. Temperature profiles at different times while heating the coated substrate with pulse 2 of Table 1. This high power/short time pulse gives larger temperature gradients in the glass than the low power/long time pulse for the same system depicted in Figure 8.

Our calculation results indicate that, in each of the experiments analyzed, the pulse which gave the larger scattering pattern had a faster cooling rate at the phase separation temperature. This agrees with literature reports that smaller domain sizes result in larger scattering patterns[13,15,16] and with our hypothesis that faster cooling rates give smaller domain sizes. Later in this paper we will demonstrate how the same cooling rates and even faster ones can be achieved in a multilayer disc with a pulsed, Gaussian-intensity-distribution laser.

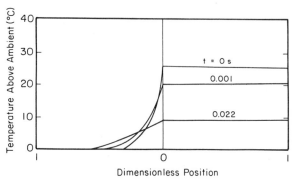

Figure 10. Transient temperature profiles versus cooling time after the termination of energy pulse 1. Free convection occurs at the sample/air interfaces in the cooling analysis.

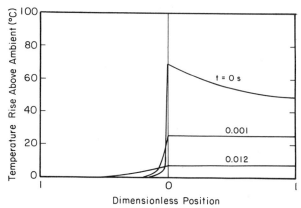

Figure 11. Transient temperature profiles versus cooling time after the termination of energy pulse 2.

Conclusion

Mixtures of 50% high molecular weight poly (methyl methacrylate) with 50% EBBA which are cooled from a single-phase state into a phase separated state have different phase-separated domain sizes, depending on the cooling rate. Mixtures which have been cooled faster have smaller domain sizes. The mixtures studied herein exhibited a range of vitrified domain sizes, indicating that phase separation was by the nucleation and growth mechanism or was by spinodal decomposition followed by phase ripening.

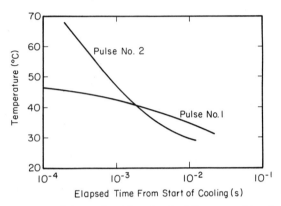

Figure 12. Temperature histories in the half-way point in the active layer following pulses 1 and 2. Pulse 2 (the short, intense pulse) leads to more rapid layer cooling after the cessation of the pulse.

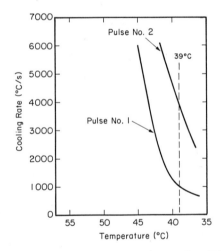

Figure 13. Cooling rate versus temperature at the film center for pulses 1 and 2. At 39°C, the phase separation temperature, pulse 1 leads to cooling at 4000°C/sec, while pulse 2 leads to cooling at 1000°/sec.

Phase separation and the resulting film morphology can be examined with an experimental set-up in which the two component mixture is spin coated onto a layer of clear, indium-tin oxide which has been applied to glass substrate. Light scattering patterns can be observed and measured while short voltage pulses are applied to the resistor. Numerical modeling of heat transfer in the active layer during and after experimental voltage pulses allows an understanding of how heat transfer rates affect the light scattering patterns which result from film morphology changes. During cooling, the temperature of a 4 μm active layer on 50 μm glass is nearly uniform in the active layer; the temperature drop is nearly entirely in the glass.

Samples with low molecular weight polymer (M_v=25,000) or high percentage of rigid rod molecule (66%) did not show a cooling rate effect on vitrified domain sizes. This could be because of increased diffusivities resulting in continued phase separation at low temperature.

Table I. Energy Pulse Characteristics
in Successful Scattering Pattern
Switches. Coatings were 50% PMMA
(M_v=620,000)/50% EBBA by Volume.

Sample	Thickness (micron)	Substrate Resistance (ohm)	Pulse	Volts	Time (sec)	Energy (joules)	Scattering Angle (max. °)
A	4	400	1	20	.01	.01	6.1
			2*	100	.0002	.005	6.9
B	6	150	3	15	.01	.015	4.3
			4	50	.0006	.01	6.1
C	8	150	5	17	.01	.02	4.1
			6	50	.001	.017	6.2

* A Cober 606P pulse generator was used for this pulse.

Table II. Cooling Rates Following Different Energy
Pulses at the Phase Transition Temperature
(39°C). Pulse Conditions Same as in Table I.

Sample	Pulse	Cooling Rate (°C/sec)	Scattering Angle (max.°)
A	1	1000	6.1
A	2	4000	6.9
C	5	200	4.1
C	6	500	6.2

PART II. HEAT TRANSFER MODELING AND PHASE EQUILIBRIA CALCULATION

Introduction

In Part II, heat transfer calculations are performed for an actual optical disc, to identify laser powers and focus radii which control the phase separation mechanism. Laser pulses are designed to give varying cooling rates such as those experienced by the polymer film deposited on the resistive glass substrate used in Part I of this study.

In addition to the heat transfer simulation, a thermodynamic model useful for chemical screening in the preparation of the optical discs is developed. The process of spin coating involves starting with a single-phase ternary mixture of solvent/liquid crystal/polymer and then evaporating solvent to leave a two-phase, binary mixture. To be feasible for optical storage, the resultant two-phase, binary mixture must become homogenized after laser heating. An understanding of ternary (and binary as a special case) phase equilibria is, therefore, important in selecting materials. A computer algorithm is presented to predict the phase diagram of solvent/liquid crystal/polymer mixtures. The thermodynamic calculations are based on the Flory lattice model, which is modified with terms to account for enthalpic effects. This computer algorithm was used in combination with cloud point experiments to study the effect of temperature on liquid crystal molecule-polymer interaction.

Literature Review

Heat Transfer Models. All of the promising write-once and reversible technologies discussed in the introduction to Part I involve laser marking by thermal effects. Therefore, an understanding of heat transfer in a multilayer optical disk during and after heating by a laser is of fundamental importance in the development of optical storage media.

Analytical solutions to the transient heat transfer problem in an optical disc have been published by Takamoto and Nakayama[24], Hill and Soong[25], Kivits et. al.[26], Chung[27], and Novotny and Alexandru[28]. These analytical solutions invoke a number of simplifications such as no disc motion, negligible radial heat transfer, and uniform light absorption or absorption only at the surface. Despite their inherent limitations, these models give estimates of the maximum disc temperature rise in limiting cases.

Numerical solutions are needed to reveal fully the physics of heat transfer in the optimal design of an optical disc. The topic was treated by Mansuripur, et.al.[29] They used the alternating direction, implicit finite difference method, and solved for a stationary multilayer disc with an impinging Gaussian laser intensity distribution. The energy absorption profile was modeled by a solution of the Maxwell equations for electric and magnetic fields. Transient results were plotted for a five layer disc of glass:manganese-bismuth alloy:glass:aluminum:polymer. Suh and Anderson[30] used the finite difference method to find transient temperature profiles in single layer films heated by Beer-Lambert absorption. Also included in their calculations was the latent heat of phase transitions. Workers at IBM have developed a two dimensional finite element model which accounts for latent heat effects[31].

Because numerical results apply only to the specific system under study, in-house heat transfer algorithms which are tailored to individual combinations of materials and parameters are often not general[32,33].

Extensive details of the computer programs and their results are often sparsely documented in the literature. We, therefore, attempted to develop a general two-dimensional finite-element model capable of handling multiple layers.

Phase equilibria. The first model for phase equilibria of rigid rod molecules in solvent was proposed by Flory in 1956[34]. Flory found that "phase separation arises as a consequence of particle asymmetry, unassisted by an energy term". He derived a partition function for rigid rod molecules with partial orientation about an axis, the partial orientation being described by a "disorientation parameter" (y) which depends on composition and rod aspect ratio. When the free energy is minimized for the whole system, a discontinuity was found in the free energy versus composition curve. This discontinuity is indicative of phase separation. The partition function was derived from a lattice model, in which a rigid rod molecule with axis ratio x, inclined at an angle ψ from the orientation axis, is divided into $y = x \sin \psi$ submolecules. Flory proposed using an additional term in the free energy of mixing expression to account for enthalpic effects:

$$\Delta H_{mxg}/kT = \chi_{12} x n_2 \upsilon_1 \qquad (6)$$

where ΔH_{mxg} = Enthalpy of mixing
χ_{12} = Solvent-rigid rod interaction parameter
x = Aspect ratio of the rigid rod
n_2 = Number of rigid rod molecules
υ_1 = Volume fraction of solvent
k = Boltzmann constant

The interaction parameter (χ_{12}) was considered to represent rod/solvent repulsions when given a positive value.

Flory extended the lattice theory in 1978 to include ternary mixtures of rigid rods, solvent, and randomly coiled polymer[9]. Enthalpic effects were neglected. An iterative procedure was detailed to find the compositions leading to separation of an isotropic phase from the anisotropic phase. Dowell studied semiflexible n-alkane molecules mixed with liquid crystal molecules composed of rigid cores and semiflexible tails[35]. A mean field, cubic lattice model was used with steric repulsions to derive the partition function. Mixing enthalpy was included in the calculation of nematic-isotropic transition curves. Predicted curves fit data qualitatively, but not quantitatively. The author ascribed the discrepancies to basic shortcomings of the model, such as the restrictions of a lattice and the approximate nature of the mean field approach.

The physical significance of the interaction parameter (χ), which Flory used in modeling rigid rod solutions and also polymer solutions, has been the subject of a number of papers[5,36,37]. It is clear that, in order for the lattice model to predict polymer solubility, χ must be more than a temperature independent parameter accounting for the enthalpy of mixing. Several forms have been proposed for χ[23]. A relatively simple semi-empirical form will be adopted in this work:[36,37]:

$$\chi = \alpha/T - \beta \qquad (7)$$

where the first term on the right-hand-side accounts for the enthalpic effects while the second accounts for entropic effects not accounted for in the lattice-derived partition function.

Model Development and Simulation Results

Heat Transfer. The objective of heat transfer modeling for an optical disc
was to find laser conditions (power, beam size, and pulse time) which can
be used to generate widely differing cooling rates in a thin layer of
polymer-liquid crystal mixture so as to repeat the successful results of
the experiment in Part I. Because a 4 μm coating worked successfully in
the experimental section, this thickness was retained in the model. To
allow the photodetector to be rigidly attached to the light source at a
fixed angle, it is proposed that the system will operate in a reflective
configuration, with a 500 Å or thicker aluminum layer serving to reflect
the beam. The aluminum layer would lie on a thick polymeric substrate
such as polycarbonate for added dimensional stability and protection.
Because the substrate is shielded from the laser by the aluminum, which
acts as a heat sink due to its high thermal diffusity (α = 0.85
cm^2/sec), the characteristics of the substrate are irrelevant to the heat
transfer model. The pertinent disc layers for the proposed storage disc
are illustrated in Figure 14.
   The top of the active layer was considered in the model to be
exposed to air, with free convection at the active layer/air interface.
This condition would exist in an "air sandwich" structure. The protected
active layer is thus insulated from direct contact with the turbulent
external air surrounding the spinning disc. To determine whether free
convective heat transfer is important versus conductive heat transfer, the
Biot number is appropriate[8]:

$$N_{Bi} = \frac{hD}{k} \tag{8}$$

where h = Upper range of typical solid-gas heat transfer
        coefficients (published in Bird, et.al.[8])
        (2.3 X 10^{-3} J/sec/$^{\circ}$C/cm^2)
      D = Characteristic dimension for heat transfer (the thickness
   of the active layer) 4 X 10^{-4} cm
      k = Thermal conductivity of polymer
        (8.6 X 10^{-4} J/sec/$^{\circ}$C/cm)

The Biot number is very small, 1 X 10^{-3}, indicating that convection is
negligible.
   An organic dye can be incorporated into the active layer to promote
absorption of the laser. Many dyes are available which absorb in the
wavelength range of semiconductor lasers and are soluble in polymers to
high concentrations[38]. A common assumption for absorption of light in a
weakly absorbing material is the Beer-Lambert law. For a laser of
Gaussian intensity distribution, this is[30]:

$$Q = \frac{P\kappa}{\pi r_o^2} \; \exp(-\kappa z) \; \exp-(r/r_o)^2 \tag{9}$$

where Q = Local heat absorption rate (watts/cm^3)
       P = Laser power (watts)
       κ = Optical extinction coefficient (cm^{-1})
       r = Radial distance from the laser spot center (cm)
       r_o= Radial position at which the laser power is less than
          the maximum by a factor of 1/e (cm)
       z = Distance from the top of the disc (cm)

Polymer/Liquid Crystal Molecule Blend with Dye (4 micron)

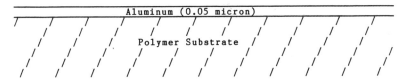

Figure 14.   Sketch of the multi-layered optical disc configuration used in
the heat transfer model.   Not drawn to scale.

In the subject system, an unpolarized laser passes through the
absorbing layer and is then reflected at the aluminum interface.   A
significant standing wave due to interference between the incident and
reflected beam could be created.   However, we will assume that the dyed
active layer is strongly absorbing and any standing wave effects are weak.
If the time-average electric fields of the incident and reflected beams
are additive at any given position in the active layer, the square root of
the intensity from the incident beam can be added to that of the reflected
beam to give the square root of the intensity of the superimposed beams,
since light intensity is proportional the the square of the electric
field.   For this study, an absorbance of $\kappa = 1000$ cm^{-1} resulted in
sufficient heating of the polymer while giving good uniformity of heating
in the 4 μm film; the heating rate varied by 4% across the film thickness.
The dye level needed for this absorbance is well within the solubility of
many dyes in common polymers[39].
    To help determine if radial heat flow predominates over axial heat
flow and vice versa, one can estimate a characteristic time for heat flow
in each direction by the square of a characteristic length for heat
transfer (l_c) divided by the thermal diffusivity (α).
    For a small diameter laser pulse in which the 1/e point (r_o) of the
intensity distribution of the laser lies at 0.35 μm, the characteristic
time in the radial direction is

$$t_c = r_o^2/\alpha = \frac{0.0000350^2 \text{ cm}^2}{9 \times 10^4 \text{ cm}^2/\text{sec}} = 1.4 \times 10^{-6} \text{ sec} \qquad (10)$$

In the axial direction, the characteristic length is the film
thickness, 4 μm, yielding a characteristic time of 1.6×10^{-4} second, two
orders of magnitude longer than that for radial flow.   Thus, for a 0.35 μm
radius laser pulse in a 4 μm film, radial heat transfer predominates over
axial heat transfer.   Because axial heat flow and convection are
negligible for a 0.35-1.0 μm radius laser, a one-dimensional heat transfer
model for radial temperature flow was used to estimate cooling rates for
small radius pulses.   The program is valid for the top of the structure

and assumes that the rate of axial heat flow is small relative to the rate of radial heat flow, as indicated by dimensional analysis.

The radial heat transfer equation which is solved in dimensionless form is:

$$\frac{\partial T}{\partial t} = \alpha \left[\frac{\partial^2 T}{\partial r^2} + \frac{1}{r} \frac{\partial T}{\partial r} \right] + \frac{Q}{\rho C_p} \tag{11}$$

where
T = Temperature (°C)
t = Time (sec)
α = Polymer thermal diffusivity (cm^2/sec)
Q = Energy absorption rate (J/cm^3/sec)
ρ = Mixture density (g/cm^3)
C_p = Mixture heat capacity (J/g/ °C)

The boundary conditions are

$$\frac{\partial T}{\partial r} (r=0) = 0 \text{ (Axi-symmetry)}$$

$$\frac{\partial T}{\partial r} (r=\infty) = 0$$

Laser pulses in the model were ramped linearly to the full power over the first 20 ns and ramped linearly down to zero power between 80 and 100 ns to simulate the switching on and off of a pulsed laser. (A 100 ns laser pulse is commonly used in optical storage.) Since writing is done on a spinning disc, longer pulses would result in a severely elongated bit and reduced storage densities.

Table III summarizes results from runs with the radial heat transfer program. Conditions were selected so as to obtain a comparison of different focus radii of a single laser and different laser powers at the same beam radius. The most important result is the cooling rate as the film passes 39°C. Temperature profiles show that this cooling rate is uniform to within 10% at all points within 0.35 μm radial distance of the laser center for even the most narrowly focussed laser. For example, Fig. 15 shows the temperature as a function of time at different radial distances from the center of the beam in run 1. The cooling rate at the center was used in preparing Table III.

Table III indicates that the major determinant of cooling rate at 39°C, and thus final film morphology, is the laser power, and not the focus radius of the laser. The explanation of this behavior in radial heat flow is that even though the laser is initially focused to a small diameter, by the time the center of the spot returns to 39°C, the heat has spread outward in the radial direction to approximate the profile which exists when the larger-radius spot of the same power laser cools past 39°C. The smaller-focus laser pulse just takes longer to reach a similar temperature profile. Due to the necessity of checking data on-line immediately after writing, larger pulses are an advantage because they take less time to reach 39°C at their center for a given power laser. A concern revealed by Fig. 15 is that the area heated by this small, low-power laser is ten times that of the original pulse size. This would have an adverse effect on storage density in optical storage. The use of a thinner active layer, so that the aluminum reflective layer indeed plays a role in heat removal by axial heat transfer, should be explored.

The model predicts that cooling rates in the active layer can be varied by two orders of magnitude, from about 10^5 to 10^7 °C/sec. This is a much greater range than that found in Part I to result in detectable differences in vitrified domain size (200 to 4000 °C/sec). Because of the

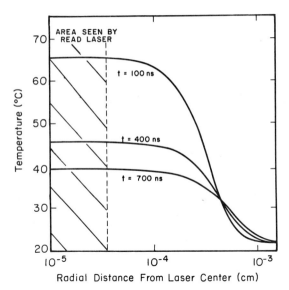

Figure 15. Calculated temperature profiles at the top surface of a polymeric optical disc after a 100 ns laser pulse. The laser pulse was 0.0004 mW and had a Gaussian distribution with an intensity at 0.35 micron of 1/e times the maximum intensity.

differences in magnitude of the cooling rates between the experimental configuration and disc configuration, further experimentation with a pulsed laser is needed in proving that the subject blends are feasible for optical storage. Our speculation is that the greatly increased cooling rates may very likely lead to much smaller domains, and thus readily detectable scattering signals.

Phase equilibria. Ternary mixtures of solvent, rigid rod molecule, and polymer form a single phase at high temperatures but separate into two phases below the critical temperature. The two-phase mixture consists of an anisotropic phase, in which polymer is almost entirely excluded, and an isotropic phase[9].

Flory used statistical thermodynamics to calculate the chemical potentials of the three components in each phase. The partition function was based on a lattice model in which each solvent molecule, each segment of a straight rigid rod molecule, and each segment of a flexible polymer has the same diameter, and occupies a lattice site. The chemical potentials for each component in both phases are given by:

Anisotropic Phase--Solvent

$$(\mu_1' - \mu_1^{\circ\prime})/RT = \ln v_1{}' + v_2{}'\ (y-1)/x_2 + c_3{}'\ (1-1/x_3) + 2/y \qquad (12)$$

Anisotropic Phase--Rigid Rod

$$(\mu_2' - \mu_2^{\circ\prime})/RT = \ln\ (v_2'/x_2) + v_2'(y-1) + v_3'x_2\ (1-1/x_3 + 2(1-\ln y) \qquad (13)$$

Anisotropic Phase--Polymer

$$(\mu_3' - \mu_3^{\circ\prime})/RT = \ln\ (v_3'/x_3) + v_2'(x_3/x_2)\ (y-1) +$$
$$v_3'(x_3-1) + 2x_3/y - \ln z_3 \qquad (14)$$

41

<u>Isotropic Phase--Solvent</u>

$$(\mu_1 - \mu_1°)/RT = \ln v_1 + v_2 (1-1/x_2) + v_3 (1-1/x_3) \tag{15}$$

<u>Isotropic Phase--Rigid Rod</u>

$$(\mu_2 - \mu_2°)/RT = \ln (v_2/x_2) + v_2(x_2-1) + v_3x_2 (1-1/x_3) - \ln x_2^2 \tag{16}$$

<u>Isotropic Phase--Polymer</u>

$$(\mu_3 - \mu_3°)/RT = \ln(v_3/x_3) + v_2x_3(1-1/x_2) + v_3(x_3-1) - \ln z^3 \tag{17}$$

where
v_1 = Segment volume fraction of solvent
v_2 = Segment volume fraction rigid rod
v_3 = Segment volume fraction of polymer
x_2 = Aspect ratio of liquid crystal molecule
x_3 = Degree of polymerization of polymer
y = Disorientation parameter = $x_2 \sin \psi_x$
z_3 = Polymer internal configuration partition function
ψ_x = Average rod inclination with respect to axis of domain in the anisotropic phase
Prime (') denotes the anisotropic phase, while no prime denotes the isotropic phase.

These equations consider only entropic combinatorials. Therefore, additional terms to the free energy are inserted based on the enthalpy of mixing[34].

$$\Delta H_{mxg} = kT (n_1 v_2 \chi_{12} + n_1 v_3 \chi_{13} + n_2 v_3 \chi_{23}) \tag{18}$$

where
ΔH_{mxg} = Heat of mixing

k = Boltzmann constant
n_i = Number of molecules of component i
v_i = Volume fraction of component i
χ_{ij} = Interaction parameter for components i and j

The relationship

$$v_i = \frac{x_1 n_1}{\left(n_1 + n_2 x_2 + n_3 x_3\right)} \tag{19}$$

is substituted into Eq. 16, giving the following enthalpic terms to be added to Eqs. 10-15:

<u>Each Phase--Solvent</u>

Component 1
$$\frac{\partial \Delta H_{mxg}}{\partial n_1 \, RT} = v_2^2 \chi_{12} + v_3 v_3 \chi_{12} + v_2 v_3 \chi_{13} + v_3^2 \chi_{13} - v_2 v_3 \chi_{23}/x_2 \tag{20}$$

<u>Each Phase--Rigid Rod</u>

component 2
$$\frac{\partial \Delta H_{mxg}}{\partial n_2 \, RT} = v_1^2 x_2 \chi_{12} + v_1 v_3 x_2 \chi_{12} - v_3^2 \chi_{23} - v_1 v_3 x_2 \chi_{13} \tag{21}$$

component 3
$$\frac{\partial \Delta H_{mxg}}{\partial n_3\, RT} = v_2{}^2 x_3 \chi_{13} + v_1 v_2 x_3 \chi_{13} + v_1 v_2 (x_3/x_2) \chi_{23} +$$

$$v_2{}^2 (x_3/x_2)\, \chi_{23} - v_1 v_2 x_3 \chi_{12} \qquad (22)$$

Three equations for the volume fractions of the components in each phase can be obtained from the above expressions by equating the chemical potential in the isotropic and the anisotropic phases for each of the three components. It is also necessary to know the disorientation index, y. The disorientation index will adopt the value which minimizes ΔG of mixing. By taking $\partial \Delta G_{mxg}/\partial y = 0$, it is determined that[9]

$$\exp(-2/y) = 1 - v_2{}' \,(1-y/x_2) \qquad (23)$$

By selecting a value for $v_2{}'$ in the anisotropic phase, unique values for v_2, v_3, and $v_3{}'$ can be found from the chemical potential equations if one uses the relations $v_1 = 1 - v_2 - v_3$ and $v_1{}' = 1 - v_2{}' - v_3{}'$. A computer program was written to solve these nonlinear equations by an iterative procedure. This program gives the resulting v_2, v_3 and $v_3{}'$ at equilibrium for selected values of $v_2{}'$.

Figure 16 shows ternary phase diagrams calculated for different rigid rod aspect ratios with all interaction parameters equal to zero. In this case, our model reduces to the purely entropic Flory model for athermal solutions. Curves for $x = x_3 = 20$ match those published by Flory[7], as expected. Tie lines are drawn for $x_2 = x_3 = 20$. Figure 17 illustrates effects on the phase behavior when the polymer interaction parameters (χ_{12} and χ_{23}) change. Again, increased rigid rod repulsion again results in a greater tendency for phase transition.

Figure 18 shows experimental data on the phase transition temperatures of ternary mixtures of toluene, EBBA, and polystyrene. Boundaries are plotted of homogeneous and two-phase regions experimentally observed at different temperatures. The samples were tested in a small, closed container with a hot stage microscope. Phase separation was visually detected. The phase diagram qualitatively agrees with the modeling results (Fig. 17). Higher temperature gives a smaller two-phase region, indicating reduced interaction parameters. Matching the experimental data quantitatively with the model allows fine tuning of interaction parameters and the assumed rigid rod aspect ratio.

The choice of x_2, the aspect ratio of the rigid rod, requires careful examination. The Flory lattice model assumes that the solvent, each segment of the randomly coiled polymer, and each segment of the rigid rod are alike geometrically and do not change upon mixing. Figure 19 shows the molecular structures of EBBA and polystyrene. With the typical published values[40] for bond radii and angles in Table IV, the length of the EBBA molecule is found to be 20 Å when the flexible ends are in the most extended conformation. Finding the molecular diameter is not as straightforward, since the widest part of the molecule is the aromatic rings, which are flat. One approach to calculating a usable rod diameter is to approximate the cross section of the aromatic rings as an ellipse of dimensions 4.7Å by 1.4Å and to use a rod diameter equal to that of a circle with equivalent area. This technique yields a rod diamter of 2.6Å and an aspect ratio of 7.7. It could also be argued that the flexible ends of EBBA are not strictly part of the rigid rod and could be neglected in calculating the apect ratio. This would indicate a molecular length of 15Å and an aspect ratio of about 5.8.

Table III. Predictions of the One-Dimensional
Model for a Laser-Heated Optical Disc.

Maximum Absorption (W/cm^3)	Beam Radius (μm)	Laser Power (W)	Maximum Temp. $(°C)$	Cooling Rate@39° $(°C/sec)$
110,000	0.35	4×10^{-7}	65	1.7×10^{7}
450,000	0.35	2×10^{-6}	210	4.2×10^{5}
55,000	1.0	2×10^{-6}	45	4.1×10^{5}
450,000	1.0	1×10^{-5}	210	1.5×10^{5}
900,000	1.0	3×10^{-5}	400	8.3×10^{4}

Table IV. Bond Radii and Angles Used in
Molecular Geometry Calculations.

Radius (angstrom)

Bond Type	H	C	N	O
Single	0.30	0.77	0.70	0.66
Double		0.67	0.60	0.56
Conjugated		0.70		

Angle Description	Value
C-C-C (Conjugated)	120°
C-C-C	109°
C-C-H	109°
C-O-C	112°
C-N=C	180°
C-C=N	120°
N=C-H	120°

Another difficulty in applying the lattice model is the size of a polystyrene segment versus that of an EBBA segment. A polystyrene segment is larger in diameter than an EBBA segment because the aromatic ring is pendant on the backbone in polystyrene, rather than in the backbone, as in EBBA. The segments are not similar in geometry, as assumed by the lattice model, because the segment diameter in polystyrene includes both the backbone and the aromatic ring, rather than just the aromatic ring. A rigid rod aspect ratio of 7.7 was assumed here, nevertheless, except where otherwise noted.

For the optical storage system under study, it is desirable to have a ternary system which starts in the one-phase region, but moves to the two-phase region after evaporation of the solvent at room temperature.

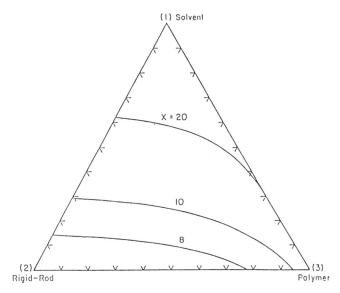

Figure 16. Ternary phase diagrams calculated with the modified lattice model for different rigid rod aspect ratios with all interaction parameters set equal to zero. The degree of polymerization was equal to the rigid rod aspect ratio. Results agree with those published by Flory[9].

Furthermore, the system should change from the two-phase to the one-phase region when heated above storage temperature. From the intersection of the binodal curves with the base line in Fig. 18, it is evident that a binary mixture of polymer and EBBA with 40-65% EBBA will change its morphology between 27°C and 50°C and will be one-phase before evaporation of toluene. Compositions in this range were thus investigated in Part I of the study.

Comparison of the series of cloud-point experiments and model predictions was made to determine the effect of temperature on the interaction between EBBA and polystyrene. First, the computer program was used to determine the critical composition for a binary polymer-liquid

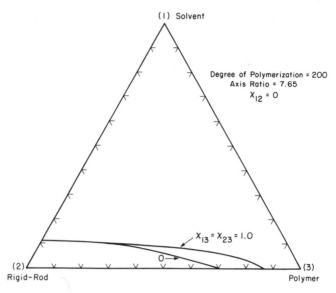

Figure 17. Effect of polymer-solvent and polymer-rigid rod interaction parameters on phase equilibria predicted by the model. The model reduces to that of Flory for the case of interaction parameters of zero.

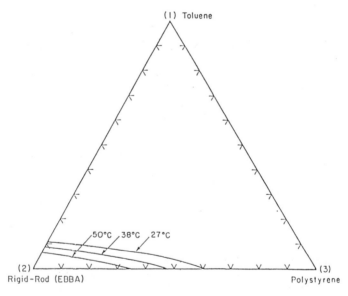

Figure 18. Ternary phase diagram for a mixture of toluene-EBBA-polystyrene at different temperatures as determined experimentally using a hot stage microscope. Model predictions agree with the experimental data if an inverse relationship of interaction parameters and temperature is assumed.

crystal molecule mixture using X_n = 7.7 and X_p = 200, and values of χ_{23} ranging from -2.4 to 0.8. The critical composition corresponds to the volume fraction of EBBA below which phase separation becomes impossible. Experiments were next performed to determine how v_2, the critical rod composition in the binary mixture, varies with phase separation temperature. EBBA and polystyrene were mixed at volume fractions of EBBA ranging from 0.35 to 0.90 (EBBA:polystyrene) and spin coated $10\mu m$ thick onto glass microscope slides using dichloroethylene as a solvent. The solvent was chosen due to its high volatility[41] in the spin coating process (boiling point = 37°C). Phase separation data were obtained using a Mettler FP82 hot stage at 1°C/min heating and cooling and a Nikon Type 104 microscope at 1500X magification. The phase separation temperatures were determined by observing the phase separated mixtures becoming a

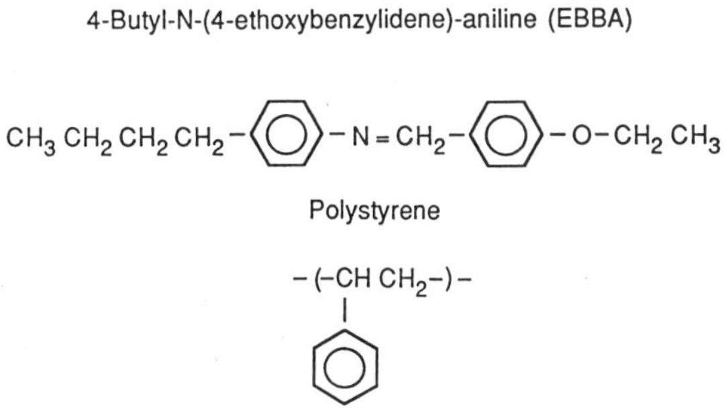

4-Butyl-N-(4-ethoxybenzylidene)-aniline (EBBA)

$$CH_3\,CH_2\,CH_2\,CH_2 - \langle\bigcirc\rangle - N = CH_2 - \langle\bigcirc\rangle - O - CH_2\,CH_3$$

Polystyrene

$$-(-CH\,CH_2-)-$$
$$|$$
$$\bigcirc$$

Figure 19. Molecular structure of EBBA and polystyrene.

single, clear phase on heating and the clear samples becoming phase separated on cooling. As noted by other researchers who have found cloud point curves by this technique[5], these temperatures typically differed by about 2°C. The phase transition temperature was taken as the average of the two, as recommended in the literature. Phase transition temperatures ranged from 22°C to 64°C as shown in Figure 20.

The v_{2crit} versus χ_{23} curve and the T versus v_{2crit} curve were combined to eliminate v_{2crit} and to prepare a plot of χ_{23} versus T^{-1} in the range $0.35 < v_2 < 0.57$, $295 < T$ (°C) < 319. A linear regression of this plot yields an expression for χ_{23} which depends inversely on temperature in agreement with Eq. 7.

$$\chi_{23} = 6540\ T(°K)^{-1} - 22.4; \qquad (r=0.990) \qquad (24)$$

In the valid temperature range, the interaction parameter ranges from -1.9 to -0.5. Negative interaction parameters correspond to energetically attractive terms upon mixing of molecules. While this is permissible within the extended framework of the theory, it is not normally found in polymer solutions. The chemistry of EBBA and polystyrene does not indicate a likelihood for strong exothermic reaction.

The phase equilibrium computer program was again run with various values of χ_{23} but with a rigid rod aspect ratio of 7.1. The above comparison procedure was employed to find the dependence of χ_{23} on T. The resulting relationship is

$$\chi_{23} = 4450 \ T(^\circ K)^{-1} - 14.5; \qquad (r=0.991) \qquad (25)$$

Here, χ_{23} ranges from -0.6 to 0.6, which is more realistic. It seems that a totally positive interaction parameter range could be obtained with an even smaller assumed aspect ratio for the EBBA. However, the model breaks down for aspect ratios below 7.1.

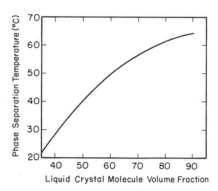

Figure 20. Phase transition temperatures for mixtures of EBBA and polystyrene ($X_p = 200$) determined by hot stage microscopy.

Conclusion

An organic mixture which is potentially useful for optical storage, when mixed with dye and subjected to laser pulses of selected focus radii and powers, can have cooling rates which vary by two orders of magnitude while avoiding excessive temperatures in the mixture.

The next step in commercial development of this phase separation system for optical storage is testing with an actual pulsed laser. These tests should look at writability and archival stability.

A Flory lattice model modified with interaction parameters with temperature dependence of the form given in equation 7 has merit and can be useful to predict whether certain mixtures of polymers and liquid crystal molecules are feasible for the phase change optical storage system. The model has the same limitations as other models which have been proposed for polymer-liquid crystal equilibria in that truly excellent prediction of equilibria requires a level of rigour which is exceedingly cumbersome.

ACKNOWLEDGEMENT

This work was supported by the National Science Foundation through CBT-8714420.

REFERENCES

1. G. Bouwhuis, J. Brant, A. Huijser, J. Pasman, G. van Rosmalen, and K. Schouhamer Immink, "Principles of Optical Disc Systems", Adam Hilger, 1985.
2. L. Alexandru, M.A. Hopper, R.O. Loutfy, J.H. Sharp, P.S. Vincett, G.E. Johnson, and K.Y. Law, "Materials for Microlithography", American Chemical Society, 435 (1984).
3. J. Drexler, J. Vac. Sci. Technol., 18, 87 (1981).
4. J.W. Cahn, Trans. Met. Soc. AIME, 242, 166 (1968).
5. O. Olabisi, L.M. Robeson, and M.T. Shaw, "Polymer-Polymer Miscibility", Academic Press, 1979.
6. E.D. Siggia, Phys. Rev. A, 20, 595 (1979).
7. I.M. Lifshitz and W. Slyozov, J. Phys. Chem. Solids, Lett. Sect., 19, 35 (1961).
8. R.B. Bird, W.E. Stewart, and E.N. Lightfoot, "Transport Phenomena", Wiley, 1960.
9. P.J. Flory, Macromolecules, 11, 1138 (1978).
10. P.J. Flory, "Principles of Polymer Chemistry", Cornell University Press, 1953.
11. A.H. Sporer, Applied Optics, 26, 1240 (1987).
12. J. Brandrup and E.H. Immergut, Eds., "Polymer Handbook", Wiley, 1974.
13. H.L. Snyder, P. Meakin, and S. Reich, Macromolecules, 16, 757 (1983).
14. T. Hashimoto, J. Kumaki, and H. Kawai, Macromolecules, 16, 641, (1983).
15. J. Gilmer, N. Goldstein, and R.S. Stein, J. Polym. Sci., 20, 2219 (1982).
16. T. Hashimoto, M. Shibayama, and H. Kawai, Kobunshi, 31, 887 (1982).
17. J.S. Huang, W.I. Goldburg, and A.W. Bjerkaas, Phys. Rev. Lett., 32, 921 (1974).
18. G.T. Feke and W. Prins, Macromolecules, 7, 527 (1974).
19. C.S. Carslaw and J.C. Jaeger, "Conduction of Heat in Solids", Clarendon Press, 104 (1959).
20. C.A.J. Fletcher, "Computational Galerkin Methods", Springer-Verlag, 1984.
21. B. Carnahan, H.A. Luther, and J.O. Wilkes, "Applied Numerical Methods", Wiley, 1969.
22. B.A. Finlayson, "Nonlinear Analysis in Chemical Engineering", McGraw-Hill, 1980.
23. W.D. McIntyre, M.S. Thesis, University of California at Berkeley, 1988.
24. K. Takamoto and S. Nakayama, Rev. Elect. Commun. Lab., 21, 647 (1973).
25. D.A. Hill and D.S. Soong, J. Appl. Phys., 61, 2132 (1987).
26. Kivits, R. deBont, and P. Zalm, Appl. Phys., 24, 273 (1981).
27. T. Chung, J. Appl. Phys., 60, 55 (1986).
28. V. Novotny and L. Alexandru, J. Appl. Phys., 50, 1215 (1979).
29. M. Mansuripur, G.A. Neville Connell, and J.W. Goodman, Appl. Opt., 21, 1106 (1982).
30. S.Y. Suh and D.L. Anderson, Appl. Opt., 23, 3965 (1984).
31. M. Chen, Personal Communication, April 20, 1988.
32. E.V. LaBudde, R.A. LaBudde, and C.M. Shevlin, Topical Meeting in Optical Data Storage (1983).
33. D.G. Howe and J.J. Wrobel, J. Vac. Sci. Tech., 18, 92 (1981).
34. P.J. Flory, Proc. R. Soc. London, Ser. A234, 60 (1956).
35. F. Dowell, J. Chem. Phys., 69, 4012 (1978).

36. R. Koningsveld and A.J. Staverman, J. Polymer Sci. Part C-1, $\underline{6}$, 1775, (1967).
37. G.T. Caneba and D.S. Soong, Macromolecules, $\underline{18}$, 2545, (1985).
38. D. Patterson, Macromolecules, $\underline{2}$, 672 (1969).
39. E.A. Guggenheim, Disc. Faraday Soc., $\underline{15}$, 24, (1953).
40. J.M. Pearson, CRC Critical Reviews in Solid State and Materials Sciences, $\underline{13}$, 1 (1986).
41. I.N. Levine, "Physical Chemistry," McGraw-Hill, 1978.
42. N. A. Lange, "Handbook of Chemistry," Handbook Publishers, 1956.

THE SIGNIFICANCE OF DIELECTRIC RELAXATION PROCESSES AND SWITCHING PROPERTIES FOR DATA STORAGE IN LIQUID CRYSTAL POLYMER MEDIA

Wolfgang Haase and Franz Josef Bormuth

Institut für Physikalische Chemie der
Technischen Hochschule Darmstadt
Petersenstraße 20, 6100 Darmstadt
West Germany

Starting from a brief introduction of basic properties of liquid crystalline polymers, we discuss some particular questions of molecular dynamics, which are related to a potential application of these materials as media for optical data storage. The principal relaxation processes as detected by dielectric spectroscopy are discussed in the light of storage properties. In a coarse model the time scales of microscopic motions and the macroscopic reorientations used for storage are compared. We come to the conclusion that the molecular reorientation time is the ultimate limit for macroscopic reorientation, which can be approached by addressing smaller areas of the probe.

1. INTRODUCTION

Liquid crystalline side chain polymers[1-3] are interesting media for data storage, because they possess some unique properties:

- the macroscopic orientation and, therefore, many anisotropic physical properties, especially in the nematic and in the fluid smectic phases, can be influenced by external forces, such as electric or magnetic field or by surface effects.

- the smectic phases of these polymers are more or less highly viscous, which implies the conservation of imprinted patterns over long periods of time without undergoing hydromechanical migration.

- furthermore, at lower temperatures there is a possibility of cooling the phases into the glassy state, where the liquid crystalline pattern is frozen in, which is of great interest for potential application.

Most effects suited for data storage make use of the interactions between an incident light beam, mostly from laser sources, and the liquid crystalline polymer matrix, which is also subjected to other external forces providing a specific preorientation. These interactions can be of different nature: the light wave can influence the liquid crystalline phase itself by temperature change, it may change the chemical structure of the compound by photoinduced reactions or, less rigorously, stimulate transitions between different conformers of the mesogenic molecules or dye molecules dissolved in the matrix. In all cases reorientations on macroscopic or submacroscopic level take place.

The aim of this paper is to describe the relationship between the dynamics on the microscopic level and some properties related to storage effects.

First, a brief summary of some principal properties of liquid crystalline phases and polymeric liquid crystalline compounds is given. Additionally some basic information on the effects which can be used for storage will be provided.

2. BASIC FEATURES OF LIQUID CRYSTALLINE POLYMERS

The crystalline state is characterized by a periodic three-dimensional long range order of the positions and orientations of individual particles, whereas the liquid state shows only incomplete short range order. The liquid crystalline state consists of different thermodynamically stable phases with variable degree of orientational and positional order[4]. The microdomains for some liquid crystalline phases (S_A, S_C, N) are shown in figure 1. The N- and S_A-phase are optically uniaxial; the S_C-phase is biaxial.

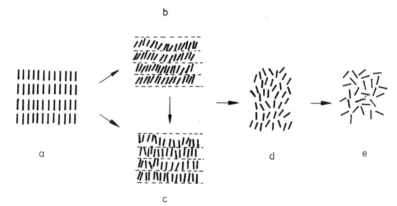

Figure 1. Microdomains in crystalline (a), smectic C (b), smectic A (c), nematic (d) and isotropic (e) phases.

The liquid crystalline state is characterized by marked anisotropies of electric, magnetic or optical properties combined with a relatively low viscosity (comparable to that of liquids, at least for nematics).

The polymeric liquid crystals form a relatively new area of materials since besides the liquid crystal properties, well known for low mass mesogenic molecules, the properties of polymers contribute to the typical mesogenic properties[1-3], [5-7]. Figure 2 shows a schematic picture of polymeric liquid crystalline phases with rod-like mesogenic units in an end-fixed spatial arrangement.

The side chain polymers show the lowest viscosity, if we assume the same polymerization degree for all of the three polymers, and therefore offer much potential as a class of functionalized polymers. The elasto-

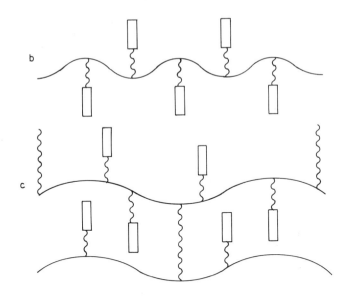

Figure 2. Principal structures of liquid crystalline polymers
a: mesogenic group b: side chain polymer c: elastomer
d: main chain polymer

$$CH_3-Si-(CH_2)_n-O-\bigcirc-R_1-\bigcirc-R_2$$
$$\underset{x}{\overset{O}{|}}$$

$$H-\underset{\underset{x}{\overset{|}{CH_2}}}{\overset{|}{C}}-\overset{O}{\overset{\parallel}{C}}\diagdown O-(CH_2)_n-O-\bigcirc-R_1-\bigcirc-R_2$$

Figure 3. Chemical composition of liquid crystalline side chain
polymers. Ia and Ib are copolymers.

compound		n	R_1	R_2	Mol.weight
Ia	92.5%	6	COO	OCH₃	38000
	7.5%	3	COO	Cl	
Ib	92.5%	6	COO	OCH₃	38000
	7.5%	3	COO	CN	
IIa		6	COO	CN	20000
IIb		6	COO	CN	3500
IIc		6	–	CN	21000
IId		6	–	CN	6000
IIe		4	–	CN	28000
IIf		6	COO	OCH₃	43000

mers, which are a relatively new variation, may be even more interesting for some special applications, e.g. in nonlinear optics[7]. Main chain polymers, on the other hand, are good candidates as structural polymers, e.g. high modulus fibres, due to their interesting mechanical properties.

In fig. 3 typical side chain liquid crystal polymers (as used in our investigations) are shown. The picture demonstrates the three elements of the molecule: the polymer backbone, often a polysiloxane or a polyacrylate unit, the flexible spacer, which decouples to a large extent the motion of the polymer main chain from that of the third element, the mesogenic unit. Specific properties of the liquid crystalline side chain polymers can be adjusted through variation in the elements, mentioned above. An important polymer-specific property is the glass temperature, which can be varied between room temperature and higher temperatures, e.g. 150 °C.

3. EXPERIMENTS RELATED TO DATA STORAGE IN LC SIDE CHAIN POLYMERS

The advantages of polymeric media, in general, for storage and recording are manifold. Some aspects are:

i) large variation of properties as a consequence of material design.

ii) reversibility is possible.

iii) integrity of the storage, due to easily accessible glassy state

iv) materials are generally easy to handle and are inexpensive

Most of these criteria are fulfilled sufficiently by liquid crystalline side chain polymers, perhaps iv) being the exception at this early stage of device development.

The preorientation of microdomains to macrodomains can be achieved by means of electric or magnetic fields or by surface effects. Different methods for the preparation of macrodomains in polymeric side chain liquid crystals have previously been described (see ref. 8). For device applications, homogeneous layers with the mesogenic groups parallel to the glass plates, or homeotropic layers with mesogenic units perpendicular to the layers have to be prepared using such methods, prior to the write-in or read-out procedure.

Three information storage mechanisms can be identified for LCPs:

- external electric or magnetic fields

- thermorecording

- photorecording

and various combinations of the above. This paper is principally concerned with E-field effects following heat pulses which allow switching between two states. Birefringence effects then generate optical contrast.

The effects of <u>external</u> <u>electric</u> <u>or</u> <u>magnetic</u> <u>field</u> have been described elsewhere[8,9,10]. The principle of the electrical two-frequency addressing is shown in figure 4; an electric field at frequencies below f_0, the frequency of dielectric isotropy, is applied to switch on the cell and a field above f_0 to switch it off.

Since switching off is driven by the field, the switch-off times are much shorter than for passive elements. At frequencies below f_0, the dielectric anisotropy $\Delta\varepsilon' = \varepsilon_{\parallel} - \varepsilon_{\perp}$, where $\parallel$ and $\perp$ refer to parallel or perpendicular to the external field direction, is positive; at frequencies above the frequency of isotropy f_0, $\Delta\varepsilon'$ becomes negative. It is clear that, for these experiments, the dielectric anisotropy in the field free state of the sample must be positive. The magnitude of $\Delta\varepsilon'$ depends on the strength of the δ-relaxation (this is the large step in figure 4) and on the level of $\varepsilon_{\perp}$, determined by the α-relaxation strength, (see section 5).

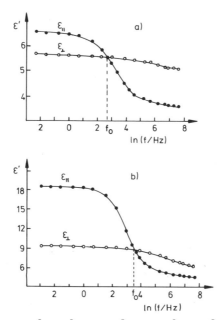

Figure 4. Frequency dependences of $\varepsilon_{\parallel}$ and $\varepsilon_{\perp}$ for two compounds
   a: methylsiloxane, Ib
   b: acrylate, IIa

The advantage of this method is that smaller switching times can be realized. The active area for one element is ≥ 1 mm². The time needed for reorientation is equivalent to the reorientation of a macrodomain.

The experimental set-up used by us was described recently[11,12]. Our data using this method will be presented in section 5 together with the intrinsic dielectric relaxation time data.

Recently some interesting results were obtained on low molecular weight liquid crystals, dispersed in polymers as droplets[13,14]. Under the influence of external fields data storage is, in principle, also possible in these materials (see also Fergason and Fang[15]).

In the method of <u>thermorecording</u>, the system is (locally) heated from the oriented glassy state with a high transmittance for light to a state with dense scattering texture, e.g. a higher temperature smectic state, the biphasic region, the nematic state or the isotropic state[16,17]. Heating is normally done by energy absorption from an external light beam in dye molecules, which may be dissolved or bound as side chains to the

LCP. This dye should have a high molar extinction coefficient at the wavelength of the laser light used. McArdle et al.[18] described for a 632.8 nm laser light a sensitivity of 12 nJ/μm^2 at 24°C for a specific liquid crystalline polysiloxane with an anthraquinone dye. Typical values for power density in the sample volume are 10^2-10^3 W/cm^2.

The method of <u>photorecording</u> is based on the local energy absorption, e.g. as heat and/or light energy, by a certain chemical group in the liquid crystal moiety or better in a dye dissolved in the liquid crystal polymer or copolymerized with it. The absorption induces,for example, a cis-trans-isomerism in the molecule. The cis-isomer can either disturb the liquid crystal order and so produce large optical effects, which is the case for the cis-trans isomerism of azobenzene[19], or it can show a different absorption coefficient (thermochromic effect, for example spiro-pyrane[20]). Photorecording on main chain LCPs has also recently been reported[21]. The advantage of photorecording is that it takes smaller power densities $\approx$ 0.1 W/cm^2, compared to thermorecording.

With respect to potential application of thermorecording or photorecording it is important to know what influence specific dyes exert on the physical properties of the mixtures. For example, the anthraquinone dye D27 dissolved in a polyacrylate with a phenylbenzoate as mesogenic group with 0.1 mol/l suppresses the clearing point by about 15 K[22]. The copolymers with dyes show analogous properties[23]. In another investigation, a concentration of two weight percent of a diazobenzene dissolved in the low molecular 4'-cyano-4-pentylbiphenyl lead to a shift of the frequency of dielectric anisotropy of 500 kHz to lower frequencies[24].

4. INTRINSIC ORIENTATION PROCESSES ON A MOLECULAR LEVEL

The macroscopic reorientation of a liquid crystalline domain is based on molecular motions on a microscopic level. Therefore it is essential in order to understand the macroscopic orientations, necessary for any kind of data storage, to study the molecular dynamics, regarding questions such as:

- Which molecular parts perform which kind of motion?

- At what temperatures or electric field frequencies are these specific motions activated?

- How do their properties depend on the chemical structure?

- What is the molecular mobility in the glass state (this is important to determine the stability of the stored information)?

There are different methods to investigate molecular dynamics, such as inelastic neutron scattering, NMR, ultrasound spectroscopy and others. A good review of these methods is given by Monnerie[25].

We and others[26] have concentrated on the use of dielectric spectroscopy , which is a very sensitive method to study motions of molecular groups containing electrically polar parts. In this case the measured physical quantity is the frequency dependent dielectric permittivity $\varepsilon^*(\omega)=\varepsilon'(\omega)-i\varepsilon''(\omega)$, with ω the angular frequency. $\varepsilon^*(\omega)$ relates the polarization P(ω) in a medium with an applied electric field E(ω):

$$P(\omega) = \varepsilon^*(\omega) \cdot E(\omega) \qquad (1)$$

In the low and radio frequency range this polarization is determined by reorientation of molecules and molecular parts along the direction of the changing field and also by an electronic contribution, which is constant up to optical frequencies.

The molecular reorientations, however, lag behind the changes of the field as their relaxation frequency is approached and cease to contribute to the dielectric permittivity well above the relaxation frequency. This behaviour shows up as a step in $\varepsilon'(\omega)$ and a maximum in $\varepsilon''(\omega)$ (figure 5).

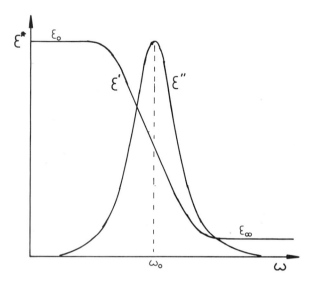

Figure 5. Principal frequency dependence of ε' and ε'' in the range of orientation polarization; ω_0 is the relaxation frequency.

The shape of the $\varepsilon(\omega)$-curve is often described by the Cole-Cole formula (eq. 2):

$$\varepsilon^*(\omega) = \varepsilon_\infty + (\varepsilon_0 - \varepsilon_\infty)/(1 + i\omega\tau)^{(1-\alpha)} \qquad (2)$$

ε_∞ is the value of ε at high frequencies, α the distribution parameter and τ the relaxation time. For liquid crystals, in general, the dielectric permittivity, which is in principle a tensor, can be decomposed into two components depending on whether the electric field is parallel to the director ($\varepsilon_\parallel(\omega)$) or perpendicular to it ($\varepsilon_\perp(\omega)$). For liquid crystals, in the rigid-rod approximation, there are in principle four molecular relaxation processes, two in each orientation[27]. A third term in both orientations was postulated by us[28] in order to take into account the main chain motions. The principal physical significance of these contributions will be discussed below.

In eq. 3, μ_l and μ_t mean the dipole moment parallel and perpendicular to the long axis of the mesogenic group and μ_h the main chain dipole moment. $A(\omega)$'s are normalized functions describing the frequency dependence, and S is the order parameter.

The original equations given by Nordio et al.[27] were supplemented by a third term, which describes the main chain motions in the liquid crystalline side chain polymers[28,29].

$$\varepsilon_{\parallel}(\omega) - \varepsilon_{\parallel}^{\infty} \sim \mu_l{}^2(1+2S) \cdot A_{oo}(\omega) + \mu_t{}^2(1-S) \cdot A_{o1}(\omega) + \mu_h{}^2 \cdot f_{\parallel}(S) \cdot A_h(\omega) \qquad (3a)$$

$$\varepsilon_{\perp}(\omega) - \varepsilon_{\perp}^{\infty} \sim \mu_l{}^2(1-S) \cdot A_{1o}(\omega) + \mu_t{}^2(1+\tfrac{1}{2}S) \cdot A_{11}(\omega) + \mu_h{}^2 \cdot f_{\perp}(S) \cdot A_h(\omega) \qquad (3b)$$

Typical dielectric spectra for a liquid crystalline side chain polymer (IIf) are shown in figure 6 for the two principal components.

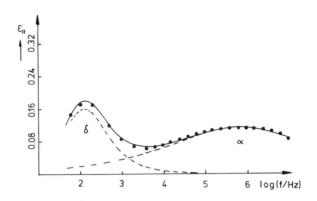

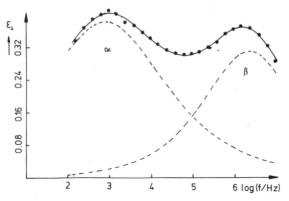

Figure 6. Dielectric permittivity ε'' in both main orientations for compound IIf; T=69.3 °C for $\parallel$ and T=47.8 °C for $\perp$.

In homeotropic orientation (E $\parallel$ n) with E the external field and n the director, we find the so-called δ-relaxation at lower frequencies. It has the smallest relaxation-time distribution (α small in eq. 2) of all the relaxations. At higher frequencies a broader relaxation regime is observed, which will be called α-process. It has a component in homogeneous orientation ($\alpha_{\parallel}$) at about the same frequency and with similar curve shape, but with a reduced relaxation strength $\varepsilon_0 - \varepsilon_{\infty}$. While these three relaxations are frozen in at the glass transition, the fourth relaxation called g_1 is still present in the glassy state. It is observed in homogeneous and, with smaller relaxation strength, in homeotropic orientation.

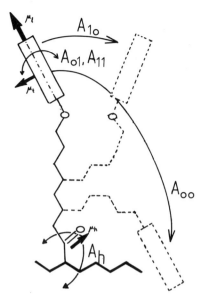

Figure 7. Molecular motions (schematic) associated with the terms of eq.
3a, 3b; thick line is the main chain, thin lines are the spacer
groups; large arrows represent the dipole components μ_l, μ_t and
μ_h.

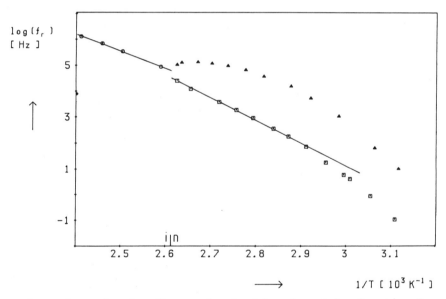

Figure 8. Activation diagram for δ– (□) and α– (▲) relaxation for
compound IIb.

In a series of measurements for oriented samples with different dipole
structure[28-30] we were able to identify the relaxations described above
with specific components of eq. 3a,b. They are schematically shown in
figure 7.

The δ–process is connected with a reorientation of the mesogenic side
chains at about approximately 180°. It is identical with the first term in
eq. 3a. This motion, which becomes dielectrically active through the

longitudinal dipole component, requires conformation changes or segmental motions in spacer and main chain.

Therefore it cannot, as in low molecular weight liquid crystals, be described as a single particle process, but the motions of neighbouring side chains are more or less correlated, leading for example to asymmetric relaxation curves and to a WLF-type[28] behaviour of the relaxation frequencies at lower temperatures (see figure 8).

The α-relaxation in homogeneous ($\alpha_\perp$) and homeotropic ($\alpha_\parallel$) orientations was assigned to superpositions of the other terms in eq. 3a, 3b, differently weighted in the two orientations and in different compounds. In some cases[28,29] a separation into a low frequency and a high-frequency part could be achieved. The high-frequency part is best interpreted by a superposition of the terms A_{01}, A_h for $\alpha_\parallel$ and A_{11}, A_h for $\alpha_\perp$, which means that it includes reorientation of the mesogenic groups around its long axis and segmental motions of the main chains. These motions are highly correlated and together form the dynamic glass transition. The low-frequency part $\alpha_\perp$ is still unclear and shall not be discussed here. The β-relaxation is probably due to an internal reorientation inside the mesogenic group.

5. MOLECULAR DYNAMICS AND STORAGE PROPERTIES

While the magnitude and sign of $\Delta\varepsilon = \varepsilon_\parallel - \varepsilon_\perp$ are preconditions for electric field switching, the switching process itself is always connected with a change of the optical axes of the cell, since reading out of information is done via the optical properties. For uniaxial systems this requires a reorientation of the director, and, in turn, is associated with reorientation of individual molecules around their short axis. This molecular motion is not exactly the δ-process, but it may be assumed, and is observed experimentally, that such reorientation is only possible at temperatures where the δ-process is already activated, that is for liquid crystalline polymers above the glass transition temperature.

The relation between switching and molecular relaxations will be clarified by the following.

The balance condition between electric and elastic forces is given by minimization of the free energy G (eq. 5).

$$G = \frac{1}{2} \int \{k \cdot (d\theta/dz)^2 - \varepsilon_0 \cdot \Delta\varepsilon \cdot E^2 \sin^2\theta)\} \, dz \qquad (5)$$

Here k is a combination of the appropriate elements of the elasticity tensor according to the arrangement: $k = k_{11}$ for Freedericksz and $k = k_{11} + \frac{1}{2}(k_{33} - k_{22})$ for TN-cell. θ is the angle between the local director and the normal to the cell walls, E is the external field strength. Eq. 5 is valid for surface-stabilized orientation, where the surface alignment is assumed to be constant (hard anchoring).

It can be assumed that in all practical cases the maximum tilt angle θ in the middle of the cell is close to 90 ° for a switched Freedericksz- or TN-cell.

We note here the important difference between a switching transition and a dielectric experiment. In case of switching, the forces that produce the orientation are overcome by the electric field term, at least in the bulk; in the case of dielectric measurements, the electric field term is

much smaller than the orienting forces. The situation is schematically
shown in figure 9. In the switching process, the whole system shifts to a
new equilibrium potential distribution; in the dielectric process, there
are two (or more) potential minima, whose depths are slightly influenced
by the electric field; only the molecules in the tail of the Boltzmann
distribution are able to overcome the energy barrier.

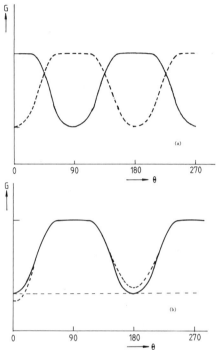

Figure 9. Potential distribution in a switching experiment for
samples with $\Delta\varepsilon>0$ in an AC-field (a) and a dielectric
experiment (b).
θ is the angle between the normal to the cell walls and
the director.
_____ : situation before applying the E-field
---------- : and after applying the field

If we presume that the reorientation during switching starts in the
middle of the cell (where the E-term dominates strongly) and proceeds by
growth of the already oriented domain on account of preoriented regions,
we have to consider carefully the movement of the transition walls between
switched and unswitched domains. In a simplified model, such a wall moves
by rotations of individual molecules about 90° (figure 10). The molecules
in the wall have indeed two potential minima and the time needed to
switch from one state to the other is approximately the δ-relaxation time
τ_δ. The velocity with which such a wall moves is then a consequence of
subsequent molecular reorientations. So it should be possible to derive a
general function:

$$\tau_d = f(\tau_\delta) \qquad\qquad (6)$$

between the active decay time τ_d, the most interesting switching time with
two-frequency addressing, and the δ-relaxation time τ_δ. Of course a re-
finement in the model is needed to achieve at least qualitative agreement
between theory and experiment.

Besides the active decay time, two other time constants play a role in switching experiments; they are defined in equations (7), where E_0 is the threshold field strength, U_0 the threshold voltage, k is the appropriate elastic constant, d the cell thickness, $\tau_d{}^0$ is the passive decay time (without field) and τ_r the active rise time.

$$E_0 = \Pi/d \cdot (k/(\varepsilon_0 \cdot \Delta\varepsilon))^{\frac{1}{2}} = U_0/d \qquad (7a)$$

$$1/\tau_d{}^0 = k/n \cdot (\Pi/d)^2 \qquad (7b)$$

$$1/\tau_r = 1/\tau_d{}^0 \cdot ((U/U_0)^2 - 1) \qquad (7c)$$

$$1/\tau_d = 1/\tau_d{}^0 \cdot ((U/U_0{}^\star)^2 - 1) \qquad (7d)$$

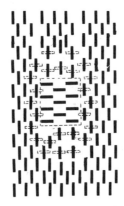

Figure 10 Schematic picture of the progression of a transition region between switched and unswitched domains in a partial volume of a homogeneously oriented macrodomain. The rectangular region in the middle includes molecules, which have already been reoriented by an external electric field. Molecules near to this region have two possible orientations, which are indicated by filled and open, dotted rectangles.

Following the interpretation for δ we want at first to relate the data obtained by macroscopic switching experiments using two-frequency addressing (see fig. 4) in a TN- or Freedericksz cell with the data obtained from dielectric experiments. This means we compare data on the macroscopic level with that of the microscopic level.

The values given in the following tables were calculated for the same cell thickness d=10μm and for the same U/U_0.

In Table I we quote data for the Freedericksz-transition. They are reduced to equal temperatures T_{red} with respect to the clearing point. Since active rise and decay times are strongly dependent on the $\Delta\varepsilon$-values at the frequency of the switching field, we have tried to make them comparable by multiplying with $\Delta\varepsilon^2$ (see eq. 7a) and also by dividing with the viscosity. Comparing these values and the passive decay times with τ_δ, we must admit that there is no general relation between them. The proportion of τ_δ for Ib compared with Ia is reflected in the $\tau_d{}^0$-values,

but only slightly in τ_δ and is even reversed in τ_r. For the pair IId/IIb
nearly equal τ_δ's are connected with significantly larger switching times
for IId. No correlation at all seems to exist on comparing methylsiloxane
and acrylate compounds.

Table I. Switching times in Freedericksz-cell.
τ_r(rise), τ_d(active decay) and $\tau_d°$(passive decay) at the same
U/U_o and same reduced temperature T_{red}=0.94 compared with the
relaxation times for δ-relaxation τ_δ and α-relaxation τ_α; η means
the viscosity.

compound	T[°C]	τ_r [s]	τ_d [s]	$\tau_d°$ [s]	$\tau_r(\Delta\varepsilon)^2/\eta$ $[1/P\cdot 10^{-3}]$	$\tau_d(\Delta\varepsilon)^2/\eta$	τ_δ [μs]	τ_α [μs]
Ib	85	0.5	0.2	10.4	3.2	6.8	247	<0.1
Ia	90	14.2	0.8	9.7	0.17	12.5	50	<0.1
IId	87	2.7	–	154	12	–	79	1.3
IIb	109	0.2	1.5	56	2.6	1.2	74	7.8

We explain this behavior by the assumption that the relevant temperature, to which the data should be reduced, is not T_{ni}, but the glass
temperature T_g. In Table II we have done this for the compounds of Table I
and some additional compounds. This procedure indeed leads to a better
correlation. It seems that proportion of τ_d and τ_δ is qualitatively similar for the siloxane compounds on one side and the acrylate compounds on
the other.

Table II. A comparison of the passive decay time in a TN-cell
with the δ-relaxation time at the same $T-T_g$, where T_g is
the glass temperature; MW is the molecular weight.

compound	T[°C]	$\tau_d°$ [s]	τ_δ [μs]	$\tau_d°/\tau_\delta\cdot 10^{-6}$	M_w
Ib	77	139	407	0.34	38000
Ia	77	44	202	0.22	38000
IId	100	30	16	1.9	6000
IIa	105	170	183	0.9	20000
IIc	77	≈320	–	–	21000
IIb	96	–	8	–	3500
IIe	90	–	250	–	28000

Two consequences stemming from our simple model outlined above shall
be discusssed with regard to Table II:

First, the limiting time for switching should be of the order of the
δ-relaxation time. This means, for example, that good candidates for fast
optical storage should have small relaxation times in the desired temperature range. If we take the glass temperature as a reference, acrylates are
better suited than methylsiloxanes (Table II). The δ-relaxation time is
shifted to smaller values when either the spacer length is increased (IIc
and IIe) or the degree of polymerization is reduced (IIb). These are some
criteria to select well-suited compounds for storage applications.

A second consequence is that one should expect much smaller reorientation times, when small areas are involved, which is the case in all
laser-addressed applications. This is qualitatively fulfilled when we
compare the passive decay times of Table II, which were measured with
areas of some mm², with those obtained by McArdle et al.[18] with areas of

10 μm². There is about a factor of 10⁶ between them. Unfortunately, to our knowledge, there are no detailed investigations on the problem of the size-dependence of switching times. We believe that more experimental and theoretical effort in this field would shed light on the microscopic mechanisms of switching and would also be a promising subject for application relevant questions.

6. CONCLUDING REMARKS

Side chain polymers are good candidates for information storage. The relevant property on the molecular level is the δ-relaxation. The relaxation times are interpreted as intrinsic times. The time scale that can be realized for macroscopic reorientations is limited by the δ-relaxation time. For the future also processes related to the faster α-relaxation may be taken into account. Moreover, it remains to be seen, if liquid crystalline polymers with ferroelectric S_c*phase offer further advantages in LCP storage media (cf. Nesrudlaev et al.[33] for non-polymeric LCs). In this case the storage effect is related to a reorientation of the dipoles around the long axis, in a way similar to the α-process.

ACKNOWLEDGEMENT

This work was supported by the Deutsche Forschungsgemeinschaft.

REFERENCES

1. V.P.Shibaev and N.A.Platé, Pure Appl. Chem.,57,1589, 1985.

2. H.Finkelmann, H.Ringsdorf, W.Siol and J.H.Wendorff, in "Mesomorphic Order in Polymers", A.Blumstein, editor, p.22, ACS Symposium series no.74, 1978.

3. C.B.McArdle, in "Side Chain Liquid Crystal Polymers", C.B.McArdle, editor, p.357, Blackie&Son Ltd., Glasgow 1988, chapter 13.

4. H.Kelker and R.Hatz,"Handbook of Liquid Crystals,"Verlag Chemie, Weinheim,1980.

5. "Polymeric Liquid Crystals", A.Blumstein, editor, Plenum Press, New York, 1985.

6. "Recent Advances in Liquid Crystalline Polymers", L.L.Chapoy, editor, Elsevier Publishing Company, New York, 1985.

7. H.Finkelmann, Angew. Chemie 99, 840, 1988.

8. W.Haase, in "Side Chain Liquid Crystal Polymers", C.B.McArdle, editor, Blackie&Son Ltd., chapter 11, pp. 309-329, Glasgow, 1989.

9. C.Noël, L.Monnerie, M.F.Achard, F.Hardouin and H.Gasparoux, Polymer 22, 578, 1988.

10. N.A.Platé, R.V.Talroze, V.P.Shibaev, Pure & Appl. Chem., 56, 403, 1984.

11. D.Pötzsch and W.Haase, Phys. Lett.,57A, 343, 1976.

12. H.Pranoto and W.Haase, Mol.Cryst.Liq.Cryst., $\underline{98}$, 99, 1983.

13. J.W.Doane, N.A.Vaz, G.G.Wu and S.Zumer, Appl.Phys.Lett., $\underline{48}$, 269, 1986.

14. W.D.McIntyre and D.S.Soane, Polymer Preprint, $\underline{29}$, 197, 1988.

15. J.Fergason, N.Fang, Japan Display '86, paper PD8, 1986.

16. V.P.Shibaev and N.A.Platé, Advances in Polymer Science, $\underline{60,61}$, Springer, Berlin 1980

17. H.J.Coles, Faraday Disc. Chem. Soc., $\underline{79}$, 201, 1985.

18. C.B.McArdle, M.G.Clark, C.M.Haws, M.C.K.Wiltshire, G.Nestor, G.W.Gray, D.Lacey and K.J.Toyne, Liq.Cryst., $\underline{2}$, 573, 1989.

19. M.Eich and H.J.Wendorff, Makromol Chem. Rapid Commun., $\underline{8}$, 467, 1987.

20. I.Cabrera, S.Yitzchaik and V.Krongauz, Polymer Preprints, $\underline{29}$, 200, 1988.

21. A.Griffen, C.Hoyle, J.Gross, K.Venkataran, D.Creed, C.B.McArdle, Makromol.Chem.Rapid Commun. $\underline{9}$, 463(1988).

22. U.Quotschalla and W.Haase, Mol. Cryst. Liq. Cryst., $\underline{157}$, 355,1988.

23. U.Quotschalla and W.Haase, Mol.Cryst. Liq. Cryst., $\underline{153}$, 83, 1987.

24. D.Bauman and W.Haase, Mol. Cryst Liq. Cryst., in print

25. L.Monnerie, Pure & Appl. Chem., $\underline{57}$, 1563, 1985.

26. C.Haws, M.Clark, G.Attard, in "Side Chain Liquid Polymers", C.B.McArdle, editor, chapter 7, p.196, Blackie&Son Ltd., Glasgow 1989, and references cited therein.

27. P.L.Nordio, G.Rigatti and U.Segré, Mol. Phys., $\underline{25}$, 129, 1973.

28. F.J.Bormuth and W.Haase, Mol. Cryst. Liq. Cryst., $\underline{153}$, 207, 1987.

29. F.J.Bormuth and W.Haase, Liquid Crystals, in print.

30. F.J.Bormuth, Ph.D. Thesis, Darmstadt, 1988

31. M.L.Williams, R.F.Landel, J.D.Ferry, J.Am.Chem.Soc., $\underline{77}$, 3701, 1955.

32. H.Pranoto, Ph.D. thesis, Darmstadt, 1984.

33. A.Nesrudlaev, A.Rabinovich, A.Sonin, Sov.Phys.Tech.Phys., 25, 1445, 1980.

PHOTO-INDUCED ISOMERIZATION --- NONLINEAR OPTICAL MATERIALS

IN NEURO-OPTICAL NETWORK

Eiichi Hanamura and Naoto Nagaosa

Department of Applied Physics
University of Tokyo
Hongo, Bunkyo-ku, Tokyo 113, Japan

From a simple model of photo-induced isomerization, we propose guiding principles to look for new nonlinear optical materials which could be used as elements of neuro-optical network. These materials have several important properties: the step-function response of transmitted light intensity against the input optical intensity, and optical learning and teaching abilities using these materials as optical neurons.Certain kinds of polydiacetylene crystals are also pointed out as good candidates for neuro-optical elements.

INTRODUCTION

A novel method for optical information processing is being sought very extensively,[1,2] which makes the best use of two advantageous characteristics of optical processing: (1) very rapid response, and (2) parallel processing. Optical neural networks are considered as one of the best candidates for optical information processing.

The simplest equations for the neural network are written in terms of the step-function $\theta(x)$ as[3]

$$x_i(t+1) = \theta \left[\sum_{j=1}^{n} T_{ij} x_j(t) + e_i - X_i \right] \quad . \tag{1}$$

Here we consider a network composed of n neurons, $x_i(t)$ describes the state of i-th neuron and takes a value 1 for the excited (firing) state and 0 otherwise, e_i and X_i are an external input field power and a threshold power for firing on the i-th neuron, respectively, and T_{ij} the coupling strength from the j-th output to the i-th neuron. The coupling matrix $\{T_{ij}\}$ and the threshold $\{X_i\}$ are given externally depending upon the problem which we want to solve. Furthermore $\{T_{ij}\}$ and $\{X_i\}$ are modified so as to represent effects of learning and teaching. A vector $\{x_i\}$, after several steps, gives us a solution to our problem.[4]

In this paper, we propose a neuro-optical network which is made up of nonlinear optical materials suffering from photo-induced isomerization. Several kinds of polydiacetylene (PDA) crystals seem to satisfy the requirements of elements in all-optical neural network.

NONLINEAR OPTICAL RESPONSE

A good crystal of bis-(2,2',5,5'-tetraxistrifluoromethylphenyl) tutadiyne (DFMP), $(C_4R_2)_n$ with R =

can be made.[5] One-photon allowed A-exciton has an excitation energy 2.23 eV and one-photon forbidden but two-photon allowed X-exciton is located at 2.74 eV. Third-order optical susceptibility $\chi^{(3)}$ was theoretically shown to be as large as 10^{-8} cgs esu at $\hbar\omega = 1.36$ eV as the X-exciton is nearly excited by two-photon processes.[6,7]

Equations for the polarizations and the populations of the PDA-DFMP are solved, being coupled with Maxwell's equation for the radiation field. Here the width of the PDA-resonator is taken to be 10 μm and the reflection constant for both sides of the resonator is fixed at $R = 0.95$. A transmitted power I_t as a function of the incident power I_i

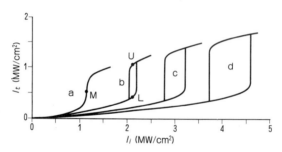

Figure 1. Optical nonlinear response (a) and optical bistable responses (b, c, d) obtained by numerical simulation using material constants of polydiacetylene-DFMP. The photon energy of the incident light is 1.361 eV and the cavity length 10 μm with the reflectivity $R = 0.95$ on both the surfaces. $\delta_0 - 2\pi m = -0.125$ (a), -0.15 (b), -0.175 (c) and -0.2 (d), where m is an integer closest to $\delta_0/2\pi$, and δ_0 is the phase difference in one round-trip in the weak limit of the incident light.

is plotted for the case around the incident photon energy $\hbar\omega = 1.361$ eV in Fig.1. This describes the step-function response and the optically bistable response depending very sensitively on the detuning between the incident frequency and the relevant cavity mode. This step-function behavior is the first characteristic required of an element of neural network. The high and low transmitting states describe the firing and non-firing states of the optical neuron, respectively. The critical input field M in Fig.1a corresponds to the threshold X_i of firing. It is noted here that these responses come from combined effects of non-linear optical response $\chi^{(3)}$, i.e., optical Kerr effect and the feedback at both surfaces of the transmitted light to the medium.

PHOTO-INDUCED ISOMERIZATION

The second requirement for a neural network element is the ability of learning and teaching, $i.e.$, self-organization. This means that the nonlinear optical materials are required to be modified in their physical properties depending on the number of applications of signal laser pulses, or the strength of externally applied laser light. These are the effects of learning and teaching, respectively. We will show here that these properties are satisfied by nonlinear optical materials such as some of polydiacetylene crystals which can show photo-induced structure change.

Some of polydiacetylenes have two kinds of isomers and a few of them will be switched optically between them.[8] Blue form (A-type) has an absorption peak at 1.9 eV so that blue light is transparent and the crystal looks blue. On the other hand, the absorption peak of red form (B-type) is located at 2.3 eV so that it looks red. The blue form of polydiacetylene-(12-8) is made to change into the red form isomer under irradiation of laser light with specified frequency, $\hbar\omega \geq 2.4$ eV. There are also preliminary experiments indicating the reversal change (from the red to the blue form) induced by irradiation of ultraviolet light.[9] Photo-polymerization yield increases as the third power of incident laser intensity in some PDA-(12-8) and more steeply in other PDA's. These photo-isomerizations can be understood from a unified point of view.[10]

We choose here the simplest model of the linear chain which is described by the following Hamiltonian:

$$H = \frac{1}{2}\sum_i Q_i^2 - \sum_{i \neq j} K_{ij}Q_iQ_j + \sum_{\ell \in E}(E_{\mathrm{FC}} - \sqrt{S}Q_\ell) \ . \tag{2}$$

Here Q_i describes the relevant displacement of the i-th molecule, K_{ij} the coupling constant between the displacements of the i-th and j-th molecules, E_{FC} the Franck-Condon excitation energy, $\sqrt{S}Q_\ell$ the Stokes shift in the excited electronic state at the ℓ-th molecule, and $\ell \in E$ means that the summation on ℓ is taken over the sites of the excited molecules. The first two terms of Equation (2) represent the lattice energy independent of the electronic ground and excited states. The photo-induced structure change can not be discussed by a conventional theory of phase transition because it is far from the thermal equilibrium. Therefore, we study relaxation mechanism of the cluster of m neighboring excited molecules. For this system, we rewrite the Hamiltonian (2) in terms of the Fourier transform $Q_\ell = N^{-1/2}\sum_k e^{ik\ell}Q_k$ and in such a way as the lattice coordinate $Q_k = Q_k^{(m)} + \Delta^{(m)}(k)$ fluctuates around the stationary displacement $\Delta^{(m)}(k) = \sqrt{S/N}\omega_k^{-2}[\sum_{\ell=1}^m \exp(ik\ell)]$:

$$H^{(m)} = \frac{1}{2}\sum_k \omega_k^2 Q_k^{(m)}Q_{-k}^{(m)} + mE_{\mathrm{FC}} - \frac{1}{2}\sum_k \omega_k^2\Delta^{(m)}(k)\Delta^{(m)}(-k) \ . \tag{3}$$

Here and hereafter we assume the nearest-neighbor coupling $K_{ij} = K\delta(j - i \pm 1)$ so that $\omega_k^2 = 1 - 4K\cos k$. The last term denotes stabilization energy due to the lattice distortion. The absolute value of this energy increases with the nearly second power of m as shown in Fig.2. Therefore under some conditions, this energy overcomes the Franck-Condon energy mE_{FC} and the cluster distorted through electronic excitation becomes more stabilized.

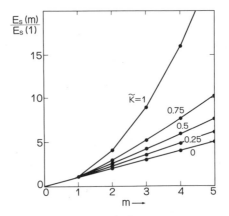

Figure 2. The distortion energy $E_s(m)$ of a cluster of m neighboring excited molecules is plotted as a function of m. $\tilde{K} = 4K$.

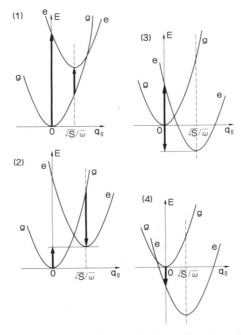

Figure 3. Energy dispersions of the 0-th molecule closest to the cluster of excited molecules: (1) The cluster of excited molecules is unstable, (2) metastable, (3) expands thermally, and (4) expands spontaneously without any additional optical pumping.

For a cluster of m neighboring molecules ($\ell = 1, 2, \cdots, m$) distorted through electronic excitation, the 0-th molecule is in the electronic ground state. Next we consider that the 0-th molecule is in the electronic excited state:

$$H_e^{(m+1)} = H^{(m)} + E_{\mathrm{FC}} - \sqrt{S}Q_0 \ .$$

Here we introduce the interaction mode q_0 at the 0-th molecule by

$$q_0 = \bar{\omega} \frac{1}{\sqrt{N}} \sum_k Q_k^{(m)} \ , \qquad \frac{1}{\bar{\omega}^2} = \frac{1}{N} \sum_k \frac{1}{\omega_k^2} \ .$$

Then the relative stability of the electronic ground and excited states at the 0-th site is classified into four types as shown in Fig.3. In the case of Fig.3(1), the minimum energy of the excited state is above the corresponding displaced ground state. Therefore, the excited state $|e\rangle$ at the 0-th molecule is unstable against radiative as well as non-radiative decay into the ground state $|g\rangle$. In the case of Fig.3(2), this vertical stability at $q_0 = \sqrt{S}/\bar{\omega}$ is reversed, but from the horizontal comparison as a function of q_0, the energy-minimum excited state is metastable while the energy-minimum lattice state of the electronic ground state is still absolutely stable. In the case of Fig.3(3), the horizontal stability of these states is exchanged. For a lattice temperature higher than the potential barrier in the ground state, the molecule at the 0-th site can jump thermally into the displaced excited state and the displaced cluster extends. Finally in the case of Fig.3(4), the ground state at the 0-th site is unstable and changes spontaneously into the excited displaced state. When we increase S/E_{FC}, K and/or m, the situation proceeds from (1) to (2), (3) and (4) in Fig.3. If we fix the size m of the displaced cluster, we have three borders as a function of $\tilde{S} \equiv S/E_{\mathrm{FC}}$ and $\tilde{K} = 4K$; e_m between (1) and (2), e_m^* between (2) and (3) and g_m between (3) and (4). These borders as a function of $\tilde{S}$ and $\tilde{K}$ are shown for several values of m in Fig.4.

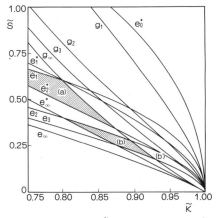

Figure 4. The phase diagram over $\tilde{S} \equiv S/E_{\mathrm{FC}}$ and $\tilde{K} = 4K$. e_m, e_m^* and g_m curves denote the stability borders of m-molecules cluster displaced through electronic excitation. e_m: the metastability, e_m^*: the absolute stability, and g_m: the spontaneous extension of the m-displaced molecules cluster.

Once we knew material constants $\tilde{S}$ and $\tilde{K}$, we can understand the photo-induced structure change from Fig.3 and Fig.4. On the other hand, we can estimate the values of $\tilde{S}$ and $\tilde{K}$ from the observed photo-induced structure change. We choose polydiacetylene crystals as an example. Many kinds of polydiacetylene crystals, *e.g.*, TCDU, ETCD, PDA-(10-8) and -(12-8), show the photo-induced isomerization from A-form to B-form. The relevant lattice mode is such displacements of carbon atoms as to change A-type into B-type, and the A and B phases correspond to the electronic ground state and the displaced electronic excited state, respectively, in the present model. First the yield of this A-to-B transition increases almost by the third power[7,8] of the incident light intensity. This means that only when 3 and larger than 3 neighboring molecules are optically excited, the displaced excited state becomes stable so that $\tilde{S}$ and $\tilde{K}$ are above e_3 curve and below e_2 curve. Second, the B-phase is absolutely stable as this phase occupies the crystal after annealing in ETCD and PDA-(12-8), and occupies the fine particles

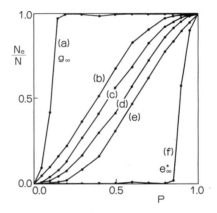

Figure 5. The final number N_e of the distorted molecules through electronic excitation divided by N as a function of the proba-
bility of the initial excitation (= proportional to the
optical intensity). $\tilde{K} = 0.7$, $\tilde{S} = 1.03$ (a), 0.80 (b), 0.66 (c),
0.55 (d), 0.48 (e) and 0.42 (f).

after laser irradiation at room temperature in PDA-(12-8). Therefore the values of $\tilde{S}$ and $\tilde{K}$ are above e^*_∞ curve. Third, the photo-isomeriza-
tion of PDA-(12-8) shows experimentally the gradual structure-change from the A- to the B-type at room temperature depending upon the incident laser power at 590 nm (= 3.7 eV).[8] These partially isomerized states exist as stable states at room temperature. We may estimate that the values of $\tilde{S}$ and $\tilde{K}$ are in the region between e^*_∞ and g_∞ curves. Therefore PDA-(12-8) corresponds to a state in the region (b) in Fig.4. Steeper structure-change as a function of the incident power was observed for other kinds of PDA's and depending upon the sample treatments. Therefore, we can choose PDA materials suitable to represent the learning and teaching effects.

ELEMENTS OF NEURO-OPTICAL CIRCUIT

When we choose a material with $\tilde{S}$ and $\tilde{K}$ in the region surrounded by e_{∞}^* and g_{∞} curves in Fig.4, we can change the crystal structure from the blue- to the red-form gradually depending upon the number of irradiated optical pulses and the pulse intensity. Crystals of polydiacetylene-(12-8) satisfy these conditions. We studied, by numerical methods, dynamics of photo-induced structure change. Figure 5 shows the degree of photo-induced structure change N_e/N as a function of the incident light intensity, and the number of applied optical pulses.[11] This structure change also brings about change in dielectric constants over wide frequency region.[12] When we use these nonlinear optical materials as the neural elements, the threshold power X_i for the step-function response and the transmittance at the exciton energies at 1.9 eV and 2.3 eV are changed depending upon the degree of the photo-isomerization. These properties can be used as the learning and teaching effects in the neuro-optical circuit.

The threshold power of firing the optical neuron depends very sensitively on the dielectric constant at the operating frequency around 1.36 eV.[6] The optical nonlinear and bistable responses are obtained as cross-points of the transmittance $I_t/I_i = 1/[1 + F\sin^2\delta]$ and the phase change of the one round-trip of the light $\delta = \delta_0 + KI_t$ as a function of incident power.[2] Here $F \equiv 4R/(1 - R)^2$ and $\delta_0 = 2\pi n_0 \ell/\lambda$. Therefore 2 % change in the linear refractive index n_0 brings about a phase change of $\Delta\delta_0$ by $\pi/2$. As a result, a small change in dielectric constants results in shifting the threshold in the step-function, depending on how often an optical neuron was fired and how strong laser light was applied to this optical neuron. The former brings about the learning effect and the latter the teaching effect.

It is noted that the optical bistable and step-function responses are most effectively realized by using the photon around 1.36 eV.[6]. Here we are making the best use of two-photon resonant excitation of the X-exciton. On the other hand, the photo-induced structure change is possible only by the laser light with the photon energy larger than 2.4 eV.

There arise a few strategies for neuro-optical circuits. First, we may use the visible laser light (> 2.4 eV) as the learning and teaching signal light and the laser light around 1.36 eV as an operating light signal. Second, when we use the operating light with photon energies 1.9 eV and 2.3 eV, the absorption coefficients at these photon energies change by a factor of 50,[8,13] so that transmitting optical intensity and attenuation can be modified considerably. In terms of this property, we can control and modify coupling constants among the optical neurons, depending upon the number of applications of the laser pulses or upon the intensity of the external laser pulses. Third, if we can use two-photon excitation of 1.36 eV for the photo-induced structure change, then a single laser beam around 1.36 eV may be used both as the operating light source and the learning as well as teaching laser light.

SUMMARY AND CONCLUSION

We have obtained the simple model to understand photo-induced structure changes from a unified point of view. Then the neuro-optical network has been proposed in which the photo-induced structure change and the associated change of refractive index are combined so as to bring about the learning and teaching effects, and the nonlinear optical

73

response, showing the step-function response of the transmitted light intensity against the change of the incident light intensity.

In conclusion, combination of a large nonlinear optical susceptibility and photo-induced structure change will make possible the neuro-optical network. Certain kinds of polydiacetylene crystals appear to be candidates. It is also interesting to study the nematic liquid crystals on an azobenzene monolayer[14] and photochromic spiropyran in Langmuir-Blodgett films[15] as the neuro-optical elements. The former shows the reversible change in alignment mode of the liquid crystals which is triggered by the photo-isomerization of the monolayered azobenzene under exposure to UV and visible light. This brings about the reversible change in the transmittance of the polarized light over the wide frequency region. The LB films of spiropyran (SP1822) and n-octadecane mixtures show very stable photochromism at room temperature. These also show the reversible structure changes under UV and visible light irradiation, and appear to be candidates for neuro-optical elements.

REFERENCES

1. For example, a special issue on "Optical Computing", IEEE Spectrum, August (1986) and a special issue on "Optical Computing", of Optics News 12, April (1986).
2. H.M. Gibbs, "Controlling Light with Light —— Optical Bistability," Academic Press, New York, 1985.
3. S. Amari, "Mathematics of Neural Networks," (in Japanese), Sangyo Tosho, Tokyo, 1978.
4. For example, M. Takeda and J.W. Goodman, Applied Optics 25, 3033 (1986), and references cited therein.
5. H. Nakanishi, H. Matsuda, S. Okada, and M. Kato, in "Proceedings of the Symposium on Nonlinear Optics of Organics and Semiconductors," held in Tokyo, 1988.
6. F. Arakawa, Master thesis submitted to Department of Applied Physics, Univ. of Tokyo, 1986.
7. H. Tanaka, M. Inoue and E. Hanamura, Solid State Commun. 63, 103 (1987).
8. Y. Tokura, T. Kanetake, K. Ishikawa and T. Koda, Synthetic Metals 18, 407 (1987).
9. A. Itsubo and T. Kojima, private communication.
10. E. Hanamura and N. Nagaosa, Solid State Commun. 62, 5 (1987); J. Phys. Soc. Jpn. 56, 2080 (1987).
11. T. Nagai, N. Nagaosa and E. Hanamura, J. Luminescence 38, 314 (1987).
12. H. Eckhardt, C.J. Eckhardt and Y.C. Yee, J. Chem. Phys. 70, 5498 (1979).
13. E. Hanamura, Y. Tokura, A. Takada and A. Itsubo, U.S. Patent 4,678,736 (July 7, 1987).
14. E. Ando, J. Miyazaki, K. Morimoto, H. Nakahara, and K. Fukuda, Thin Solid Films 133, 21 (1985).
15. K. Ichimura, Y. Suzuki, T. Seki, A. Hosoki, and K. Aoki, Langmuir, 4, 1214 (1988).

PART II.   PHYSICOCHEMICAL CONSIDERATIONS IN OPTICAL RECORDING

POLYMERIC DATA MEMORIES AND POLYMERIC SUBSTRATE

MATERIALS FOR INFORMATION STORAGE DEVICES

G. Kaempf, W. Siebourg, H. Loewer, and N. Lazear

BAYER AG, Department of Technical Development
and Applied Physics, 5090 Leverkusen, F.R.G.
MOBAY Corp., Plastics and Rubber Division
Pittsburgh, PA 15205-9741

Starting with a review of optical data memories,
this paper then goes into detail on the specific
uses of polymers as substrate materials and/or
active memory layers in optical data memories. The
polymer developments underway for these applications are compared and discussed on the basis of
selected examples. In this connection , reference
is made to calculation methods for determining the
birefringence of polymers and an explanation is
given of improvements in measurement and evaluation techniques in the determination of the
birefringence of polymers.

Of particular interest are recent developments
relating to the use of ferroelectric materials,
especially ferroelectric polymers, as active
memory material for high-density main and mass
memories. This is explained on the basis of following
examples:

- Ferroelectric materials as active memory layers
  in conjunction with conventional main memories
  based on silicon semiconductor technology
  ("non-volatile RAM").
- Ferroelectric materials as high-density, erasable, re-writable mass memories: writing and
  reading the information by means of an electron
  beam or by means of a modified scanning tunneling microscope.

1. INTRODUCTION

Technological developments in the 1990s will be determined by
progress in the field of microelectronics, particularly with
regard to information processing and data storage. Progress

in information storage is directly related to the ability to store and recall large amounts of data in the smallest possible space, preferably in erasable form. With respect to data storage units or memories, a distinction is made between

- main memories for moderate data volumes (currently - $\leq 4 \times 10^6$ bit per storage unit) and extremely short access time ($< 10^{-7}$ s); and

- mass memories for very large data volumes ($> 10^6$ bit per storage unit) and moderate access time ($> 10^{-4}$ s).

Today, main memories are exclusively silicon semiconductor memories of the MOS (metal oxide semiconductor) type or, to a growing extent, CMOS (complementary MOS) types. In standard semiconductor main memories, polymeric materials are only of secondary importance. In contrast, polymers are of great importance in the manufacture of main memories - not discussed here - based on semiconductor technology, e.g. as positive or negative resists in UV, X-ray and electron beam lithography.

The most important mass memories are magnetic storage units with packing densities of approx. 10^7 bit cm^{-2}, for instance magnetic tapes and magnetic disks (floppy or hard disk). Polymers are widely used for these data memories: films made of polyethylene terephthalate (PET) are used as the base material for magnetic tape memories, while PVC is used, as well as PET, as a substrate material for floppy disks. Polymeric binders based on PVC/PVA or polyurethane (PUR) are used for the top coating of magnetic tapes and disks, filled with finely dispersed magnetic storage materials (e.g. γ -Fe$_2$O$_3$, CrO$_2$, Co-doped Fe$_3$O$_4$).

Optical mass memories with packing densities up to approximately 10^8 bit cm^{-2} are currently being developed and launched. In these optical mass memories, polymers are of great importance as substrate materials and active memory layers (e.g. CD-ROM) or as substrate materials (WORM and erasable systems).

This paper reviews the state of the art and the future potentials of polymers on the one hand as data storage units for main memories and for mass memories, and, on the other hand, as substrate materials for optical mass memories.

2. FERROEELCTRIC POLYMERS AS ACTIVE STORAGE MATERIALS

Ferrimagnetic and ferromagnetic storage units (tapes, disks, drums) are used almost exclusively as the mass memories of current computer systems. Analogies between the behavior of ferromagnetic materials in a magnetic field and that of ferromagnetic materials in an electric field suggest that ferroelectric materials could conceivably be used for storage purposes. The relationship between the electric polarization P and the electric field strength E of ferroelectric materials is described by a hysteresis curve (Fig. 1)[1]. The remanent polarization P_R and the coercive force pEc are characteristic parameters of the ferroelectric material. The reversibility of remanent polarization makes ferroelectric materials suitable for use as analog or binary storage media.

In the 1950s and 1960s, the storage capabilities of inorganic ferroelectric materials, such as $BaTiO_3$, were intensively researched and numerous memory versions were tested. However, these proposals proved to be inferior to magnetic storage units, since inorganic ferroelectric materials do not usually have a unique coercive force and are also subject to fatigue after many read/write cycles.

In the early 1970s, it was found that certain polymers possess suitable piezoelectric and pyroelectric properties; ferroelectric properties were also demonstrated in some polymers. This is true for polymers whose monomers have large dipole moments and are structured in such a way that they eliminate internal compensation of the dipole moments; examples include polyvinylidene fluoride (PVDF: $(CH_2 = CF_2)_n$); poly(vinylfluoride) (PVF: $(CH_2 = CHF)_n$); etc. Today the most extensively researched polymers with excellent ferroelectric properties are PVDF and PVDF-copolymers, the beta-phase of PVDF having a remanent polarization P_R of 5 $\mu C/cm^2$ and a coercive force of pEc of 5 MV/cm.

In our experiments, a copolymer comprising 65 (60) parts PVDF and 35 (40) parts PTrFE (polytrifluoroethylene) proved to be a particularly suitable ferroelectric polymeric material for reversible data storage[3]. Figure 2 shows the typical, almost rectangular hysteresis curve of such a material[2], while Figure 3 provides information as to the switching times of this copolymer as a function of the electric field strength and temperature[3].

In ferroelectric materials, the writing process (that is the polarization of domains) is accomplished by applying a strong electric field of the order of MV/cm (Fig. 4)[1]. For reading, a distinction must be made between reading processes that destroy the information and processes that preserve it. In destructive reading, the originally written information is read out by determining the polarization current to be applied. In non-destructive reading, either the piezoelectric effect or the pyroelectric effect is used, which means an electric charge corresponding to the permanent polarization is liberated by applying heat or pressure to the storage medium. This charge can be detected by appropriate methods.

The following read and write processes have been described in the literature[1]:

- reading and writing using electrically-conductive electrodes (Fig. 5 a),

- reading and writing using a metal tip (Fig. 5 b),

- reading and writing using a laser beam (Fig. 5 c) and

- reading and writing using an electron beam (Fig. 5 d).

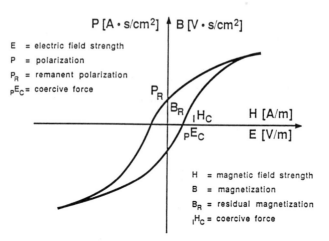

Fig. 1. Typical hysteresis curve of a ferromagnetic or ferroelectric material.[1]

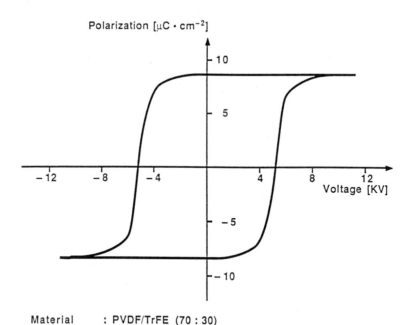

Material : PVDF/TrFE (70 : 30)
Thickness : 146 µm
produced by Atochem (France)

Fig. 2. Typical hysteresis curve of a PVDF/TrFE-copolymer.[2]

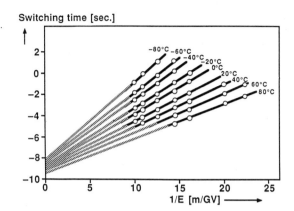

Fig. 3. Plots of the log of switching time against the reciprocal of applied electric field for a 65/35 PVDF/TrFE copolymer.

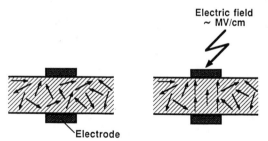

Fig. 4. Ferroelectric polymer: Local polarization by alignment of the electric dipoles in the electric field.

81

In our laboratories, we structured a purely polymeric ferroelectric data memory of low bit density for experimental and
demonstration purposes (Fig. 6)[32]: wordlines and bitlines
were vapor-deposited onto both sides of a suitably polarized
PVDF or PVDF/ TrFE film (β-modification) in the form of
crosswise arrangement of strip electrodes[4]. This experimental
polymeric ferroelectric storage unit meets the usual requirements regarding readability, writability, access time and
extrapolated lifetime of the stored information. Expansion to
a high density memory, however, is not possible because of
basic physical limitations (capacity attenuation of the read
signal).

3. MAIN MEMORIES

Non-volatile RAM with Polymeric Ferroelectric Memory Layer

At the moment, the main memories used in computer engineering
are exclusively RAMs (random access memories) based on Si
technology, either in the form of dynamic RAMs (DRAM) or
static RAMs (SRAM). In terms of storage density (e.g. max. 4
Mbit) and access time (< 100 ns), these memories leave very
little to be desired, but they have the major disadvantage of
being volatile, i.e. if the power fails or is switched off,
the stored information is irreversibly lost.

Consequently, there has been no lack of attempts in the past
to avoid this decisive drawback of volatility without impairing the other characteristics of a RAM to any major degree.
Examples deserving mention in this context include the EEPROM
(electrically erasable programmable read-only memory) and the
combination of a RAM with a battery as emergency power
supply, the NVRAM (non-volatile RAM)[5]. However, in relation
to the original RAM, all these developments led to decisive
disadvantages in terms of storage density and/or write speed
and/or number of write cycles and/or cost; thus, they are
only used in certain special areas.

In the late 70s, the proposal was made to directly integrate
a ferroelectric memory layer based on $KMnO_3$ with a silicon
semiconductor driver module[6]. More recently, two US companies
(Ramtron and Krysalis), working on the basis of Cook's
patents, have begun to combine ferroelectric memory layers
based on PZT (lead zirconium titanate) with silicon semiconductor driver modules[7]. In Spring 1988, they announced a
development in which a silicon semiconductor RAM is combined
with a sputtered ferroelectric memory layer based on PZT[8]. No
information was provided as to how the problem of possible
contamination with the elements of the ferroelectric ceramic
material was solved on the Si process line operating under
clean-room conditions. In addition, it is known that ceramic
ferroelectric materials such as PZT do not have a defined
coercive force and also that they only allow a limited number
of write cycles due to uncontrollable ageing effects.

From the very outset, we did not consider a ferroelectric
memory layer with a driver module based on silicon semiconductor technology, but intensified our efforts to develop a
conventional (volatile) RAM (e.g. DRAM) combined with a
non-volatile memory (in accordance with the pick-a-pack
principle) based on polymeric ferroelectric materials.

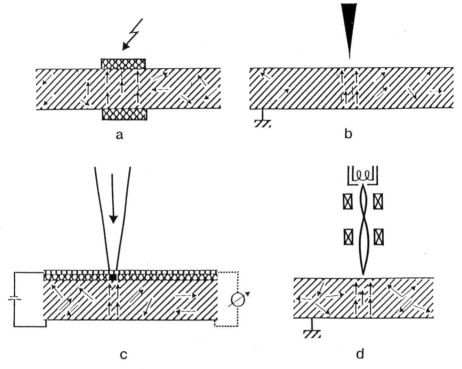

Fig. 5. Ferroelectric polymer: Local polarization by means of
a) metal electrodes, b) metal tip, c) laser beam,
d) electron beam.

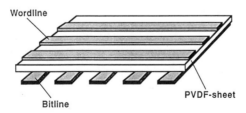

Fig. 6. Experimental ferroelectric polymer data memory
(with very low bit density).

This data memory, referred to by us as FERAM (ferroelectric RAM), combines two storage technologies: during normal operation, only the first memory area is activated, i.e. the FERAM acts like a conventional RAM (DRAM or SRAM), thus having the desired property profile of a RAM in relation to write/read speed, free addressability and storage density. Only if certain events occur (computer power-down, power failure, STORE command) then the information present in the first memory area (RAM) is transferred to the second (ferro-electric) memory area, where it is stored permanently. When the power supply is restored, or following a special RECALL command, the information can be transferred back to the first part, the RAM part[9].

To this end, a ferroelectrically polarizable layer (e.g. PVDF/ TrFE copolymer or - to fulfill processing conditions in regard to postbake temperature (130°C) - PVDCN/VAc copolymer) is applied to the semiconductor module (RAM) in pick-a-pack form; like the semiconductor memory, its upper and lower sides are provided with word and bit lines in the form of strip electrodes. Here, the strip electrode system on the underside of the ferroelectric layer forms at the same time the word and bit line system of the semiconductor memory facing the surface. In this way, each semiconductor memory cell is assigned a non-volatile ferroelectric memory cell with permanent storage properties (Fig. 7).

4. MASS MEMORIES

4.1 Optical Mass Memories (Review)

Alongside magnetic mass memories, optical mass memories with storage densities up to approx. 10^8 bit cm^{-2} will become increasingly important in future. For example, non-erasable optical data memories are already being launched in the market, where the data are either already recorded by the manufacturer (so-called prerecorded optical ROMs based on the Compact Disc principle = CD-ROM) or can be written by the user with a laser (WORM = write once, read multiple). Era-sable optical data memories (E-DRAW = erasable direct read after write) on the basis of magneto-optical recording (MOR) or phase change recording (PCR) are currently still under development or soon to be launched[1].

CD-ROMs are made entirely of polymeric material (except for metal reflection layer), since substrate and memory layers are a single entity. At the moment, CD-ROMs are made exclusively of polycarbonate (PC), which adequately fulfills all the requirements imposed on both the substrate and memory layers (e.g. polycarbonate Makrolon® CD 2000 from Bayer AG, F.R.G., and Mobay Corp., Pittsburgh, USA). In contrast, only the substrate of WORM and E-DRAW systems is made of polymers (currently preference for polycarbonate) or - for experimental purposes - glass.[*)]

The different types of optical mass memory, their characte-ristics, the structure of the memory layer and corresponding analogous semiconductor devices are compiled in Table I.

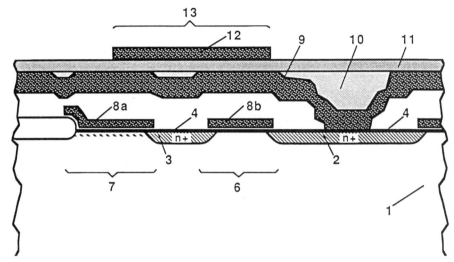

Fig. 7. Non-volatile RAM: Drawing of a memory cell of a
dynamic NMOS-based semiconductor DRAM with associated polymeric ferroelectric memory layer
(pick-a-pack principle)
1   p-doped Si substrate
2, 3   n-doped regions
4   gate oxide (SiO_2-layer)
6   selection transistor
7   memory capacitor
8a, 8b   word lines
9   bit lines
10   passivation layer (inert material, e.g.
           polyimide)
11   ferroelectric layer (e.g. PVDF/PTrFE
           copolymer)
12   strip electrode
13   ferroelectric memory cell (backup memory).

4.2 Potential High-density Polymeric Data Storage Systems of the Future

The following possible developments in polymeric data memories are currently being debated[1]:

a) Memories with 2-dimensional data storage:

- photopolymers
- liquid crystalline polymers
- ferroelectric polymers

b) Memories with 3-dimensional data storage:

- volume holograms in polymers
- photochemical "hole burning" in polymers

These developments, the state of the art and possible future applications have already been described in detail elsewhere[1].

At this point, further information will be given on more recent work which may possibly form the basis for the development of polymer-based high-density mass memories with storage densities $> 10^8$ bit cm^{-2}.

4.3. Writing and Reading Information in Thin Ferroelectric Polymer Films by Means of an Electron Beam

In relation to mass memories, modern computer engineering demands the availability of ever-increasing amounts of data in the smallest possible volume or area. The efforts of the manufacturers of mass memories are thus aimed at increasing storage density while retaining very short access times and other relevant parameters.

The limiting values for magnetic mass memories are around 3×10^8 bit cm^{-2} (due to finite flux quantization during the reading process) and calculated for the use of UV light – about 10^8 to 10^9 bit cm^{-2} for optical mass memories (due to diffraction phenomena)[1].

Consequently, a significant increase in storage densities can only be achieved through the use of new technologies. The use of an electron beam is one particularly good possibility, since beam cross-sections down to as little as 5 nm can be achieved under practical conditions. In purely mathematical terms, data densities up to approx. 10^{12} bit cm^{-2} could be obtained in this way. This value has already been reached on an experimental model of a non-erasable data memory (burning of tiny holes in β-aluminium by means of an electron beam).

---

*) Further alternative polymers as substrate materials for erasable optical data memories are described in more detail in Section 5.

Table I. Optical Memories: Different Types, Characteristics, Memory Layer and Designations of Analogous Semiconductor Memory Devices.

Type	Characteristics	Memory Layer (structure)	Analogous Semiconductor Device
Prerecorded O-ROM[a] (CD-ROM)	Read only Non-volatile Replicable DRAW[b] not essential	Polymer (Polycarbonate)	ROM[g]
Write-once O-ROM (WORM)[c]	User write/read only Non-volatile DRAW needed	Thin metal layers, Polymeric layers containing light-absorbing additives (dyes, pigments, carbon black)	PROM[g]
Optical RAM (EOD)[d]	User write/read/erase selective erase Non-volatile Rapid access	Amorphous layers with reversible magnetization (MOR)[e], Phase transition amorphous/crystalline polymer or metal or metal oxide layers (PCR)[f], Polymer layers containing photochromic dyes	EEPROM[g], RAM[g]

a O-ROM = Optical Read Only Memory
b DRAW = Direct Read After Write
c WORM = Write Once, Read Multiple
d EOD   = Erasable Optical Disk

e MOR   = Magneto Optical Recording
f PCR   = Phase Change Recording
g ROM   = Read Only Memory
  PROM   = Programmable ROM
  EEPROM = Electr. Erasable and Programmable ROM
  RAM   = Random Access Memory

The development of reversible data storage with electron beams faces far greater difficulties, particularly in regard to the reading process. Various concepts have been proposed in the literature and in patent applications[11], in which electrically remanently polarizable data media (e.g. PVDF films) are used as the storage media. These concepts could not be implemented in a practical manner, since the necessary requirements in relation to signal-to-noise ratio, lifetime of the stored information, lateral resolution, etc., could not be fulfilled.

In cooperation with our laboratories, Prof. K. Dransfeld of the University of Konstanz (F.R.G.) developed a method for writing and reading data in a ferroelectric layer (e.g. PVDF or PVDF/TrFE polymer film) by means of an electron beam which, with a lateral resolution of < 1 μm, allows data transfer rates of $> 10^6$ bit s^{-1} with a high signal-to-noise ratio[12]. Since the lateral resolution of the structures in ferroelectric polymers polarized by a focused electron beam is limited by the interaction volume of the primary electrons (PE) in the material, extremely thin ferroelectric polymer films must be used to achieve a high lateral resolution (= high effective data density) in order to shift the major part of the interaction volume of the PE into the substrate[13].

A high signal-to-noise ratio during reading is achieved by high frequency modulation of the secondary electrons via periodic exposure of the data medium to pulsed electromagnetic radiation (e.g. IR laser) or by using high-frequency ultrasound. Figure 8 outlines the fundamental arrangement for reading polarized domains in a ferroelectric layer by means of electron beam with pyroelectric activation. A detailed description of the reading technique with pyroelectric (IR laser) or piezoelectric (ultrasonic) activation can be found in ref. 12. In order to achieve the high data transfer rates required, the modulation frequency must be at least one order of magnitude higher than the data transfer rate. At a data transfer rate of 10^6 bit s^{-1}, this is easily ensured by an ultrasound frequency of 1 GHz and a laser modulation frequency of 10 MHz. The reading operation is always reversible and the polarization state of the data medium is not changed even after repeated reading, since the method selected involves only extremely slight interaction between the electron beam and the ferroelectric data medium[12].

4.4 Writing and Reading Information in Thin Ferroelectric Polymer Films by Means of a Scanning Tunneling Microscope

Another method for a mass memory with a storage density significantly above 10^8 bit cm^{-2} is writing and reading by means of a metal tip in thin ferroelectric layers (Fig. 5 b). In comparison with the technique of remanent polarization by means of a tip electrode, described for the first time by Pulvari[11] in the 1950s, and in comparison with the limited lateral resolution when writing by means of an electron beam due to the finite interaction volume of the primary electrons in the ferroelectric memory layer (see Section 4.3)[13], the scanning tunneling microscope (STM)[14] developed by Binnig and

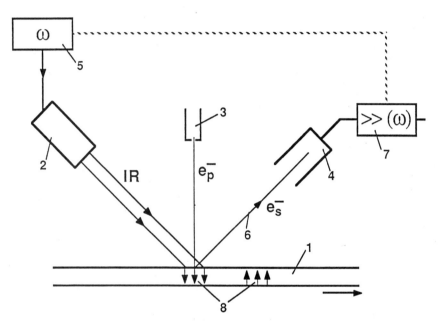

Fig. 8. Reading of polarized domains in a ferroelectric
         layer by means of an electron beam with pyroelectric
         activation (elementary diagram)[12]
         1 Electically polarisable layer (e.g. PVDF/TrFE-
            copolymer)
         2 IR-laser
         3 Electron gun
         4 Collector electrode (multiplier)
         5 HF-generator (modulating frequency)
         6 Scattered/secondary electrons
         7 Narrow-band highly sensitive amplifier
         8 Ferroelectric domains.

Rohrer offers a lateral resolution down to the range of < 10 nm. Exploiting the tunneling effect, the STM involves passing an extremely fine-tipped metallic needle over the object to be examined at very close range (typically < 1 μm). The more recent development of the STM[15] also allows operation in normal air and at room temperature with scanning rates in the kHz range[16]. Smaller scale and simplified STM models are extremely handy (measuring approx. 10 x 10 cm), but have a correspondingly lower resolution. This means that the wide ranging use of the STM for tasks other than fundamental examinations of the microstructure of solids is possible; for example, the usefulness of the STM for lithographic purposes has been tested[17].

In cooperation with K. Dransfeld, we have started to develop an STM-based high-density data memory, where the STM is used to write and read information into and out of a ferroelectric memory layer (ceramic material, e.g. PZT, or polymer, e.g. PVDF/TrFE copolymer) in the form of polarized domains[18]. Using an etched molybdenum needle and an applied voltage > 1 volt (corresponding to field strengths of up to 100 MV m^{-1}), initial trials proved it possible to write electrically polarized domains of approximately 100 nm^2 in very thin PVDF/TrFE layers (< 100 nm) applied to aluminium-vaporized silicon wafers by spin-coating; in other words, this method can achieve storage densities of > 10^{10} bit cm^{-2}. In this context, it is essential to maintain a constant distance between write/read tip and ferroelectric layer. In our system, this is achieved by means of an electrically decoupled twin tip which adjusts the STM with the aid of an electrically conductive coating on the ferroelectric memory layer (Fig. 9)[18].

The stored information can be read either by "destructive reading" or by means of pyroelectric activation of the memory layer; the signal is recorded by means of an electrometer probe with extremely high resolution. The previously customary data transfer rates of approx. 1 kHz are totally inadequate, but can be increased to the GHz range by means of a specially developed technique[19].

5. DEVELOPMENT OF SPECIAL POLYMERS AS SUBSTRATES FOR OPTICAL MASS MEMORIES (review)

Substrate materials of non-erasable and especially of erasable optical memories have to meet stringent requirements for the level of

- Birefringence   (Δn)
- Impact Strength   (IS)
- Bending modulus   (BM)
- Heat-Deflection-Temperature   (HDT)
- Density (Weight)   (Dens.)
- Water-Absorption   (WA)
- Cost   (Cost)
- Processability   (Proc.)

In Fig. 10[20,21] different polymers (Polycarbonate PC, Polymethylmethacrylate PMMA, Cyclic Polyolefin CPO, modified Duro-

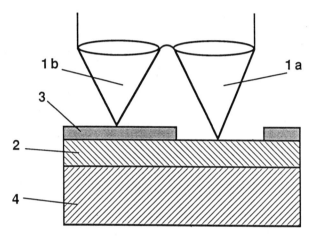

Fig. 9. High-density data storage in a ferroelectric layer,
          using a scanning tunneling microscope (elementary
          diagram)

          1 a: Twin tip for setting a constant gap between
               1 b and 3
          1 b: Write/read tip
          2 : Si wafer
          3 : Ferroelectric layer (e.g. PVDF/TrFE copolymer)
          4 : Al substrate.

plast MDP (e.g. epoxy resin), polycarbonate/polystyrene blend PC/PS) and inorganic glass are qualitatively compared for the above-mentioned important properties.

Up to now, all CD-ROM disks are made from high-flow polycarbonate, as this material offers the best property balance to meet the standards and material specifications. The development of modified polycarbonate grades, e.g. Makrolon® CD-2000 from BAYER AG (FRG) or from MOBAY Corp. (USA), and the introduction of modified manufacturing techniques for CD-disks and CD-ROMs with low molecular orientation have allowed successful mass production of PC disks with acceptable birefringence level (Δn< 20 nm).

The benefit of PMMA having low sensitivity to orientation birefringence is negated by the disadvantage of high water absorption. On optical data disks recorded on one side only, excessive water absorption results in unacceptable warping. Copolymers made of MMA/CHMA (methylmethacrylate/cyclohexylmethacrylate) or MMA/ST (methylmethacrylate/styrene) display a considerably lower water absorption, but the material strength is substantially reduced (MMA/CHMA) and/or the birefringence is increased (MMA/ ST)[1, 21].

The disadvantage of polycarbonate is its relatively high intrinsic birefringence. Very low birefringence is particularly important when writing data in WORM discs respectively writing/reading data in/from E-DRAW discs. Excessive birefringence will, on the one hand, deteriorate exact focussing of the laser beam during writing and, on the other hand, will change the rotation angle of the plane of polarization during reading resulting in a relatively low signal to noise ratio. Furthermore, birefringence causes a certain amount of light intensity to fall back to the diode laser causing the laser to oscillate. This so called Laser Feedback Power (LFP) has different limits depending on the geometry of the focus lens. The American National Standard for Information Systems (ANSI) proposes a limit of 2.5 % for the LFP for a given Numerical Aperture of NA = 0.52 for a standard optical drive focus lens[31].

Therefore, efforts are underway to reduce the birefringence level of polycarbonate and at the same time maintain typical polycarbonate properties. One approach uses extruded polycarbonate sheet (e.g. Makrolon® OD of ROEHM GmbH, FRG) requiring the 2P process or the printing procedure to add the information pits or the pre-grooves[21]. Another way to reduce birefringence is to substitute polycarbonate, by e.g CF_3- or phenyl-substituted PC, (also see Section 5.1)[22].

The most effective, but certainly also the most difficult approach towards obtaining birefringence-free substrate materials is to blend polymers with positive birefringence (e.g. PC, but also PVC, PET, PVDF etc.) with polymers with negative birefringence (e.g. PMMA, PST, PAN etc.). Sumitomo Chemical is working on such blends of PC with modified PST (Fig. 11)[23]. Also, a blend of PMMA and PVDF is particularly interesting, since, if blended in an appropriate ratio, it is not only birefringence free (Fig. 12), but also displays considerably reduced water absorption in comparison with pure PMMA[21].

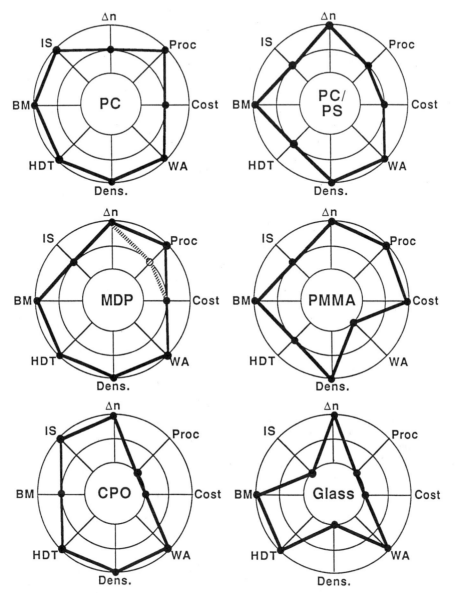

Fig. 10. Qualitative comparison of substrate materials for optical disks.

Increasing radius means more favourable properties (based on [20,21], modified and supplemented)

PC = Polycarbonate, PMMA = Polymethylmethacrylate, CPO = Cyclic Polyolefin, MDP = Duroplast based on high reactive resins, PC/PS = Polycarbonate/Polystyrene blend, Glass = Inorganic Glass.

An interesting development is reported by formulating a cyclic polyolefin (CPO) which combines transparency, sufficient mechanical properties and high heat resistance with low birefringence and low water absorption (see Fig. 10)[20].

According to a recent publication of 3M, St. Paul (USA)[24] polycarbonate grades currently used for CD-ROM substrates (e.g. Makrolon® CD 200) generally meet the requirements of erasable optical memories (magneto-optical systems of up 130 mm diameter).

According to measurements at Bayer AG on polycarbonate substrates for data storage media made by various manufacturers and different moulding techniques[25], none were found showing a lower LFP than 4 %. From this conflicting information, ANSI requirement and 3M results, we conclude that the ANSI requirement is either too stringent or the manufactring of substrates from polycarbonate eventually requires[30] annealing of the substrates to release internal stress and reduce birefringence levels to guarantee good yield ratios in a production environment.

To meet stringent requirements in birefringence, especially for substrate diameters of up to 300 mm, Bayer AG and Mobay developed trial products based on thermoplastic blends, as well as duroplasts (crosslinked polymers) based on high reactive resins with short cycle time, which meet all the requirements. The measured birefringence for the above-mentioned thermoplastic blend is presented by way of example in Fig. 13.

5.1 Calculation of Birefringence for a Single Linear Polymer Chain

The major disadvantage of polycarbonate as substrate material for optical disks is its relatively high intrinsic birefringence. Very low birefringence is particularly important when writing/ reading data on E-DRAW disks, especially E-DRAW disks of the MOR type.

The birefringence of a cooled substrate disk is caused by two factors during manufacture: orientation in the melt (orientation birefringence) and internal stresses (stress birefringence). Orientation birefringence occurs in the melt and depends on the magnitude and sign of the rheo-optical constants of the polymer, among other process parameters. Stress birefringence occurs during the cooling phase, predominantly at temperatures below the glass transition temperature of the polymer, and depends not only on the process parameters, but also on the magnitude and sign of the stress-optical constants of the polymer.

The intrinsic birefringence of a polymer (= difference in the refractive indices for light oscillating parallel and vertical to the chain direction) is based on a corresponding difference in the polarizability $\Delta\alpha$ of the monomer for the two directions: $\Delta\alpha = \alpha_{\parallel} - \alpha_{\perp}$. The birefringence Δn of the

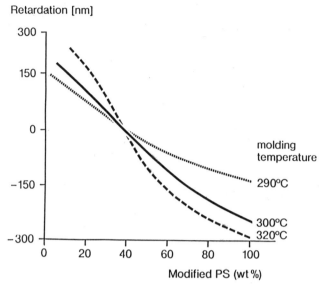

Fig. 11. Effect of blend ratio on birefringence of a PC/
        modif. PST blend (after SUMITOMO Chem. Co. Ltd.)[23]

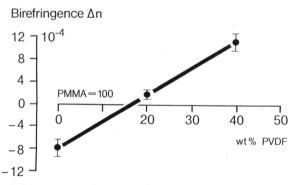

Fig. 12. Birefringence of uniaxially stretched samples of
        PMMA/PVDF-blends as a function of the composition[21]

polymer ensemble thus disappears if the orientation distribution of the polymer chains is completely random in the relaxed state (i.e. in the stress-free state). If the internal stresses are ignored (or if the polymer is relieved of stresses by heat treatment), the macroscopically measured birefringence Δn of a polymer substrate disk is determined by the orientation distribution of the polymer chains F_{Or} and by the difference in the optical polarizabilities of the monomer unit $\Delta\alpha$ as[21]

$$\Delta n = K \cdot F_{Or} \cdot \Delta\alpha \qquad \text{with } \Delta\alpha = \alpha_{\parallel} - \alpha_{\perp}$$

$$F_{Or} = (3 \cos^2 \upsilon - 1) \qquad \text{"Herman's orientation function"}$$

$$K = \frac{4}{3} \frac{\varsigma \cdot N_L}{M_O} \frac{(n_o + 2)^2}{n_o} \qquad \varsigma = \text{density}$$

With complete orientation of the polymer chains, $F_{Or} = 1$, and the maximum possible magnitude for the birefringence of a given polymer Δn_{max} is measured in this case.

Therefore, in order to obtain low-birefringence disk substrates, it is necessary to set the most random possible orientation of the chain segments (task of process engineering) and/or to develop polymers with a very low or negligible difference in the polarizabilities of the individual chains (task of the polymer manufacturer).

Processes have already been developed for low-orientation processing (e.g. injection compression, injection coining, etc.) and they are being improved all the time[25].

Suitable computational methods were generated in our laboratories, or known methods adapted, for specific development of polymers with the lowest possible intrinsic polarizability difference[26]:

- Determination of the conformation of the chains of polymers of interest as substrate materials via model calculations using the data from X-ray structural investigations; from there, determination of bond spacings, as well as valency angles and torsion angles was done.

- Calculation of the polarizability tensor from the additive contributions of the individual covalent bonds and the polarizability components of the monomer units as a function of the conformation for the three directions in space, and also of the arithmetic mean of $\alpha = \bar{\alpha}$ by averaging across all possible bond directions (diagonal tensor):

$$\bar{\alpha} = (\alpha_{\parallel} + \alpha_n + \alpha_t)/3$$
$$\Delta\alpha = \alpha_{\parallel} - (\alpha_n + \alpha_t)/2$$

where
$\alpha_{\parallel}$ = Polarizability parallel to chain direction
α_n = Polarizability perpendicular to chain direction
α_t = Polarizability transverse to chain direction

- Calculation of the mean refractive index n_o and the birefringence Δn.

The values obtained with our computational method[*] correlate very well with the experimental data, where available, and provide valuable assistance in the development and optimization of polymers for optical substrate materials.

Considerations relating to reducing the optical anisotropy of bisphenol-A polycarbonate by way of substitution without impairing the other outstanding properties, such as transparency, toughness, heat deflection temperature, etc., are particularly interesting. Werumeus Buning et al.[22] pointed out that incorporation of a lateral phenyl ring in the isopropylidene group (3-nucleus bisphenol-A PC) can greatly reduce the birefringence. Our calculations show that complete compensation of the birefringence only occurs in purely mathematical terms when the three phenyl rings stand vertically one on top of the other. If the ring plane of the third aromatic is also in the paper plane (Fig. 14), the values obtained for $\Delta \alpha$, depending on the reference plane used or on the measuring direction, are $+ 3.8 \overset{3}{\text{Å}}$ or $+ 7.6 \text{ Å}^3$ (comparison value for bisphenol-A PC: 4.8 Å^3)[26].

By way of example, Table II shows the values calculated for the refractive index n and the birefringence Δn_{max} for various substituted polycarbonates (with complete orientation of the chain molecules).

5.2 Birefringence Measurements of Polymer Substrates

Exact measurement of the birefringence and the most informative presentation possible of the results is of great importance in the development of new polymers, in the industrial testing of the polymeric starting materials and in controlling the production quality of Compact Discs, CD-ROMs and disk substrates for optical data memories.

Because of the large aperture of the write and read optical system in the conventional hardware of optical data memories, the optical path differences (birefringence) must be measured in all these directions in space M_1 to M_3 for exact material characterization (see Fig. 15). For control measurements, it is sufficient to measure the optical path difference vertical to the disk plane (M_1), which makes the decisive contribution towards the total birefringence of a disk substrate in practical operations, and to record the results in a well-arranged form.

Simple test set-ups were built in our laboratories for this purpose. The schematic beam path of the customary version of such an instrument with crossed polarizers ($P_1 = 0°$, $P_2 = 90°$) is shown in Fig. 16 a and the evaluation characteristic for transmission $T = f$ (path difference, PD) in Fig. 16 b.

---

[*] A detailed presentation and discussion of the results will follow elsewhere at a later date.

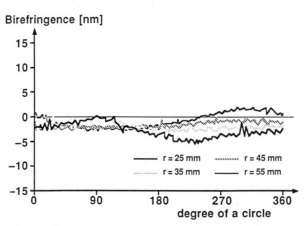

Fig. 13. Birefringence of optical disk substrates as a function of radius (special thermoplast blend of BAYER AG and MOBAY Corp.).

The parabolic curve profile entails considerable difficulties: low measuring sensitivity with small path differences
and no unequivocal sign.

The following applies for the set-up shown:

$$T = \sin^2 (\delta/2)$$ δ = phase difference

where $\delta = 2\pi \frac{PD}{\lambda}$

For small PDs, the following approximation applies:

$$T \sim (\delta/2)^2$$

However, if the original equation is developed by selecting
an operating point around $\delta_o = \pi/2$, the result is

$$T = 1/2\ (1 + \sin\delta),\ \text{corresp.}\ T \sim 1/2\ (1 + \delta)$$

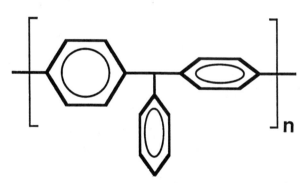

Fig. 14   Structural diagram of a 3-nucleus bisphenal A
polycarbonate.

In terms of apparatus, this possibility for evaluation can be
implemented with P_1 = 0° and P_2 = 45°. The resultant characteristic (Fig. 16 c) allows accurate measurement of the
magnitude and sign of the path differences, even with small
values of PD, in the range of interest -50 nm $\leq$ PD $\leq$ +50 nm.
By connecting the birefringence meter to a transient recorder, it is possible to measure the birefringence over a
complete circumference with a given disk radius (see Fig. 13
for an example); thus, the birefringence can be determined at
all points on the entire substrate disk substrate surface[27].

Table II. Computation of Density, Refractive Index and Orientation Birefringence for different substituted Polycarbonates.

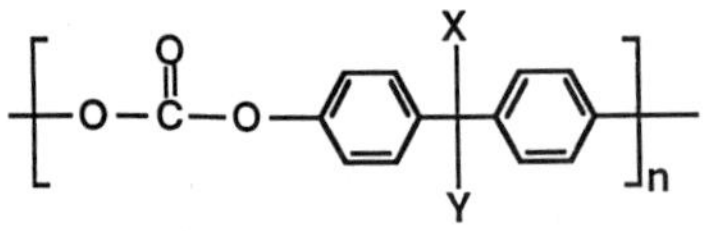

with: X,Y = Me
= CF_3
= Ph

Results:

X	Y	Density g cm⁻³	Refractive Index n	Birefringence (at max. molecular orientation) Δn_{max}
Me	Me	1,18	1,58	+ 0,12 (PC, reference)
CF_3	CF_3	1,50	1,51	+ 0,12
Me	Ph	1,20	1,60	+ 0,03 ... + 0,06
Ph	Ph	1,21	1,61	+ 0,01

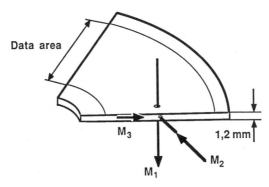

Fig. 15.   Measurement of the birefringence of disk substrates. Principal measuring directions are M_1, M_2, M_3.

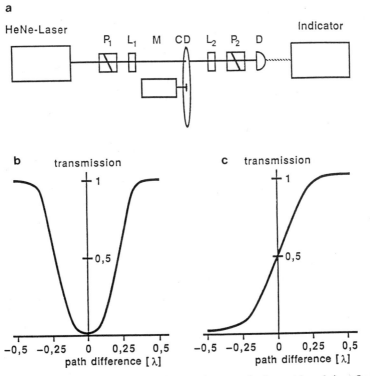

Fig. 16.   Simple test set-up for determining the birefringence of disk substrates vertical to the disk
plane
a) Elementary block diagram of the test set-up
     P_1, P_2 = Polarizers, L_1, L_2 = λ/4-disks,
     M = Motor, CD = Compact Disc, D = Photodiode
b) Evaluation characteristic, conventional technique
c) Evaluation characteristic, improved technique.

In order to present all the individual results for different disk radii together, we use, among other things, a graphic plotter to reproduce the birefringence curves for the selected radii in a spatially separated form. As an example, Fig. 17 shows the evaluation results for a trial product with low birefringence[28]. This type of result presentation allows substrates to be judged at a glance.

6. OUTLOOK ON FUTURE DEVELOPMENTS OF MAGNETIC AND OPTICAL DATA STORAGE SYSTEMS

At present, the advantage of optical mass memories in comparision with magnetic storage media is the 10 to 100 times greater storage density. This leads to a cost advantage per MByte for optical systems of the same order of magnitude[1]. In addition, optical data memories are insensitive to dust and fingermarks, and the hardware shows reduced susceptibility to failure. It should be remembered in this context that the gap between the magnetic disk and the read/write head in a Winchester drive is only 0.3 µm.

The major advantage of magnetic storage systems is the fact that this technology is well established in the industry worldwide. It can be seen here that the information industry is conservative in this respect and unwilling to give up time-proven technologies unless there are special advantages in favour of a new one.

Furthermore, there is ongoing development with regard to improved magnetic media (e.g. rice grain pigments, barium ferrites, improved Co-doped magnetites, pure iron pigments and thin-film memories) and improved magnetic recording techniques (e.g. perpendicular recording technique). In the next couple of years, an increase of storage density to 2 to 5×10^7 bit cm^{-2} will be reached, with storage costs declining proportionally. On the other hand, optical memories have reached their limit storage density of approx. 6×10^7 bit cm^{-2}; a further increase would only be possible after the development and introduction of inexpensive diode lasers operating in the visible or near UV range[1].

If this increase in magnetic storage density occurs, it is unlikely that erasable optical media will replace magnetic techniques to a great extent, unless it becomes possible to develop economical techniques that provide erasable optical media with high signal-to-noise ratio. On the other hand, non-erasable optical storage systems will have an increasingly high impact in numerous special application fields, such as catalogues, car repair manuals, computer operating systems, etc., as well as in those applications where high data security, including non-erasability, is required (e.g. libraries, accounting in large companies, spectrum documentation, etc.).

These considerations are also taken into account in the MackIntosh study "High-Capacity Information Storage"[29], which does not forecast a breakthrough in erasable optical memories until 1993, but a steadily increasingly market share of non-erasable optical data storage systems in the same period, at least in Western Europe (Fig. 18)[1].

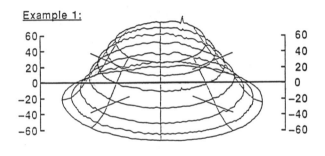

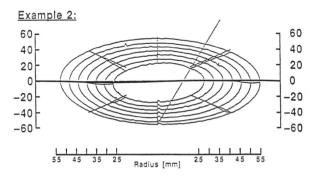

Fig. 17.   Presentation of birefringence profiles (single
path) for selected radii in 3 D. Example 1: Thermoplast; Example 2: Special thermoplast blend (BAYER
AG and MOBAY Corp.)

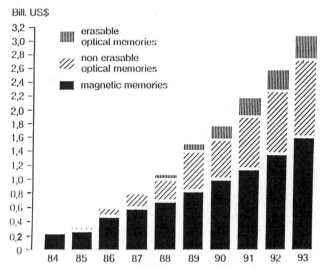

Fig. 18.   Data Storage Media of High Capacity in Western
Europe (1984 - 1993).
(Source: MackIntosh   Study "High Capacity Informa-
tion Storage")

REFERENCES

1. G. Kaempf, Polymer J. 19, 257 - 268 (1987)
   G. Kaempf, H. Loewer and M.W. Witman, Polymer Eng. Sci. 27, 1421 - 1435 (1987)
2. F. Bauer, K. Harnischmacher and J. Rika, Proc. 5th Int. Symp. on Electrets ISES, 924 - 929 (1985)
3. T. Furukawa et al, J. Appl. Phys. 56 (5), 1481 - 1486 (1987)
4. W.J. Merz and J.R. Anderson, Bell Lab. Record 33, 335 - 342 (1955)
5. Elektronik 25, 82 - 86 (1983)
6. R.C. Cook, US-Patent No. 4 149 301 and 4 149 302
7. G.A. Rohrer, US-Patent 4 707 897
   L. McMillan, Int. Patent Appl. WO 86/04447 and WO 86/06752, US-Patent 4 713 157 and other
8. Electronics, 32 (February 4, 1988) and 91 - 94, 95 (February 18, 1988)
9. R. Pott, A. Eiling, G. Kaempf, Europ.Patent EP 0 236 696, US Serial No. 7048/1987
10. D. Wohlleben, R. Pott, and A. Eiling, J. Appl. Phys. 61, 2999 - 3010 (1987)
11. C.F. Pulvari, US-Patent 2 698 928
    D.W.G. Byatt, US-Patent 4 059 827
12. K. Dransfeld, Europ. Patent EP 0 186 813;
    US-Patent 6 806 246
13. K. Dransfeld and D. Schilling. in press
14. G. Binnig and H. Rohrer, Europ. Patent EP 0 027 517
15. B. Drake et al, Rev. Sci. Instrum. 57, 441 - 445 (1986)
16. A. Bryant, C.F. Quate et al, Appl. Phys. Lett. 48, 832 - 834 (1986)
17. M. Ringger et al, Appl. Phys. Lett. 46, 832 - 834 (1985)
18. G. Kaempf, K. Dransfeld and A. Pott, Europ. Patent No. EP 0 275 881
19. K. Dransfeld, Europ. Patent No. 3 812 684.2
20. A. Todo and H. Kajiura, Japan Plastics 38, No. 7, 41 - 44 (1987)
21. J. Hennig, Proc. Int. Symp. on Optical Memory ISOM (1987) Jap. J. Appl. Phys. 26, 9 - 14 (1987)
22. G.H. Werumeus Buning et al, Proc. Int. Symp. on Optical Memory ISOM (1987), Poster WC 23
23. M. Isobe and S. Imai, Sumitomo Chem. Co., Ltd., Chemical Research Lab. (1986)
24. R.N. Gardner et al, SPIE Optical Mass Data Storage II, Vol. 695, 48 - 55 (1986)
25. M. Witman, H. Loewer and G. Kaempf, SPI/SPE Plastics-West, 40 - 44 (1987)
26. F.P. Hoever, H. Schmid and E. Kops, BAYER AG Leverkusen and Uerdingen (FRG), internal report (1988)
27. F.-M. Rateike, BAYER AG Leverkusen (FRG), internal report (1988)
28. K. Sommer, BAYER AG Leverkusen (FRG), internal report (1988)
29. MackIntosh-Study "High Capacity Information Storage", Markt + Technik, Nr. 41 v. 11.10.85, 14
30. M. Okada et al, NEC Corp., Proc. of The 4th Topical Meeting on Optical Data Storage, 123 - 126 (1987)
31. American National Standards for Information Systems ANSI X3B11/87 - 166R1
32. R. Pott, A. Eiling and G. Kaempf, BAYER AG Leverkusen (FRG), internal report (1985)

THERMALLY IRREVERSIBLE PHOTOCHROMIC MATERIALS FOR ERASABLE OPTICAL DATA

STORAGE MEDIA

M. Irie

Institute of Advanced Material Study
Kyushu University
Kasuga-Koen 6-1, Kasuga, Fukuoka 816, Japan

Organic compounds which possess the photochromic property
have attracted a significant amount of attention from the viewpoint of using them as optical memory media. Despite favorable
conditions provided by the recent development of laser
technology, the compounds have found little applications in
optical information storage. Among various reasons for this,
the most important one is the lack of thermal stability. The
present study proposed a guiding principle for molecular design
of the thermally irreversible diarylethene type photochromic
compounds on the basis of molecular orbital calculation of
state correlation diagrams. According to the theoretical
prediction, a new type of thermally irreversible photochromic
compounds, diarylethene derivatives having heterocyclic rings,
were developed. The compounds had no thermochromic property
and the colored state was stable for more than three months at
80°C.

INTRODUCTION

Optical data storage systems, that use write-once organic recording
media, are now beginning to become available. Organic media have the
following advantages as compared to inorganic ones.
1) They are less subject to degradation by air and moisture.
2) They are generally less toxic.
3) They can be prepared by spin coating, leading to lower fabrication
   costs.

Because of these advantages, there is an increasing interest in using
organic materials not only for write-once recording media but also for
rewritable media. By far the most extensively studied rewritable media
are inorganic compounds, which use the magneto-optic effect or phase change
as the basis of optical recording. Although several organic rewritable
media, which include pit-forming media, bump-forming media, and organic
phase changing media, were reported, the bump-forming dye-polymer is the
only media directed toward commercial use. Organic media have not yet been
fully explored.

All of these media are based on a heat-mode optical recording method. In order to utilize the versatile function of light fully, photon-mode recording should be ultimately superior to the heat-mode system. Photon-mode recording will be advantageous in the sense of resolution, speed of writing, and multiplex recording. One of the candidates for the photon-mode and organic erasable recording media is a photochromic material.

Photochromism is defined as a reversible change in a chemical species between two forms having different absorption spectra,

$$A \; \underset{h\nu',\, \Delta}{\overset{h\nu}{\rightleftarrows}} \; B.$$

The instant image forming property without processing has led to the consideration of their use in rewritable direct read after write systems[1]. Despite favorable conditions provided by the recent development of laser technology, organic photochromic compounds have found little application in optical information storage. The requirements for a reversible optical recording medium are as follows:
1) Archival storage capability (thermal stability).
2) Sensitivity at diode laser wavelength.
3) Low fatigue (can be cycled many times without significant loss of performance).
4) Non-destructive read out capability.

The limitation of the application is due to the lack of suitable compounds which fulfill the above requirements. Among the requirements, the most important property is the thermal stability of both A and B forms of the above scheme. Although 3-furyl-fulgide was reported to have the thermal stability,[2] the open ring E form, A, was rather unstable and showed thermochromism above 130°C.

In this paper, we propose a new guiding principle for molecular design of thermally irreversible diarylethene type photochromic compounds based on MNDO calculation of state correlation diagrams. Based on this principle, we have synthesized several thermally irreversible photochromic compounds, which have no thermochromic property and the colored forms are stable for more than 3 months at 80°C.

RESULTS AND DISCUSSION

Theoretical Study

We chose a 1,3,5-hexatriene to cyclohexadiene type reaction as the model photochromic system. In order to come up with a guiding principle for molecular design of thermally irreversible compounds, we carried out a theoretical study of the system.

According to the Woodward-Hoffmann rule based on the pi orbital symmetries[3] for 1,3,5-hexatriene, a conrotatory cyclization reaction to cyclohexadiene is brought out by light, and a disrotatory cyclization by heat.

The cycloreversion reaction is also allowed both photochemically in the conrotatory mode and thermally in the disrotatory mode. From the simple symmetry consideration of the hexatriene framework, we might not expect the thermal irreversibility of the reaction.

Typical compounds of the 1,3,5-hexatriene framework are diarylethene derivatives having phenyl or heterocyclic rings. Semiempirical MNDO calculations[4] were carried out for several diarylethene derivatives, such as 1,2-diphenylethene(1a), 1,2-di(3-furyl)ethene(3a), 1,2-di(3-thienyl)ethene(3b), and 1,2-di(3-pyrrolyl)ethene together with their closed forms. Geometrical optimizations were fully performed, except that the aromatic rings were assumed to be planar.

$$ 1 \quad \overset{\cdot h\nu}{\underset{h\nu'}{\rightleftharpoons}} \quad 2 $$

1a, R1=R2=H
1b, R1=R2=CH3

2a, R1=R2=H
2b, R1=R2=CH3

$$ 3 \quad \overset{h\nu}{\underset{h\nu'}{\rightleftharpoons}} \quad 4 $$

3a, X=O, R1=R2=R3=H
3b, X=S, R1=R2=R3=H
3c, X=S, R1=R2=CH3, R3=H
3d, X=O, R1=R2=CH3, R3=H
3e, X=S, R1=CN, R2=R3=CH3

4a, X=O, R1=R2=R3=H
4b, X=S, R1=R2=R3=H
4c, X=S, R1=R2=CH3, R3=H
4d, X=O, R1=R2=CH3, R3=H
4e, X=S, R1=CN, R2=R3=CH3

Figure 1 and 2 show the state correlation diagrams[5] for the reactions from 3a to 4a and from 1a to 2a in disrotatory and conrotatory modes, respectively. The relative ground state energies of the ring closed forms are given in Table I. The two heterocyclic rings were assumed to be in the parallel orientation for the disrotatory reaction and in the anti-parallel orientation for the conrotatory reaction, respectively. According to the correlation diagrams of Fig. 1, orbital symmetry allows the disrotatory cyclizations in the ground state, in both cases under Cs symmetry, from 3a to 4a and from 1a to 2a. The relative ground state energies of the products are, however, 41.8 and 27.0 kcal/mol higher than the respective energies of the reactants. This indicates that thermal cyclization practically does not take place in both cases.

Orbital symmetry forbids the conrotatory cyclizations in the ground state from 3a to 4a and from 1a to 2a, which proceed under C2 symmetry, because each So open ring form correlates with a highly excited state of the ring closed form, as shown by dotted lines in Fig. 2. Because of a non-crossing rule, the actual correlations are those given by the full lines. Even in this case, the large barrier prohibits the conrotatory cyclization in the ground state. The thermal cyclization reaction is not allowed in both disrotatory and conrotatory modes. This means that the open ring forms of the diarylethenes having phenyl or heterocyclic rings are thermally stable and do not show any thermochromic property.

No large energy barrier exists in the cyclization processes in the S2 state for 3a and S1 state for 1a, as shown in Fig. 2. This indicates that the cyclizations of both 1,2-di(3-furyl)ethene and 1,2-diphenylethene are allowed in the photochemically excited states.

107

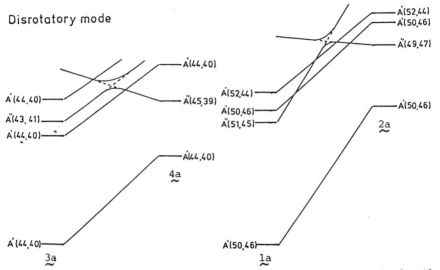

Figure 1. The state correlation diagrams in disrotatory mode for the reactions from 3a to 4a and from 1a to 2a.

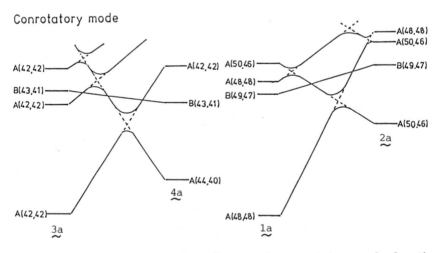

Figure 2. The state correlation diagrams in conrotatory mode for the reactions from 3a to 4a and from 1a to 2a.

According to the state energy calculation given in Table I, the ground state energy difference between the open ring and the closed ring forms strongly depends on the aryl groups. The large energy difference, 27.3 kcal/mol, of 1,2-diphenylethene decreases to 9.2 kcal/mol by replacing the phenyl rings by furyl groups. When the phenyl rings are replaced by thienyl groups, the energy difference further decreases and the closed ring form becomes more stable than the open ring form.

The energy difference is considered to play an important role in the thermal cycloreversion reaction. Full lines of Fig. 2 suggest that the energy barrier of the cycloreversion reaction correlates with the ground state energy difference. When the energy difference is large, as in the case of 1,2-diphenylethene, the reaction energy barrier becomes small and the cycloreversion reaction is expected to take place readily. On the other hand, the reaction barrier becomes large when the energy difference becomes small as shown for 1,2-di(3-furyl)ethene. In this case, the reaction is hardly expected to occur. The energy barrier, which correlates with the ground state energy difference between the open ring form and the closed ring one, controls the thermal stability of the closed ring form. The closed ring form becomes thermally stable, when the ground state energy of the closed ring form is close to that of the open ring one.

The next question is what kind of molecular property causes the difference in the ground state energy of the two isomers. First, we compared the strain energy of the six membered rings formed by the con-rotatory reaction. The optimized geometries of 1,2-di(3-furyl)ethene and 1,2-diphenylethene ,however, show almost identical molecular framework of the six-membered ring. Ring strain can not explain the energy difference.

Next, we examined the aromaticity change from the open ring form to the closed ring one. The difference in the energy between the following right and left side groups was calculated, as shown in Table II. The energy difference is considered to correspond to the difference in aromaticity as a result of the conjugated electron migration.

Table I. Relative Ground State Energy Difference between the Open and Closed Ring Forms.

compd	disrotatory kcal/mol	conrotatory kcal/mol	half lifetime		
1,2-diphenylethene	41.8	27.3		1.5	min(2b)
1,2-di(3-pyrrolyl)ethene	32.3	15.5	32	min	
1,2-di(3-furyl)ethene	27.0	9.2	12	h	(4d)
1,2-di(3-thienyl)ethene	12.1	-3.3	12	h	(4c)

Table II. Aromatic Stabilization Energy Difference.

group	Energy, kcal/mol
phenyl	27.7
pyrrolyl	13.8
furyl	9.1

The highest energy difference was calculated for the phenyl group and the lowest one for the furyl group. Destabilization due to the destruction of the aromatic rings during the course of cyclization increases the ground state energy of the ring closed form. The aromaticity explains well the trend of the relative stability.

From the above considerations, we can conclude that the thermal stability of both isomers of the diarylethene type photochromic compounds can be improved by introducing aryl groups, which have low aromatic stabilization energy.

Synthetic Studies

Based on the above theoretical prediction, we synthesized several diarylethenes having phenyl or heterocyclic rings with various aromatic stabilization energies, in the hope to obtain thermally irreversible photochromic compounds. In the synthesis, we paid attention to the following. When R2 of compounds 1 and 3 is hydrogen, the photo-generated dihydro-compounds become unstable to convert further to dehydrogenated forms[6]. Therefore, R2 groups were substituted by methyl groups.

At first, cis-2,3-dimesityl-2-butene(1b) was synthesized and the photochemical reaction was followed. Upon ultraviolet irradiation (λ=289nm), a yellow color appeared at 445 nm. According to the assignment of Muszkat and Fischer[7], the absorption is attributable to the dihydro derivative of 9,10-dimethylphenanthrene. In the dark, the yellow color disappeared with half lifetime of 1.5 min. at 20°C. The ring-closed form reverts quickly to the open ring form.

Both the molecular orbital calculation and the above experimental finding indicate that the closed ring form of diarylethenes having phenyl rings with high aromatic stabilization energy is thermally unstable. We replaced the phenyl groups by heterocyclic rings. 2,3-bis(2,5-dimethyl-3-thienyl)-2-butene(3c) and 2,3-bis(2,5-dimethyl-3-furyl)-2-butene(3d) having methyl groups at R1 and R2 positions were synthesized and the photochemistry was followed.

Ultraviolet irradiation(λ=313nm) of the degassed benzene solution of 3c and 3d led to the formation of yellow color. The absorption maxima were observed at 431nm for 4c and at 391nm for 4d. Exposure of the solution to visible light(λ>390nm) led to rapid disappearance of the yellow color. The color could be regenerated by irradiation with 313 nm light. The yellow color is attributable to the closed ring forms 4c and 4d. Exposure to the air caused no decrease of the yellow color. Oxygen did not convert the dihydro form into the condensed ring because of the presence of methyl groups. The absorption bands of 4c and 4d lie, however, in the wavelength range shorter than 500 nm and the photochromic reactions cannot be efficiently induced by conventional laser light such as Ar ion laser(λ =488, 514nm) or He-Ne laser(λ =633nm). In order to shift the absorption maximum to longer wavelength, we synthesized cyano and acid anhydride derivatives.

Figure 3 shows the absorption spectral change of CCl_4 solution of 3e by photoirradiation with 405nm light. Irradiation of the solution with 405nm light led to the decrease of the absorption at 297nm and the formation of a red solution, in which a visible absorption at 512 nm was observed. The introduction of cyano groups shifted the absorption maximum to a longer wavelength by as much as 81nm. Isosbestic points were observed at 277, 320, 377, and 410 nm. No indication of the formation of the trans form was discerned from the absorption spectrum. Upon visible irradiation (λ >500nm) the red color disappeared along with the ultra-violet absorption at 355nm and the initial cis form absorption was restored.

In order to shift the absorption band further to longer wavelength, and to prohibit the cis to trans isomerization completely, the cyano groups were converted to an acid anhydride group. Upon irradiation with 405nm light, the benzene solution containing the acid anhydride derivative became brown and a new absorption peak appeared at 560nm (Fig. 4).

The absorption maximum shifts to longer wavelength by 48 nm in comparison with the dicyano derivative(4c). The new band is ascribable to the closed ring dihydro form. Upon exposure of the brown solution to the visible light (λ=520 nm), the solution again became yellow and the initial absorption was restored.

The photogenerated closed ring forms of 4e and 6 were stable even at elevated temperature. The absorption spectra at 512 nm(4e) and 560 nm (6) remained constant for more than three months at 80°C in toluene in the absence of oxygen, as shown in Fig. 5. The figure also shows that the photochromic property does not change even after the three months storage. The thermal stability of the closed ring form of the chromo-phores dispersed in polystyrene film was also measured. The polystyrene film was prepared by casting the toluene solution containing polystyrene and the compound (5) on glass plate. After being dried in vacuo, the film was irradiated with 405 nm light and the decay of the absorption at 560 nm(6) in the dark was followed, as shown in Fig. 6. Any significant decay of the color is not observed even after one year storage in the dark in the presence of air. In addition, neither 3e nor 5 showed any thermochromic reaction even at 300°C. Compound (5) melted at 161.5°C without changing the initial yellow color. The experimental results support the theoretical prediction that the closed ring forms of diaryl-ethenes having heterocyclic rings with low aromatic stabilization energy are thermally stable.

The theoretical prediction was further confirmed by measuring the lifetime of the closed ring form of 1,2-di(5-cyano-2-methyl-3-pyrrolyl)-2-butene, which has pyrrole rings. The aromatic stabilization energy of the pyrrole ring lies in between those of phenyl and furyl rings. The lifetime was observed to be 32 min. at 20°C, which is longer than that of

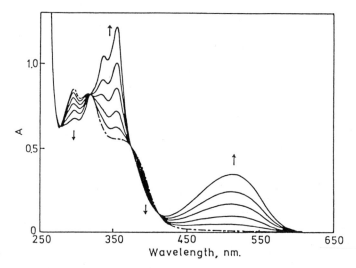

Figure 3. Absorption spectral changes of CCl_4 solution of 3e(—·—) by irradiation with 405 nm light.

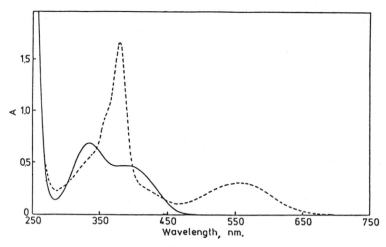

Figure 4. Absorption spectra of 5 (——) and in the photostationary state under irradiation with 405 nm light (----).

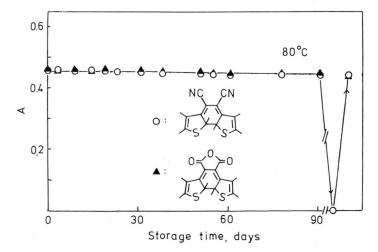

Figure 5. Thermal stability of 4e and 6 in toluene solution.

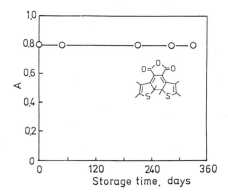

Figure 6. Thermal stability of 6 in polystyrene film.

Table III. Fatigue-Resistant Property of Diarylethene Derivatives(a).

compd	irradiation wavelength, nm		repeatable cycle no.(b)
	coloration	bleaching	
3c	313	435	3
3d	313	405	3
3e	405	546	10
5	405	546	70
5(c)	405	546	> 100
Furyl fulgide	366	546	21

(a):Irradiation was carried out in the presence of air.
(b):The cycle number when the colored intensity decreased to 80%
    of the first cycle.
(c):In the absence of air.

Table IV. Quantum Yields of Coloration and Bleaching.

	excitation wavelength, nm	CCl_4	benzene
$5 \rightarrow 6$	366	0.11	
	405	0.09	0.07
$6 \rightarrow 5$	546	0.15	0.12
	577	0.09	0.07
	633	0.04	

cis-2,3-dimesityl-butene but shorter than that of 2,3-bis(2,5-dimethyl-
3-furyl)-2-butene. Table I summarizes the lifetimes of 2b, 4d and
4c along with above compound and the relative ground state energy
differences between the open ring form and the closed ring one of diarylethenes synthesized. The lifetimes of the colored forms measured
experimentally correlate well with the energy differences calculated for
the conrotatory cycloreversion reaction.

It is difficult to compare quantitatively the fatigue resistance,
i.e., how many times coloration and decoloration cycles can be repeated
without permanent side product formation, because the property strongly
depends on the environmental conditions, such as solvent purity or
matrix properties. We tried to measure the fatigue of 3c, 3d, 3e, and 5
in benzene, as shown in Table III. 3-Furyl-fulgide was also measured
under the same conditions. The repeatable cycle number indicates the
cycle number when the colored intensity decreases to 80 % of the first
cycle. The acid anhydride derivative shows the best fatigue resistance,
which is superior to that of 3-furyl-fulgide under the present conditions.
Elimination of oxygen from the solution prolongs the cycles more than
100 times.

The quantum yields of coloration and decoloration reaction were
measured in benzene and CCl_4. Mercury lines, which were isolated with
a monochromator, were used to induce the reaction. Table IV summarizes
the results. The coloration quantum yield is around 0.1. A slight
increase in the quantum yield with shorter wavelength excitation is
observed. The decoloration quantum yield strongly depends on the wavelength of excitation.

Polymers with Pendant Diarylethene Groups

A vinyl monomer having dithienylethene chromophore was synthesized by the reaction of compound(5) with p-aminostyrene.

7

A homopolymer of the compound (7) and copolymers with styrene were prepared by using 2,2-azoisobutyronitrile as initiator at 60°C in benzene. Molecular weight of the homopolymer was 3×10^5. Upon irradiation with 405 nm light, the benzene solution containing the homopolymer turned brown, the absorption maximum being observed at 550 nm. The photogenerated band at 550 nm is due to the closed ring dihydro form. The copolymer film also showed a similar spectral change upon photoirradiation. In the film, the conversion to the closed ring form under the photostationary state depended on the photoirradiation temperature. In contrast to thermally unstable photochromic compounds, conversion of the thermally irreversible compound increased at higher temperature. The conversion at 120°C was twice larger than the conversion at 20°C. Increase in the flexibility of the polymer chain at higher temperature is considered to assist the photocyclization reaction. The reverse reaction also showed similar temperature dependence.

CONCLUSION

The MNDO calculation of state correlation diagrams suggested that the thermal stability of both isomers of diarylethene type photochromic compounds could be attained by introducing aryl groups, which have low aromatic stabilization energy. The theoretical prediction was confirmed by the synthesis of several diarylethene derivatives with various aromatic stabilization energies. In order to obtain a thermally irreversible photochromic compound, which has sensitivity at longer wavelength and no side reactions such as cis to trans isomerization, 2,3-bis(2,3,5-trimethyl-3-thienyl)maleic anhydride was synthesized. The compound had no thermochromic property and the colored state was stable for more than 3 months at 80°C. The colored form was converted to the colorless form with a He-Ne laser(633 nm). In addition, the coloration and decoloration were repeated more than 100 times.

REFERENCES

1. a) Y. Hirshberg, J. Am. Chem. Soc., 78, 2304 (1965).
   b) H.G. Heller, IEEE Proceeding, 130, 209 (1983).
2. P.J. Darcy, H.G. Heller, P.J. Strydom, J. Whittall, J. Chem. Soc., Perkin I, 202 (1981).
3. R.B. Woodward, and R. Hoffmann, "The Conservation of Orbital Symmetry", Verlag Chemie GmbH, Weinheim, 1970

4. M.J.S. Dewar, and W. Thiel, J. Am. Chem. Soc., $\underline{99}$, 4899 (1977).
5. H.C. Longuett-Higgins, and E.W. Abrahamson, J. Am. Chem. Soc., $\underline{87}$, 2045 (1965).
6. R.M. Kellog, M.B. Groen and H.J. Wynberg, J. Org. Chem. $\underline{32}$, 3093 (1967).
7. K.A. Muszkat, and E.J. Fischer, J. Chem. Soc., B, 662 (1967).

REWRITABLE DYE-POLYMER OPTICAL STORAGE MEDIA

N. E. Iwamoto and J. M. Halter

Optical Data, Inc.
9400 S.W. Gemini Drive
Beaverton, OR   97005

A unique rewritable dye-polymer, bump forming media has been
developed for high density optical data. The media structure
consists of two active layers, an elastomeric layer and a
plastic layer, which are dyed to selectively absorb at different
wavelengths. Writing and erasing is accomplished through the
viscoelastic response of the heated layer. The marks are created
mechanically by thermal expansion of the elastomeric layer. The
subsequent deformation of the surface is captured by the plastic
layer upon cooling. Erasure of the mark is caused when the plastic
layer is relaxed with heat.

Material requirement trends have been identified which change
performance and lead to media structures which create marks with
higher read contrast and which write and erase with lower energy.
The selection of materials is based upon polymer structure. When
combined with layer thickness and dye concentration, the media
system response can be adjusted.

The results obtained for several material systems will be reported
along with observations. Mechanical modeling data which confirm
these trends will be included.

INTRODUCTION

The patented[1] rewritable dye-in-polymer optical media which has been
developed by Optical Data, Inc. consists of two active layers each of
which is dyed to selectively absorb at different wavelengths.   The
bilayer structure consists of an elastomeric layer called the expansion
layer and a plastic layer called the retention layer; writing and erasing
is accomplished through the viscoelastic response of the heated layers.

Media operation is illustrated in Figure 1. During the write process,
the expansion layer is heated by direct absorption of the write beam.
The expansion layer thermally expands against and deforms the retention
layer to create the written mark. During the erase process, the
retention layer is heated past its Tg, and viscoelastically relaxes to
its original shape aided by the elastic reversibility of the expansion
layer.

Examples of the physical properties for expansion and retention layer
materials are given in Figure 2 and Table I. The expansion layer

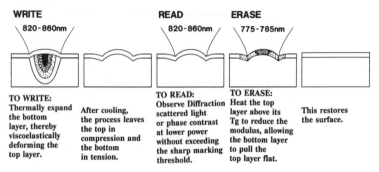

| WRITE | READ | ERASE |
| 820-860nm | 820-860nm | 775-785nm |

TO WRITE:
Thermally expand the bottom layer, thereby viscoelastically deforming the top layer.

After cooling, the process leaves the top in compression and the bottom in tension.

TO READ:
Observe Diffraction scattered light or phase contrast at lower power without exceeding the sharp marking threshold.

TO ERASE:
Heat the top layer above its Tg to reduce the modulus, allowing the bottom layer to pull the top layer flat.

This restores the surface.

Figure 1. Optical Data, Inc. Process for Write, Read, Erase.

materials operate in their rubbery phase and almost entirely in the linear elastic range. It has been estimated by finite element analysis that strains in excess of 50% can be reached during the write process.

The retention layer is required to traverse a wide range of properties controlled by temperature for both the write and erase processes. During the write process, the retention layer is heated primarily by conduction. This serves to lower the modulus of the retention layer enough so that the expansion layer can effectively mold the deformation into the retention layer. Relatively high Tg retention layer materials with high room temperature moduli have shown to yield media with high contrast deformations which erase well. Because of this "molding" process which occurs during the write process, the thermal/strain histories become very important. During the erase process, the retention layer attains temperatures in excess of its Tg, and is thought to be working close to a rubber-like phase. This ensures reversibility and cyclability of the media.

Mechanical modeling has been used in conjunction with the optimization of polymer formulation to establish trends for better media performance; results can be found in a previous paper[2] The media modeling uses the Finite Element Analysis (FEA) technique. The bump profiles of media systems modeled with FEA are consistent with observations of actual media photographed with SEM (Figure 3). Mechanical modeling has been shown to be useful in elucidating the role played by individual layer strengths and thicknesses in mark contrast trends. Typical polymer properties used in the modeling are given in Table I. The size of the deformation is about one micron.

THEORY

The questions which arise most frequently concerning the performance of the bump-forming process are: 1. What is the process whereby the bumps can maintain and release a deformation? and 2. Can the processes be made fast enough in order to improve media performance? There are many studies which help to establish the theory behind the types of molecular structures or processes which may be present during bump formation and erasure. These studies use techniques such as acoustic relaxation[3,4], dielectric relaxation[5,6] ESR[7,8], fluorescence labeling[9,10], neutron scattering[11], NMR[5,6,12-17] and x-ray scattering[18-20] in order to determine polymer molecular structure in solution and in the solid state at varying temperatures. Solid state NMR is especially helpful in understanding high frequency transitions that occur in bulk polymers. These studies confirm that main chain rotational conformational freedom, and therefore molecular structure, is important in determining main chain flexibility, and that chain flexibility varies with temperature.

118

Such studies lend support to theories of conformational transitions of polymers in solution and in bulk based upon Brownian dynamics[21-30]. They also contribute understanding to molecular structure-property polymer theories[31-33,59-61], viscoelastic behavior[34-38], and molecular theories of plastic flow[39-55] such as those considered by Argon[43-45], Bowden[47-51], and Robertson[46] which explain plastic flow at different temperatures based upon monomer unit size and shape and the resulting inter/intramolecular forces.

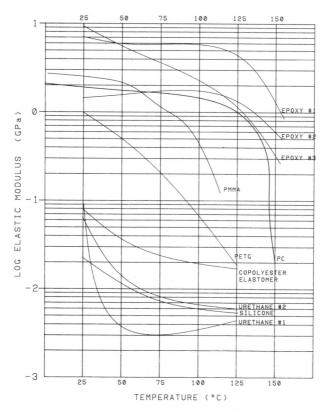

Figure 2.   Typical Material Properties: Effect of Temperature on Modulus.

Use of the principles contained in these studies and theories has been helpful in determining polymer variables which may change media response. Factors which affect the ability of the polymer to undergo conformational transitions (besides the main chain chemical structure) include: crosslink density, side group size and shape, stereochemistry, matrix viscosity, temperature, pressure, stress, molecular weight, packing density, domain structure, crystallinity, and effects such as hydrogen-bonding. A common characteristic of these factors is that they all affect molecular movement by physical constraint. This can be done either by affecting the room available within which the polymer has to move (temperature, viscosity, pressure, crosslink density, molecular weight, stress, side group and steric considerations, polar effects, packing density, crystallinity, or domain structure), or by affecting the intrinsic flexibility of the chain (main chain chemical structure, crosslink structure, side group size and shape, stereochemistry, temperature). For example, the properties of polymers made up of hard-block and soft-block segments vary depending upon the distribution of domains. Some polymers contain micro-domains of short range order which also contribute to the overall flexibility of the polymer and may be thought of as pseudocrosslinks. They have been estimated as small as 3-18nm in size[14,20] in some blends and polymer systems.

TABLE I. Material Parameters Utilized in the Models with Typical Material Property Ranges.

	Retention Layer	Expansion Layer	Substrate (Polycarbonate)
Elastic Modulus (GPa)			
@ 20C	1.75	0.028	2.225
@ 70C	0.52	0.011	
@ 120C	0.14	0.006	
Typical Ranges			
@ 20C	10.0–1.0	0.1–0.01	
@ 70C	5.0–0.5	0.05–0.005	
@ 120C	5.0–0.01	0.056–0.005	
Thermal Expansion ($^{\circ}$C)	0.15×10^{-3}	0.25×10^{-3}	0.15×10^{-3}
Typical Ranges	$0.1–0.3 \times 10^{-3}$	$0.20–0.5 \times 10^{-3}$	
Poisson Ratio			
@ 20C	0.34	0.45	0.35
@120C	0.499		
Typical Ranges			
@ 25C	0.32–0.45		
@ Tg	0.499		
Specific Heat (J/g C)	1.8	1.7	1.17
Typical Ranges	0.9–2.0	1.5–2.5	
Thermal Conductivity (J/m sec C)	0.2	0.20	0.19
Typical Ranges	0.1–0.5	0.1–0.3	
Density (g/cm^3)	1.05	1.05	1.20
Typical Ranges	0.9–2.0	0.9–1.9	

Therefore, if the correct temperature/pressure profile and history is used, deformations can be frozen in place by a change in conformational population and distribution as well as by a change in the micro-domain structure and distribution. Consideration of these states helps to explain why write bumps can be made in expansion layer elastomers (when no retention layer is utilized in the structure) which depend upon domain or microcrystalline states for their elasticity, even though the marks are annealed out by low temperature bulk erase of these materials. Similar processes are likely involved in density changes during temperature quenching studies of epoxy materials[61].

120

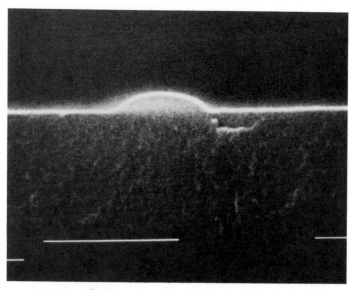

Figure 3.   SEM Micrograph of Optical Data, Inc. Air Incident Media
Showing Cross Section of a Mark (40,000X Magnification, at
90°, 1.0μm Marker).

Writing and erasing in any material may be easy to understand with such
considerations; however, the speed of the process is a difficult property
to assess. The speed of main chain rotation/translation will be
ultimately determined by the temperatures traversed which affect the
conformational states of the main chain and by the steric influences and
rotational states of the side groups. CMAS T1 studies indicate that
correlation times (approximately the time required for a rotation or
translation of a segment or group) for the diffusion of main chain
carbons (main chain motion) can be less than a nanosecond in solution at
room temperature and in the micro to nanosecond range in solids[13].
Temperatures attained in the heated bump have been estimated by FEA[2] (a
typical energy deposition was calculated from a square pulse applied at
10mW for 100ns) to be over 700°C, and can be over 300°C in portions of
the retention layer during the write process. At such high temperatures,
high frequency energy rotations should become more common as energy
barriers are surpassed and viscosity factors drop markedly, especially if
a phase change (e.g. Tg) has also occurred. If such correlation times
are considered as an upper estimate of the speed of material response
(the nanosecond range) during the heating phase, then the limiting factor
in bump formation/erasure becomes the speed at which thermal repopulation
of "normal" room temperature conformational states is resumed during the
cooling phase and also whether or not a disruption of chain packing,
crystallinity or order (or micro-domains) has resulted from the process.

Therefore, for elastomers whose chains exhibit fairly unhindered rotation
with little or no steric interferences, anchor groups or hard blocks
present, the speed of conformational repopulation could be as fast as the
cooling process. These effects are demonstrated in various expansion
layer materials. Expansion layer materials made of elastomers of low
modulus whose structures exhibit free rotation around single bonds do not
retain write deformations by themselves; whereas expansion layer
elastomers, which consist of hard and soft block domains, can be made to
retain a small deformation with the write beam.

121

Another way of looking at bump formation and erase considers effects of mobility and conformational changes on free volume. If free volume is decreased, then viscosity increases[10]. An increase in viscosity is associated with hindered movement of the matrix, the matrix in this case being the polymer. A decrease in viscosity is associated with larger freedom of movement and larger free volume. Factors which affect the freedom of movement and, therefore, the free volume in the condensed polymer include polymer molecular structure (intrinsic flexibility), secondary order such as packing, and temperature and pressure as mentioned above. It is a logical extrapolation that effects which alter free volume, and therefore the molecular movement in a polymer will also affect glass transition which is a reflection of the backbone motion of a polymer[21]. Studies of Tl relaxation at high pressures indicate that high pressures lead to an increase in the glass transition point[5] as qualitatively predicted by consideration of the effect of pressure on molecular motion.

MECHANISMS

During the write process, several transformations within the bump are taking place according to thermo-mechanical modeling[2]. High temperatures are generated in the expansion layer corresponding to an energy deposition based upon laser energy profile, media thickness, dye loading and time. Thermal expansion occurs which is a direct consequence of thermal energy exchange from dye to polymer linked through molecular motion[13,56-58]. After beam shut off, the energy dissipates and the hot spot migrates toward the surface of the retention layer. A portion of the retention layer material attains temperatures above its glass transition temperature, and is thought to be working close to its rubbery phase. This implies low viscosity, high rotational conformational freedom and translation, and increased free volume. High pressure areas at the top and sides of the bump created by radial compressive stresses are generated within the retention layer. High compressive stresses are also generated in the expansion layer. As the generation of stress and pressure continues, at the molecular level the polymer is redistributing its conformational population; variables such as packing effects can become important. At the higher pressure areas, internal rotation and net translation is hindered at a faster rate than at areas of low pressure[5], leading to a non-isotropic change in free volume throughout the bump retention layer.

Therefore, it is those areas of high strain, temperature and pressure in the retention layer which correspond to the maximum amount of rotational and conformational changes and which lead to the changes in polymer properties in reference to the rest of the retention layer material in the bump. Any other factors such as aromatic rings, polar entities, or hydrogen bonding capabilities, which can influence localized order, can only enhance this process; groups subject to steric influence will also change the molecular spacing and free volume in these areas. Because these are areas of increased inhibition of rotational and translational movement, it is a logical sequence that these areas will pass through their glass transition state before the rest of the retention layer material during cooling. Below the glass transition point there are no net changes in free volume in these areas. The rest of the material, still above its glass transition, is allowed to cool following its own separate relaxation process. At room temperature the discontinuity of properties is maintained and a net deformation has resulted. The high thermal quench rate on the order of 10^{6}°C/second allow little time for viscoelastic based restoration forces to return the material toward its initial configuration.

Changes in pressure and strain are also occurring in the expansion layer. But because these materials are usually chosen for high chain flexibility (few bulky substituents and crystallizable groups), there is less concern over creating lasting property discontinuities within the material. However, if the elastomer is of insufficient molecular weight and crosslink density, there still remains the danger of creating permanent deformation due to processes such as chain disentanglement or other relaxation processes.

During the erase process, the retention layer is heated past its glass transition again to a state of high motional freedom and increased free volume (back to its rubbery regime); however, this time pressure and stress differentials across the heated area are reduced. The previously deformed area is given time to relax out to a material more like the surrounding heated retention layer material. This can be accomplished by varying factors such as dye loading, a larger beam diameter, longer erase irradiation and lower laser energies, in addition to changing the polymer structure. In this way the bump can cool more isotopically resulting in continuous properties across the bump area of the retention layer which leads to an erased bump. The same type of effect is used in bulk erase where the entire disk is heated, resulting in more continuous properties across the retention layer. The expansion layer which is not affected by discontinuities as much as the retention layer, because of its high chain flexibility, aids bump erasure by helping to pull the softened surface retention layer back with it as both layers cool; this also serves to speed the process of erasure.

MATERIAL TRENDS

Many of the early correlations at Optical Data, Inc., of material vs. media response involved measurements of bulk materials properties such as tensile modulus or glass transition as indices of performance. For example, write contrast and cycling ability are particularly sensitive to the type of retention layer material used. In general, higher modulus, high Tg materials give the best write contrast and high cycling ability. However, it was later found that the types of measurements made were not adequate enough to use as predictive tools of media response. A finer index was needed, preferably a predictive tool which could more directly link performance with formulation or polymer structure.

Because of the subtle nature of temperature[61] and pressure effects on material changes, property variations could be easily overlooked if an insufficient or insensitive analysis technique is employed. Therefore, at present, the link between polymer structure, macroscopic bulk properties and media performance is still unclear.

However, certain trends have emerged which link polymer structure and media performance when considered from a molecular viewpoint. The principles contained in other polymer studies such as solid state NMR have given support to the rationale of media response dependence upon changes in internal polymer structure. Thus, without further measurements taken, observed molecular structure trends could be used to estimate probable media performances, just as correlations of structure to polymer bulk properties are being made today[31-33,59-61]. Some of the trends observed in expansion layer materials have already been mentioned in previous sections; other trends will be discussed in this section. Unfortunately, because of the proprietary nature of the polymers involved, only general trends and generic names can be used.

Both the expansion and retention layer materials need to be crosslinked in order for the media to write and erase with short irradiation times and maintain write/erase cycling characteristics. Although crosslink density measurements have not been made, measurements such as performance vs. stoichiometry suggests that a relatively high crosslink density is necessary for both layers. Such trends suggest that at least part of the retention layer is working in a temperature regime in excess of its glass transition where crosslink density becomes more important. As is predictable from considerations of modulus versus crosslink density where the strength of the material maximizes with increased crosslink density, the retention layer dependence upon crosslink density is demonstrated by write performance (signal) which maximizes stoichiometry (Figure 4).

This also agrees with write/erase cycling studies of uncrosslinked retention layers: although the media retains high contrast and sensitive write deformations, the erase and cycling ability of these polymers is poor in contrast to crosslinked systems because of the growth of a residual deformation caused by the erase beam upon erasure. The residual growth suggests a mechanism which can be affected by crosslinking.

On a molecular scale, the need for crosslinking suggests the need to control long range movement (and translation). The need for crosslinking also suggests that systems which already consist of segments of polymer with restricted conformational freedom due to chain structure, upon crosslinking should be even more susceptible to high temperature effects and pressure differentials created within the bump. This is generally consistent with T1 studies which indicate that motional freedom (and therefore free volume changes and packing effects) is altered by the chain length, molecular weight and crosslink density. This is also true for expansion layer materials. Higher crosslinked materials and higher modulus elastomers appeared to perform slightly better than lower crosslinked materials, which agrees with trends found for the retention layer.

However, it was also found that the type of crosslink used is just as important as the density of the crosslink. The correct tradeoff between the two has become important to the understanding of media performance.

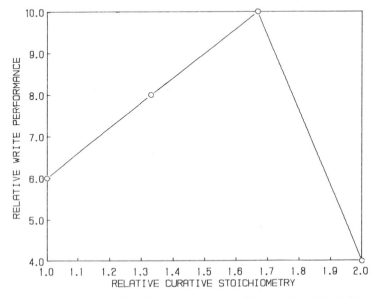

Figure 4. Relative Write Performance vs. Increasing Stoichiometry.

The molecular influence to the write/erase cycling of the media was first observed in static tests (microscopically evaluated) of media with retention layer systems which varied curative type while keeping the resin constant (Table II). Those resins crosslinked with curatives of low bond flexibility, such as those with aromatic character, generally produced media which wrote with high contrast bumps, exhibited intermediate write/erase cycling characteristics at medium to high write powers. Those resins crosslinked with highly flexible curatives with high aliphatic character generally wrote with low energies, with low contrast bumps and exhibited good write/erase cycling characteristics. Those resins crosslinked with large steric groups present exhibited high contrast bumps at medium write energies with good write/erase cycling.

Using the preceding arguments of molecular effects on the conformational freedom in polymers, these effects can be rationalized. Due to the ability of the aromatic groups to undergo only limited motion, the aromatic groups contribute an anchor effect (sterically induced and induced through packing) at low temperatures which hinders motion of the chain and causes a higher temperature to be necessary in order to surpass energy barriers for segmental rotation or translation; they also contribute to changes in the polymer conformational structure during the write process. In contrast, the aliphatic curatives contribute a high amount of flexibility to the chain. A lower amount of heat is necessary for segmental motion, leading to an increase in sensitivity. By the same token, the high flexibility makes the polymer matrix harder to set up conformational differentials, and the contrast and signal is lower. Because of the high flexibility of these groups, erasure and write/erase cycling is easier as the tendency of these groups to relax out deformations is increased. The curative using bulky substituents adds steric inhibition to the already inhibited motion of the chain due to crosslinking. These substituents may be thought of as acting as physical anchors due to size (and the sterically induced inability of these groups to undergo rotation or translation) instead of through packing effects or bond type. They also add to the free volume available for main chain conformational changes, again due to the large size of these groups relative to other curative types which contributes to the ability of other segments to rotate and reorient. In comparison to the other crosslink types, these groups create a balance of adding to main chain flexibility by increasing local free volume (which contributes to sensitivity, erasability and cyclability), and decreasing total flexibility through the steric anchors present (which contributes to the contrast or signal strength of the mark) and are especially important considering the high temperatures attained.

A similar trend was seen when resins were varied in the retention layer and the curative was kept constant. When resins of high flexibility and low flexibility bond character were compared, the better performing resins were generally of the latter type.

The resins also exhibited mixtures of lower erase performance or lower environmental stability depending upon the curative. Aliphatic resins cured with high steric curatives generally led to media which had higher write and erase energies and those cured with flexible curatives produced media with lower environmental stability, when compared with resins of aromatic character. In addition, aliphatic resins with higher aliphatic content overall did not perform as well as those with lower aliphatic content. This suggests the importance of steric control as well as the control of flexibility and intermolecular interaction (chain packing) or micro-domain structure within a retention layer.

TABLE II. Relative Media Performance Based Upon Curative For Retention
Layer Material.

Curative	A	B	C*	D	E
		In order of decreasing flexibility			
Resin #1: high steric, low flexibility					
Relative write energy	L	L	M	M	M-H
Write contrast	L	L	M	M	L
Relative erase energy	L	M	M	M	H
Rewrite contrast	M	M	M	L-M	L
Resin #2: med steric, med flexibility					
Relative write energy	--	--	L-M	--	M
Write contrast	--	--	M	--	L
Relative erase energy	--	--	H	--	H
Rewrite contrast	--	--	L	--	L
Resin #3: low steric, high flexibility					
Relative write energy	L	--	L-M	--	M
Write contrast	L	--	L-M	--	L-M
Relative erase energy	M	--	H	--	H
Rewrite contrast	M	--	--	--	L
*High Steric Curative		L= low; M= medium; H= high			

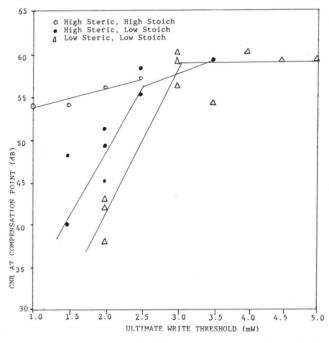

Figure 5. CNR vs. Write Threshold as a Function of Stoichiometry and
Steric Hindrance.

In addition, these trends were confirmed when disks were produced and
dynamically tested to obtain carrier-to-noise ratios (CNR) vs. write
energy data (Figures 5, 6, 8). These figures relate performance
parameters (such as signal or sensitivity to polymer structural factors
such as crosslink density (stoichiometry) and steric affects (group
size). In general, the high steric resin and curative combinations which
were highly crosslinked produced disks with better signal (higher CNR) as
shown in Figures 5 and 8 and with better sensitivity (lower write
energies)as shown in Figures 5 and 6.

Further evidence of the importance of molecular aspects was seen in
studies which used additives such as plasticizers as part of the
formulation. In general, an increase in the amount of plasticizer
brought about a slight decrease in bump contrast. The molecular
mechanisms could be rationalized as follows. The plasticizer is able to
move more easily at lower temperatures than the polymer. In addition,
the presence of the plasticizer interrupts packing, domain formation or
crystallization, increasing the ability of the chain to move. An
increase in motion, which may be thought of as a decrease in matrix
viscosity, leads to a subsequent increase in main chain motion. The
polymer is less constrained and is able to flex with a slightly higher
range of motion and frequency. The tendency for the retention layer to
set up conformational differentials is decreased; therefore the bump
would be expected to be smaller.

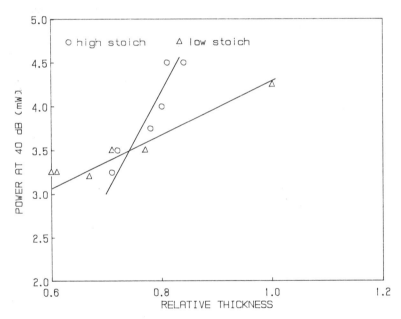

Figure 6.   Sensitivity vs. Increasing Relative Thickness of Retention
             Layer.

A fine balance exists between the thickness of the material, its modulus
as a dependence of temperature, and the temperatures achieved. For
example, both finite element analysis[2] and experimental work has shown
that retention layer thickness plays a key role in the bump height, and
therefore the signal obtained from the media (Figures 7, 8). In both
these studies, an optimal bump shape or signal was obtained. The working
retention layer is predictable from a mechanical standpoint. If the
layer is too thin then mechanically it cannot support any internally
induced stresses and a smaller deformation results; if the layer is too

thick, it becomes harder for the expansion layer to aid deformation of the retention layer and a smaller deformation results. Because of the existing balance between thickness and structure effects, care must be exercised in evaluating materials for improved media performance. Many materials were initially discarded when the thickness of the layer used in evaluation masked the ultimate performance capabilities of the polymer. In some cases, the effect of a curative which added too much inflexibility to a chain was moderated by a mixture of layer thickness decrease and dye content increase so that higher temperatures were attained in the retention layer. On a molecular scale, higher temperatures would imply larger free volume, lower viscosity, and higher internal pressures created within the layer which would lead to the increase in sensitivity needed for the stiffer material.

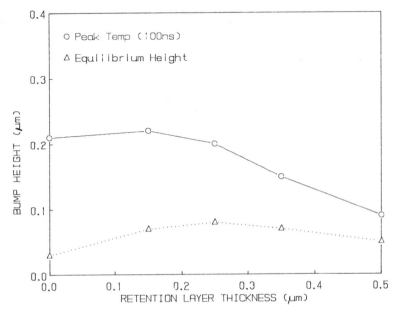

Figure 7. Modeling Results Showing Bump Height at Beam Shut-off (peak media temperature), and Bump Height Upon Cooling Media to Equilibrium Temperature (room temperature).

The dependence of media performance on both thickness and stoichiometry is also demonstrated in Figures 6 and 8. The thinner the retention layer, the more sensitive the media. The higher the curative stoichiometry of the retention layer (the higher relative modulus of the retention layer), the higher the ultimate CNR at the lower thicknesses which are required for increased sensitivity. Thus, for a combination of both increased sensitivity and signal, the retention layer must be both thinner and stronger.

CONCLUSIONS

Various studies have shown that molecular motion exists in polymers in the condensed state at varying frequency ranges. These studies indicate that the motional range and frequency are dependent upon lattice energy; i.e., temperature. Therefore, it is a natural extrapolation that there exists a continuum of motional frequencies and ranges which vary with temperature and depend upon the chemical make-up of the molecule. It is the structure of the polymer which determines motional freedom through the sum of the conformational states available.

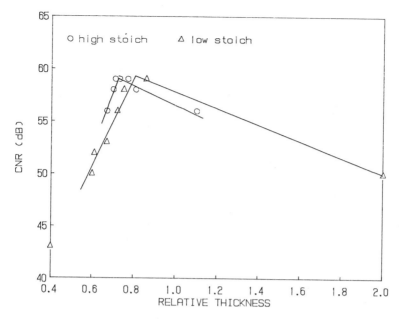

Figure 8. Ultimate CNR vs. Increasing Relative Retention Layer Thickness

The media performance is directly linked to controlling the type of conformational motions (or the motional continuum) existing at any time during the write and erase process. What we have tried to show experimentally is that there indeed exists a direct relationship of the polymer structure to the performance of the media. (Table III)

TABLE III. Typical Effects of Structural Changes on Media Performance Parameters.

Structural Changes	Performance Parameter	
	Signal	Sensitivity
Crosslink Density	+	+/-
Steric Group Content	+	+
Aromatic Content	+	-
Aliphatic Content	-	+

(+ = positive influence on parameter,
 - = negative influence on parameter)

Because the dye-in-polymer media must be both written and erased at high speeds, the laser energy profiles and thermal histories cannot be easily tailored to the polymer for maximum signal or erase cycling, rather the polymer must be matched to the high speed of the required write and erase process. Understanding of the impact of the molecular processes involved will make it easier to choose materials which respond to the fast processes which are required and make it possible to estimate ultimate performance limits of the systems chosen.

ACKNOWLEDGEMENTS

The help of many individuals has enabled this materials effort at ODI to take place. They include Renne Harris and John Swanson for their useful insights and assistance in sample preparation; Robert Marx, Sonia Sachdev, Geoffrey Russell and W. Eugene Skiens for material property measurements and discussions; Brian Holloway and Joseph Monek for their tireless media testing; and Michael A. Lind and John Hartman for useful discussion and encouragement. Appreciation is also extended to George H. Johnson of Philips & Du Pont Optical for obtaining SEM photographs of our media. Finally, without the help of Benita Harper and Jim Pylant, this manuscript would still be an ink blot on a napkin.

REFERENCES

1.   Optical Data, Inc.; U.S. Patent 4,719,615; 1988.
2.   J.M. Halter and N.E. Iwamoto, SPIE Symposium Proceedings No. 899, January 1988.
3.   Michael A. Cochran, John H. Dunbar, Alastair M. North, and Richard A. Pethrick, JCS Faraday II, $\underline{70}$, 215 (1974).
4.   Alastair M. North, Richard A. Pethrick and Ian Rhoney, JCS Faraday II, $\underline{70}$, 223 (1974).
5.   W. P. Slichter; NMR (1971) v.4, p. 209 ff; ed. P. Diehl, E. Fluck, R. Kosfeld, 1971.
6.   David W. McCall, Acct. Chem. Res., $\underline{4}$, 223 (1971).
7.   K. Shimada and M. Szwarc, J. Amer. Chem. Soc., $\underline{97}$, 3313 (1975).
8.   N.M. Atherton, Chem. Phys. Lett., $\underline{23}$, 454 (1975).
9.   Curtis W. Frank and Larry A. Harrah, J. Chem. Phys., $\underline{61}$, 1526 (1974).
10.   Rafik O. Lautfy, Pure and Appl. Chem., $\underline{58}$, 1239 (1986).
11.   Geoffrey Allen, Julia S. Higgins and Christopher J. Wright, JCS Faraday II, $\underline{70}$, 348 (1974).
12.   Francoise Laupretre, Lucien Monnerie and Joseph Virlet, Macromolecules, $\underline{17}$, 1397 (1984).
13.   F. A. Bovey and L. W. Jelinski, J. Phys. Chem., $\underline{89}$, 571 (1985).
14.   Brigitte Albert, Robert Jerome, Philippe Teyssie, Gerard Smyth and Vincent J. McBrierty, Macromolecules, $\underline{17}$, 2552 (1984).
15.   Jacob Schaefer, M.D. Sefcik, E. O. Stejskal, R. A. McKay , W. Thomas Dixon and R.E. Cais, Macromolecules, $\underline{17}$, 1107 (1984).
16.   Gary E. Maciel, Science, 226, 282 (1984).
17.   James C. Randall, ed., ACS Symposium Series No. 247, 1984
18.   H.H.M. Balyuzi and R. E. Burge, Biopolymers, $\underline{10}$, 777 (1971).
19.   R. Adams, H.H.M. Balyuzi and R. E. Burge, J. Mater. Sci., $\underline{13}$, 391 (1978).
20.   H.H.M. Balyuzi and R. E. Burge, Nature, $\underline{227}$, 489 (1970).
21.   Eugene Helfand, J. Chem. Phys, $\underline{54}$, 4651 (1971).
22.   H. A. Kramers, Physica VII no. 4, 284 (1940).
23.   Herbert Morawetz, Acct. Chem. Res., $\underline{3}$, 354 (1970).
24.   Gerald Wilemski and Marshall Fixman, J. Chem. Phys., $\underline{58}$, 4009 (1973).
25.   R. Cerf, Chem. Phys. Lett., $\underline{22}$, 613 (1973).
26.   Eugene Helfand, J. Chem. Phys., $\underline{69}$, 1010 (1978).
27.   Eugene Helfand, Z. R. Wasserman and Thomas A. Weber, J. Chem. Phys., $\underline{70}$, 2016 (1979).
28.   Jeffrey Skolnick and Eugene Helfand, J. Chem. Phys. $\underline{72}$, 5489 (1980).
29.   Elisha Haas, Ephraim Katchalski-Katzir and Izchak Z. Steinberg, Biopolymers, $\underline{17}$, 11 (1978).
30.   D. A. Rees and R. J. Skerrett, Carbohyd. Res., $\underline{7}$, 334 (1968).
31.   George C. Derringer and Richard L. Markham, Organic Coatings and Applied Polymer Science Proceedings, $\underline{46}$, 237(1981).

32. D. H. Kaelble, Organic Coatings and Applied Polymer Science Proc., 46, 241(1981).
33. R. J. Young, ACS Symposium Series No. 337, Chapter 20, 1987
34. Malcolm L. Williams, Robert F. Landel and John D. Ferry, J. Amer. Chem. Soc., 77, 3701 (1955).
35. F. Bueche, J. Chem. Phys., 22, 603 (1954).
36. K. W. Scott and R. S. Stein, J. Chem. Phys., 21, 1281 (1953).
37. Prince E. Rouse, Jr., J. Chem. Phys., 21, 1272 (1953).
38. Bruno H. Zimm, J. Chem. Phys., 24, 269 (1956).
39. A. Peterlin, J. Mater. Sci., 6, 490 (1971).
40. S. Rabinowitz, I. M. Ward and J.S.C. Parry, J. Mater. Sci., 5, 29 (1970).
41. R. A. Duckett, S. Rabinowitz, and I.M. Ward, J. Mater. Sci., 5, 909 (1970).
42. Salim Yamini and Robert J. Young, J. Mat Sci., 14, 1609 (1979).
43. A. S. Argon, Phil. Mag., 28, 839 (1973).
44. A. S. Argon and M. I. Bessonov, Phil. Mag., 35, 917, (1977).
45. A. S. Argon, R. D. Andrews, J. A. Godrick and W. Whitney, J. Appl. Phys., 39, 1899 (1968).
46. Richard E. Robertson, J. Chem. Phys., 44, 3950 (1966).
47. P. B. Bowden and S. Raha, Phil. Mag., 29, 149 (1974).
48. P. B. Bowden and J.A. Jukes, J. Mater. Sci., 7, 52 (1972).
49. P. B. Bowden, Phil. Mag., 22, 455 (1970).
50 P. B. Bowden and S. Raha, Phil. Mag., 22, 463 (1970).
51. A. Thierry, R. J. Oxborough and P. B. Bowden, Phil. Mag., 30, 527 (1974).
52. J.B.C. Wu and J.C.M. Li, J. Mater. Sci., 11, 434 (1976).
53. Salim Yamini and Robert J. Young, J. Mater. Sci., 15, 1814 (1980).
54. I. M. Ward, J. Mater. Sci., 6, 451 (1971).
55. A. Keller and D.P. Pope, J. Mater. Sci., 6, 453 (1971).
56. G. Hirsch and G. Rehage, Angew. Chem. Int. Ed., 8, 385 (1969).
57. S. W. Benson, F. R. Cruickshank, D. M. Haugen, H. E. O'Neal, A. S. Rodgers, R. Shaw and R. Walsh, Chem. Rev., 69, 279 (1969).
58. Sidney W. Benson and Jerry H. Buss, J. Chem. Phys., 29, 546 (1958).
59. J. H. Coates and Charles E. Carraher, Jr., Polymer News, 9, 77 (1983).
60. James E. Mark, Acc. Chem. Res., 18, 202 (1985).
61. Eric S.W. Kong, in ACS Symposium Series No. 243, S.S. Labana and R.A. Dickie, editors, Ch. 9, American Chemical Society, Washington D.C., 1984

REWRITABLE DYE-POLYMER OPTICAL STORAGE MEDIUM: DYNAMIC PERFORMANCE CHARACTERISTICS

W. E. Skiens and G. A. Russell

Optical Data, Inc.
9400 S.W. Gemini Drive
Beaverton, OR   97005

The fundamental mechanism of storing digital information in the form of thermally-induced mechanical deformations in glassy organic polymer films has been briefly described in the preceding companion paper. A mechanism for erasure is also given. This paper discusses the practical structures and materials configuration that are used to record and erase information. The dynamic optical performance of the medium as a function of carrier to noise ratio, mark length characteristics and rewritability are discussed. The environmental stability for certain configurations of the medium is also shown.

INTRODUCTION

A rewritable optical data storage medium has been developed by Optical Data, Inc. This medium is totally organic containing only polymers and dyes (no metal or metal oxide compounds). It has been demonstrated to have high density write/read characteristics, is capable of being erased and rewritten and has excellent environmental stability over a relatively wide range of operating temperature and humidity. The mechanism of operation of this solvent-coated medium was described in the previous paper[1] together with some of the thermomechanical properties which influence its performance. This medium, consisting of two or more layers, has been coated on a variety of substrates. Its unique mark forming mechanism produces deformations (bumps) which are optically written and readable, yet can be erased locally by laser irradiation or universally by thermal treatment. Typical laser written bumps are shown in Figure 1, SEM micrograph taken at a 45° angle. These deformations are generally a little less than 1μm in diameter. Figure 2 illustrates the adaptability of this medium to run length encoding, showing variable length marks written on the media.

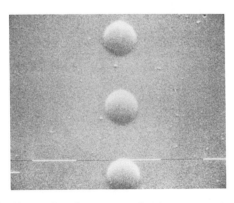

Figure 1. SEM (45° Tilt) of Write Bumps @ 20kx magnification, 1.0μm markers.

MEDIA STRUCTURE

The exact type of structure and mechanism of operation used in this technology has been described in a previous paper[1], however, it is possible to prepare the medium in more than one configuration. Figure 3 shows two significantly different configurations in which this medium may be formed and demonstrated. These two types are termed "air-incident (AI)" and "substrate-incident (SI)" media. These names typify the method in which these media are written, read and erased by incident laser beams. As shown in Figure 3a, the laser beam is directly incident upon the active layers at the air interface, while in Figure 3b the laser beam passes through the clear substrate, upon which the active layers are coated, before interaction with the dye-polymer structure. In either instance, a non-vesicular bump is formed which may be read as low power laser light is scattered from it.

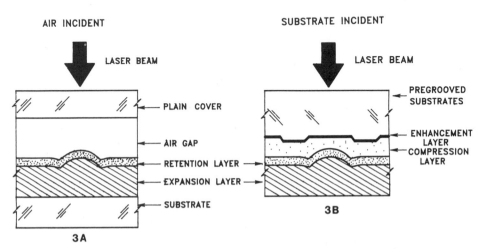

Figure 2.   Run Length Code on ODI Media Produced by Varying Laser Pulse Time. Smallest Deformations are about 0.8μm diameter.

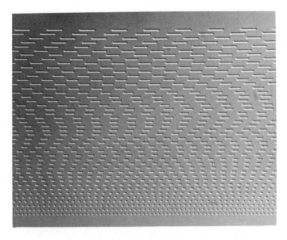

Figure 3.   Air Incident and Substrate Incident Media Configurations.

134

Both configurations use the same type of active expansion and retention layers in which the bump is formed and retained as explained previously. The AI structure is the simpler of the two configurations consisting of just two dyed layers, an elastomeric expansion layer and a glassy polymer retention layer. The expansion layer typically is applied by spin coating from a dye-polymer/solvent formulation onto a polycarbonate optical disk blank to produce an optically flat coating. Depending upon the polymer system used, this coating may be cured, either by thermal or radiation induced polymerization, producing a film a few μm thick. The glassy retention layer formulation then is similarly coated over the expansion layer. Again, it may be cured as was the expansion layer to provide a final structure with a reflectivity in the 12 to 30% range. It is usually desirable to control the thickness of the retention layer, not only to optimize mechanical properties as discussed in the previous paper[1], but also from optical considerations to obtain maximum reflectivity. To protect the media surface from abrasion and also to act as a dust defocusing structure for the disk, a clear polycarbonate cover is attached to the substrate over the coatings using adhesive stand-off spacers at the periphery and inside the coating area.

SI medium is somewhat more complex and typically consists of four layers which are sequentially coated and cured (where necessary) on a pre-grooved or formatted polycarbonate substrate. The first coating is a groove reflectivity enhancement layer which may consist of a material of substantially higher refractive index, either organic or metal/metal oxide, than the substrate. Due to this index mismatch some reflection occurs at the formatted substrate interface allowing the disk drive servo to track on the disk. The major portion of light, however, passes through this layer to the subsequent layers. The second coating is a clear elastomer of relatively low refractive index, titled the compression layer. The thickness of the layer is carefully controlled to permit simultaneous reading of both the format structure and the bumps generated in the third layer, the retention layer. The thickness of the compression layer, as an optical cavity, is also controlled to provide maximum reflectance from the structure. The effect of thickness of the compression layer on the reflectance of a SI four layer structure is shown in Figure 4. The next coating, the retention layer, contains a narrow band-width dye which strongly absorbs 780nm diode laser light, but is primarily transparent to 840nm light. The refractive index of the retention layer is substantially higher than that of the compression layer. This layer thickness is also carefully controlled to provide both mechanical strength for bump retention and constructive interference with light reflected from both the retention and enhancement layers. The final active coating, the expansion layer, is formulated with a dye absorbing strongly at 840nm, which upon heating expands to generate the bump in the retention layer. Two SI disks may be assembled together to form a double-sided assembly with the substrate furnishing a protective and dust defocusing cover.

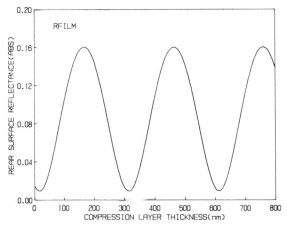

Figure 4.   Effect of Compression Layer Thickness on Rear Surface Reflectance from SI Quad Layer.

The AI medium does not require a transparent substrate and has been coated on many types of materials: aluminum, glass, plastic, silicon, paper, etc., while the SI configuration has been constrained to use glass and plastic substrates.

The spectral properties of typical expansion and retention layer near-IR dyes are shown in Figure 5. Note that the retention layer dye is highly transmissive of the 840nm write/read laser, but highly absorptive for the 780nm erase beam. The bottom curve shows the reflectance of the retention layer coating alone.

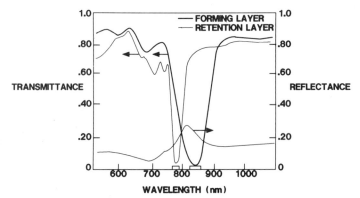

Figure 5.   Spectral Curves of Retention Layer and Expansion Layer Coatings. Narrow Band-width at 780nm is important for proper erase medium.

DYNAMIC OPTICAL PERFORMANCE

Carrier to Noise Evaluations

Carrier-to-noise ratio (CNR) is one of the most frequently shown evaluations of optical media. This ratio of the carrier level to the noise level in the same channel at increasing laser power, if used judiciously, can be a means for comparing signal contrast of similar media. The CNR curve shown for AI media in Figure 6 provides an initial screening test for performance. The signal is expressed in decibels and gives a value of about 60dB above a very quiet noise floor. The second harmonic for this medium is also shown with a compensation point (minimum second harmonic content) between 10 and 11mW. Figure 7 reveals similar curves for a SI medium with the exception that this shows slightly lower maximum CNR values (about 55dB compared to 60dB for AI). The compensation point for the SI medium is at about 8mW indicating somewhat larger marks than those observed for AI medium with both written at 1.5µm bit length (1MHz signal). The mark onset thresholds (write power at which a CNR signal is first observed) for both types of media are very sharp and similar in value.

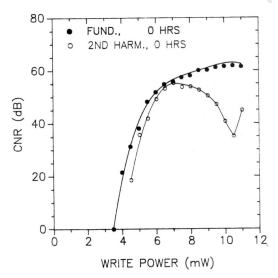

Figure 6.   CNR vs Power for AI Media. Linear Velocity 3.0m/s; Write and Read Wavelength 840nm; Duty Cycle - 50%.

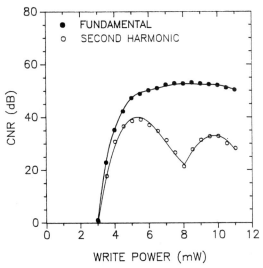

Figure 7.   CNR vs Power for SI Media. Grooved Media; Linear Velocity 3.0m/s; Write and Read Wavelength 840nm; Duty Cycle - 50%.

The quality of the substrate used in SI media can be of critical importance in optical recording.[2] The materials used in the substrates can affect such characteristics as birefringence, and the molding and stamping (formatting) techniques also influence greatly the final quality of a disk. As an illustration of this, Figure 8 shows CNR curves for two different "good quality" polycarbonate substrates coated with the same formulations and prepared under identical conditions. At 40dB and above, the CNR of the two disks vary by up to 10dB. It has been found that the noise floor levels of carefully finished media are determined primarily by the noise floor of an uncoated disk. Clearly then, the substrate quality can influence CNR measurements very significantly.

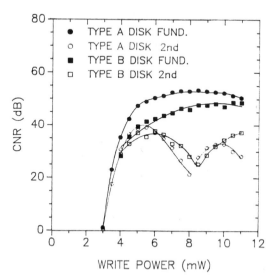

Figure 8.   CNR vs Power for Identical Coatings and Differing Substrates.

The CNR may also be shown as a function of linear velocity of the media. In general, the media discussed here has been optimized to perform in the 3-5m/s linear velocity range and this is illustrated by the data in Figure 9 in which the CNR is shown as a function of write power at various linear velocities. Above about 40dB, the contrast is observed to scale with linear velocity. Using the ratio of the velocities, the CNR data at a particular linear velocity and write power can be used to estimate the power necessary to provide this contrast at another velocity. Thus the marking is a linear function of the absorbed energy.

Mark Length Characteristics

Both AI and SI media have very similar mark length characteristics. Figure 10 illustrates the write mark length versus read mark length at two write powers. Mark length effects are considered with the bit cell length held at 10.0μm. Little or no write compensation is necessary to yield appropriate mark lengths down to the diffraction limited spot size of the laser. At the compensation power, the marking process is essentially linear. This indicates that the medium is very suitable for use with run length encoding schemes.

The usefulness of the medium is also shown in Figure 11 which plots mark length uniformity (counts versus a time base sample) using MFM code at three marking frequencies, with very good mark separation at 0.2μm. The very clear distinction revealed in the mark length histograms also shows full-width bit edge jitter with values of ~0.10μm on AI media and with SI media nearly as good at values of 0.15μm

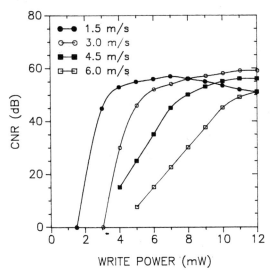

Figure 9.   CNR vs Power for Increasing Linear Velocities on SI Disk.

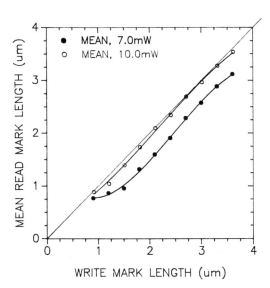

Figure 10.   Read Mark Length as a Function of Write Mark Length for 10μm bit cell.

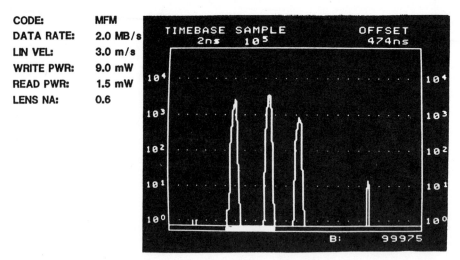

CODE:	MFM
DATA RATE:	2.0 MB/s
LIN VEL:	3.0 m/s
WRITE PWR:	9.0 mW
READ PWR:	1.5 mW
LENS NA:	0.6

TIMEBASE SAMPLE   OFFSET
2ns   10⁵   474ns

Figure 11.   Counts vs Time Base for MFM Encoding - Data Rate 2.0 MB/s; Linear Velocity 3.0m/s, Write Power 9.0mW; Read Power 1.5mW.

Erasability (Rewritability)

Another distinctive feature of the ODI media is its rewritability. As described in the previous paper[1], heating of the retention layer with a laser of a different wavelength erases the bumps. This occurs as a 780nm laser pulse heats the retention layer above its glass temperature (Tg) permitting the elastomeric expansion layer, which has been held in tension, to relax, thus helping to erase the bump. A write and erase CNR versus power curve is shown in Figure 12 in which bumps written at 7.5mW with an 840nm laser are erased using a 780nm beam. This action results in the 57dB carrier being reduced to about 22dB - 35dB signal differential. Computer

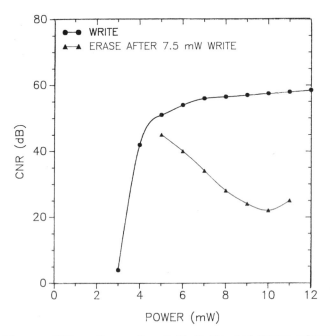

Figure 12.   CNR vs Power for Both Write and Erase After Write.

modeling studies[3] indicate that relatively high instantaneous temperatures
are likely in these systems (500° to 800°C are estimated). The results
shown are all single pass erasure. Even better erasure has been observed
recently as the quality of the erase beam on the tester is improved.

Write/erase data are also illustrated in Figure 13, which plots 1000 erase
cycles for an AI sample. Note that during the 1000 cycles, the erase
performance actually appears to improve as the CNR for the erase cycle
drops from about 25 to 5dB. The reason for this effect is not known and
may be tester related, but no significant changes in either bit edge jitter
or mark length were observed in these tests. The write and erase
effectiveness begins to degrade significantly after a few thousand cycles
on the AI medium and the less mature SI medium loses contrast after a few
hundred to 1000 cycles. Work to extend the erase cycling to 10^6 or more
cycles is being carried out.

It is also of interest to note that erasure may be carried out by a
different method than bit erasure using a laser beam at a specific
wavelength. This other method, termed "bulk erasure", is done by simply
placing the media in an oven and raising the temperature above the Tg of
the retention layer. This allows the bumps to relax and all erasable data
on the disk are deleted. This technique has been done only manually, and
as a result only a few cycles have been carried out, but it appears
reversible over the small number of cycles.

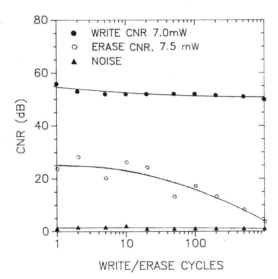

Figure 13.   CNR Before and After Erase vs No. of Write/Erase Cycles.
              Noise Floor for 1000 Cycles.

Environmental Stability

A mechanism for the erase process has been described in the previous
paper[1]. The thermally induced deformations in the polymer retention layer
depend for their integrity on relaxation processes in amorphous polymer
glasses. Excellent environmental stability for this media has been
demonstrated. This stability appears to be outstanding when compared with
that observed in other types of optical recording media utilizing metallic
or inorganic films[4]. ODI's dye-polymer medium has been evaluated in a
number of types of stability tests. These include: reread stability,
shelf life studies at elevated temperatures and humidities and archival
studies both at ambient and elevated temperature and humidity.

Reread studies of data shown in Figure 14 indicate that little change occurs at 10^6 reads either in CNR or more particularly in mark length. No change in mark length over this number of cycles indicates that the medium is not being further deformed by the low energy ($\sim$1.0mW) 840nm read beam.

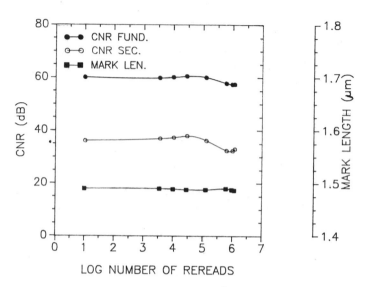

Figure 14.   Bump (data) Stability for 10^6 Reread Cycles on AI Media.

Media have been exposed to temperatures of 60°C and 85% relative humidity for periods of more than 6000 hours with periodic rewrite/read (CNR) information being taken on the exposed disks. The data illustrated in Figure 15 indicate that no change has occurred under these conditions over approximately 3/4 of a year.

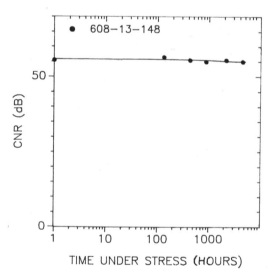

Figure 15.   CNR After Environmental Exposure of AI Media to 60° and 85% Relative Humidity for $\sim$ 6000 Hours.

Further environmental testing using a cyclic testing regimen (2 cycles/day between 25°C and 65°C at 93% relative humidity) indicated no degradation in sensitivity over a 30 day period, as shown in Figure 16.

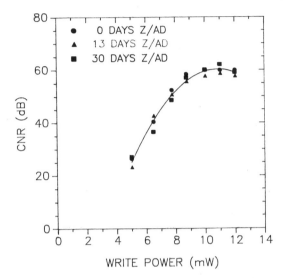

Figure 16.   CNR After Cyclic Environmental Exposure of AI Media to 25°/65°C and 93% Relative Humidity.

SUMMARY

Both AI and SI dye-polymer structures readily support high data densities and have good environmental stability. These structures are potentially "tunable" to other wavelengths of interaction with dyes as new, better and/or shorter wavelength diode lasers appear on the market. It is also very possible to switch the system to a 780nm write and 840nm erase by changing the dye system in the active layers.

The "characteristics" of the media are summarized as follows:

o Can be similarly produced in either air or substrate incident configuration.

o Reflectance of the media in the 12-30% range can be determined to a significant degree by the dyes and by control of layer thicknesses.

o Media have very sharp write thresholds.

o Low noise floors with good substrates, and CNRs to 60dB.

o Erasable, either locally by laser or in bulk, thermally.

o Media tunable to other write/erase wavelengths by use of appropriate dyes.

o Run length encoding readily performed due to linear mark length/pulse length relation.

o Bit edge jitter of $\sim$ 0.1μm will support high density marking (codes).

o Environmental stability is very good as shown by accelerated shelf life, archival life and Z/AD cycling tests.

REFERENCES

1. N. E. Iwamoto and J. M. Halter, Polymer Preprints, <u>29(2)</u>, 217 (1988)

2. T. W. Smith, J. Vac. Sci. Technology, <u>18</u>, 100 (1981)

3. J. M. Halter and N. E. Iwamoto, Proc. of SPIE, <u>899</u>, 201 (1988)

4. Unpublished data - Optical Data, Inc., Test Lab Results

5. M. A. Lind and J. S. Hartman, Proc. of SPIE, <u>899</u>, 211 (1988).

6. C. D. Feyrer, N. R. Gordon and W. E. Skiens, U. S. Patent 4,719,615, 1988

PHOTOCHEMICAL HOLE BURNING OF QUINIZARIN AND TETRAPHENYLPORPHIN IN

MAIN-CHAIN AROMATIC POLYMERS

Kazuyuki Horie[1], Kazuo Kuroki[1], Itaru Mita[1] and
Akira Furusawa[2]

[1]Research Center for Advanced Science and
Technology, University of Tokyo, 4-6-1, Komaba,
Meguro-ku, Tokyo 153, and [2]Research Laboratory,
Nikon Corp., 6-3-1, Nishi-ohi, Shinagawa-ku, Tokyo
140, Japan

The efficiency of hole formation and the temperature
stability of holes burnt at 4.2 K by using argon-ion laser,
helium-neon laser, or ring dye laser are studied for
quinizarin (Q) and tetraphenylporphin (TPP) in various
main-chain aromatic polymer matrices. The efficiency of
hole formation for TPP in phenoxy resin (PhR) is higher
than those in poly(ethylene terephthalate) (PET),
polyimide (PI), poly(methyl methacrylate) (PMMA), and liquid
crystalline polymer (LCP). The temperature stability of the
holes was evaluated by cycle annealing experiments. In the
case of TPP in phenoxy resin (PhR), the hole burnt at 4.2 K
could be detected even at liquid nitrogen temperature (80 K)
after 30 min annealing at the temperature. The hole burnt
at 4.2 K partially recovers when cooled again to and
measured at 4.2 K after cycle annealing up to 110 K for TPP
in PMMA, 120 K for TPP in PhR, and 130 K for TPP in PI. The
first experiment of photochemical hole burning (PHB) without
using liquid helium as a refrigerant was carried out for TPP
in PhR system, where we could burn and observe clear holes
even at 80 and 100 K.

INTRODUCTION

The phenomenon of photochemical hole burning (PHB) has recently
attracted considerable interest not only as a tool for high-resolution
solid state spectroscopy[1-3] but also as a means for frequency-domain high
density optical storage[4-6]. The PHB consists of burning very narrow and
persistent holes into the absorption bands of guest molecules molecularly
dispersed in amorphous solids by the narrow band excitation with a laser
beam at very low temperatures.
The temperature dependence of hole profiles burnt at liquid helium
temperature is one of the important aspects of PHB phenomenon. The
photochemical hole of free-base phthalocyanine (H_2Pc) in poly(methyl
methacrylate) (PMMA) burnt at 4.2 K was reported to become difficult to
measure above 50 K by hole broadening, but the hole was recovered[4] when

it was cooled to 4.2 K after the annealing at 80 K. When the hole burnt in $H_2Pc/PMMA$ system was elevated to 100 K, the hole was totally destroyed and did not reappear[4] even if the system was recooled to 4.2 K. The hole was observed for quinizarin (Q) in amorphous silica ($a-SiO_2$) system up to 34 K, and a partial hole recovery at 4.7 K was observed[7] after cycle annealing up to 67 K.

In order to increase the thermal stability of photochemical holes, we introduced as matrices main-chain aromatic polymers composed of stiff aromatic main chains without side chains[8]. Phenoxy resin (aromatic polyhydroxyether) proved to be an effective matrix for the efficiency of hole formation and for the temperature stability of the hole of free-base tetraphenylporphin (TPP)[9]. In the present paper, the efficiency of hole formation and the dependence of hole stability in PHB of quinizarin (Q) and free-base tetraphenylporphin (TPP) are studied in polycarbonate (PC), polysulfone (PSF), polyethersulfone (PESF), aromatic polyimide (PI), phenoxy resin (PhR), poly(ethylene terephthalate) (PET), liquid crystalline polymer (LCP), as well as in PMMA and poly(vinyl alcohol) (PVA), by using Ar ion laser, He-Ne laser, or ring dye laser.

EXPERIMENTAL

Materials

Various polymer films containing quinizarin (1,4-dihydroxyanthra-quinone) ($4-10\times10^{-4}$ mol/ℓ) with thickness of 0.1-2.0 mm were prepared by solvent casting (s) or by hot press casting (p) of solvent-cast polymers containing molecularly dispersed quinizarin. The films of PMMA containing 2.2×10^{-2} mol/ℓ of TPP and side-chain liquid crystalline polyacrylate (LCP) with 1 mol % of chemically bound TPP group[10] were prepared by solvent cast method and annealed. The phenoxy resin films containing $(0.4-2.2)\times10^{-2}$ mol/ℓ of TPP (0.1-0.66 mm thickness) were solvent cast from chloroform solution and then hot pressed and annealed. The PET film containing 8×10^{-3} mol/ℓ of TPP[11] was solvent cast, extruded, hot pressed, and then cold drawn. Aromatic polyimide (PI) films composed

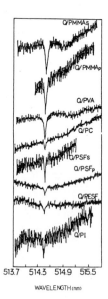

513.7   514.3   514.9   515.5
WAVELENGTH (nm)

Figure 1.   Typical PHB spectra at 4.2 K of quinizarin in various polymer matrices indicated beside the profiles[8]. Laser intensity: 0.15-1.5 mW/cm², irradiation time: 5-30 min.

of biphenyltetracarboxylic dianhydride and p-phenylenediamine containing quinizarin or TPP were prepared by the solvent cast of the precursor polyamic acid with quinizarin or TPP and successive thermal imidization and annealing.

PHB Measurements

The hole burning was performed with a CW argon ion laser (NEC,GL3200) at 514.5 nm with a linewidth of 0.3 Å for quinizarin, and

with a CW helium-neon laser (NEC,GL5800) at 632.8 nm with a linewidth of 0.045 Å or a ring dye laser (Coherent 699-05) at about 647 nm for TPP. The optical absorption spectra of the sample films set in a continuous flow type liquid helium cryostat (Oxford, CF1204) were measured with a 1-m monochromator (Jasco, CT100C) with a resolution of 0.1 Å. The transmitted light was detected by a photomultiplier (Hamamatsu R934-02), lock-in amplified (Jasco LA126W), digitized by a transient-time converter (Riken Denshi, TCDC 12-4000), and data processed with a desk-top computer (NEC,PC9801vm2). The sample temperature was monitored with a AuFe/Chromel thermocouple.

PHB OF QUINIZARIN IN MAIN-CHAIN AROMATIC POLYMERS

The PHB measurements were carried out for quinizarin (Q) in PMMA,[8] PVA, PC, PSF, PESF, and PI by using argon ion laser at 514.5 nm. Typical holes burnt in these polymer matrices are illustrated in Fig. 1.

The photochemical holes of quinizarin in PC, PSF, PESF, and PI are observed in this study for the first time. The increases in hole depth, $\Delta A/A_0$, where ΔA is the difference in absorbance produced by hole formation and A_0 is the absorbance before irradiation, are plotted against irradiation time for all cases in Fig. 2. The irradiation intensity of argon ion laser was 0.15 mW/cm^2 at the beginning, and after 30 minutes it was raised to 1.5 mW/cm^2 when the efficiency of hole formation was very low. The samples are divided into two groups according to the efficiency of hole formation. In PMMA and PVA the holes were burnt very effectively and the relative hole depth, $\Delta A/A_0$, reached about 20 % in the maximum case of Q/PMMA; while the efficiency of hole formation was diminished markedly in PC, PSF, PESF, and PI, where the existence of aromatic hydrocarbon groups might weaken the hydrogen bonding ability of these polymers with hydroxy group of quinizarin. There are two possible mechanisms for the PHB reaction of

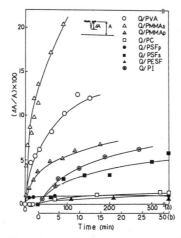

Figure 2. Irradiation time dependence of hole depth, $\Delta A/A_0$, for PHB of quinizarin(Q)[8]. Intensity: (a) 0.15 mW/cm^2, (b) 1.5 mW/cm^2.

quinizarin, i.e., tautomeric isomerization with hydrogen atom transfer and the change in intramolecular hydrogen bonds to intermolecular ones with the matrix. Graf et al.[12] have suggested the second mechanism with intermolecular hydrogen bond formation based on the fact that PHB of quinizarin was not observed with detectable optical yield in nonhydrogen bonded matrices. The small efficiency of hole formation of quinizarin in PC, PSF, PESF, and PI in the present experiments supports the mechanism involving the formation of intermolecular hydrogen bonds, and the introduction of aromatic rings in the main chain aiming to increase the temperature stability of hole by using the stiff matrix is supposed to bring about the decrease in the extent of intermolecular hydrogen bonding.

PHB OF TETRAPHENYLPORPHIN IN PhR, PI, PET, LCP, AND PMMA

The change in guest molecule from quinizarin to free-base tetraphenylporphin (TPP) whose hole-burning phototautomeric reaction is of intramolecular nature[13] suggests the possibility of studying the effect of a stiff polymer matrix on the temperature stability of photochemical holes.

The PHB for TPP in PMMA, PI[14], PhR[9], or PET[11] film, and the PHB for TPP bonded to LCP[10] were measured by using a helium-neon laser (632.8 nm) or a ring dye laser (about 647 nm) pumped by an argon-ion laser. Typical hole profiles are shown in Fig. 3. The hole width of TPP burnt at 4.2 K was initially about 0.6 cm^{-1} and gradually increased during the irradiation, but no dependence on the nature of polymer matrices was observed. Photochemical hole of TPP in PhR burnt at 30 K showed[9] the hole width of 3.3 cm^{-1}.

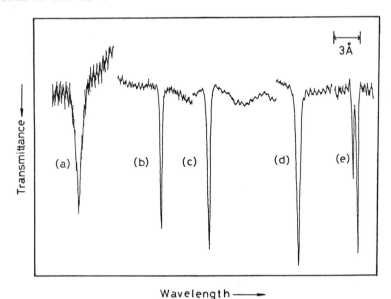

Figure 3. Typical PHB spectra of tetraphenylporphin (TPP) in various polymer matrices at 4.2 K. (a) PMMA, (b) PET, (c) PI, (d) PhR, (e) LCP.

The changes in hole depth, $\Delta A/A_0$, during 0.36 or 3.6 mW/cm^2 helium-neon laser irradiation at 632.8 nm are shown in Fig. 4 for TPP in PhR and PI at 4.2 K together with the results for TPP in PMMA. The efficiency of hole formation in polyimide is almost the same as in PMMA, but the efficiency in the phenoxy resin is much higher than that in PMMA. The purification of the phenoxy resin markedly improves the efficiency of hole formation, resulting in the hole depth of more than 30 % of initial absorbance for TPP in reprecipitated PhR. The annealing of the PhR sample films at a temperature a little below the glass transition temperature, T_g, gave additional effect on the efficiency of hole formation of TPP at 4.2 K. The changes in hole depth, $\Delta A/A_0$, during 2.8 mW/cm^2 ring dye laser irradiation at about 647 nm are shown in Fig. 5 for TPP in PhR, PI, PET and for TPP bonded to LCP. Though the rate of increase in hole depth for TPP in PhR seems to be much larger in Fig. 5 than the corresponding rate in Fig. 4 due to the difference in molar extinction coefficient at different wavelengths, the quantum yield of hole formation, Φ, proved to be the same ($\Phi = 1.2\times10^{-3}$ at 632.8 nm and at 647 nm for TPP in PhR). The TPP in PI and LCP gave lower efficiency

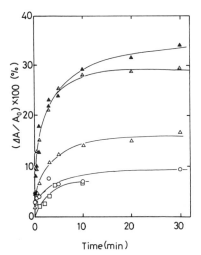

Figure 4. Irradiation time dependence of hole depth, $\Delta A/A_0$, for PHB of TPP in PhR ($\triangle$: as received, $\triangle$: reprecipitated, $\blacktriangle$: reprecipitated and annealed), PI ($\square$), and PMMA ($\bigcirc$) at 632.8 nm (He-Ne laser, 3.6 mW/cm^2) at 4.2 K.

hole formation ($\Phi = 7.0\times10^{-4}$ for PI and $\Phi = 1.9\times10^{-4}$ for LCP) than TPP in PhR. The TPP bound to LCP showed rather narrow hole width (1.0 cm^{-1}) and no phonon side holes were observed even after 30 min irradiation. Adjacent holes separated only by 0.086 nm (2.0 cm^{-1}) could be burnt for TPP bound to LCP (Fig. 3 (d)). The initial efficiency of hole formation for TPP in PET ($\Phi = 6.2\times10^{-4}$) is about a half of that in PhR and does not change by five-time drawing of the sample. Though we do not exclude the possibility of obtaining higher values of Φ than the present ones by improving the sensitivity of the apparatus, it is clear that efficient hole formation of TPP can be realized in phenoxy resin.

The formation of satellite holes in PHB is illustrated in Fig. 6 for TPP in phenoxy resin. By laser irradiation at 632.8 nm, several satellite holes in both higher and lower energy regions could be observed. The energy differences given in cm^{-1} in Fig. 6 correspond to vibrational frequencies in excited singlet state of TPP, and can be compared to those of chlorin and porphin obtained from fluorescence excitation spectra[15].

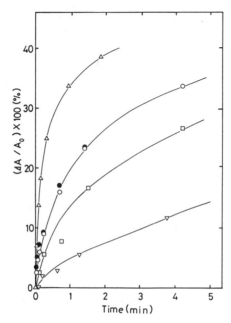

Figure 5. Irradiation time dependence of hole depth, $\Delta A/A_0$, for PHB of TPP in PhR ($\triangle$), PET ($\bigcirc$: undrawn, $\bullet$: 5 times drawn), PI ($\square$), and LCP ($\triangledown$) at 647 nm (ring dye laser, 1.5 mW/cm^2) at 4.2 K.

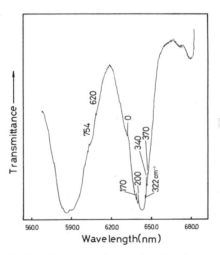

Figure 6. Satellite holes in PHB of TPP in PhR burnt with He-Ne laser at 632.8 nm at 4.2 K.

TEMPERATURE STABILITY OF HOLES FOR TPP IN PMMA, PI, AND PhR

The temperature stability of hole profiles was studied by cycle annealing experiments. Typical examples are shown in Fig. 7 for TPP in PMMA. The hole was burnt at 4.2 K by 16 min irradiation by helium-neon laser (3.8 mW/cm^2). The hole spectrum was measured at a certain temperature after annealing at that temperature for 30 min, and the spectrum was measured again after cooling to 4 K. The annealing was extended to higher temperatures, and the measurements of hole profile were carried out at each annealed temperature and each cooled temperature (4.2 K). The relative hole depth, $\Delta A/\Delta A_0$, at annealed temperature decreased with increasing temperature and disappeared at 60 K for TPP in PMMA, but 60 % of the initial hole depth was recovered again by cooling to and measuring at 4.2 K. The reappearance of hole after cycle annealing was reported[4] up to 80 K but not at 100 K for free-base phthalocyanine in PMMA. Fig. 7 shows, however, the hole recovery up to 110 K for TPP in PMMA.

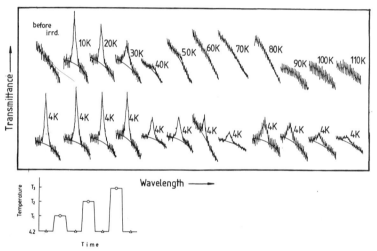

Figure 7.   Typical photochemical holes of TPP in PMMA during annealing cycle. Irradiation: He-Ne laser, 3.8 mW/cm^2, 16 min. The upper row corresponds to the measurements at the annealed temperature shown beside the spectra after 30 min annealing; the lower row corresponds to spectra measured at 4.2 K after cooling from the annealed temperature.

The cycle annealing experiments were also carried out for TPP in aromatic polyimide (PI) and in phenoxy resin (PhR). The hole for TPP in PI measured at annealed temperature disappeared at 60 K as in PMMA, but the hole recovery when measured at 4.2 K could be observed after annealing for 30 min up to 130 K. This means that the informations written in at 4.2 K are archivally stored at 130 K and can be read out again at 4.2 K.   The advantage of TPP in PI system would be due to the rigid molecular structure and the absence of dangling side group in this aromatic polyimide prepared from biphenyltetracarboxylic anhydride and p-phenylenediamine.

The changes in hole profiles during cycle annealing of TPP in PhR are given in Fig. 3 of Ref. 9. It is remarkable that the photochemical hole of TPP in phenoxy resin (PhR) burnt at 4.2 K for 30 min with 3.6 mW/cm^2 helium-neon laser was clearly observed at 60 K ($\Delta\omega_h \cong 12$ cm^{-1}) after 30 min annealing at the temperature, and the hole could be detected even at 80 K (above liquid nitrogen temperature) though its profile became broad and shallow. Two factors, i.e., small dipole interaction between excited state of guest molecule and matrix polymer and the suppression of molecular relaxation by introduction of a stiff matrix are supposed to be important for restricting the dephasing process leading to the observation of photochemical holes at higher temperatures. Phenoxy resin composed of aromatic main-chain without bulky side group and with rather low dielectric constant, compared to PMMA and PI, would satisfy the above factors.

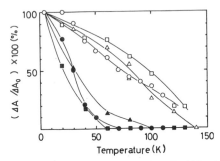

Figure 8. Changes in relative hole depth, $\Delta A/(\Delta A_0)$, during annealing cycle measured at the annealed temperatures ($\blacktriangle$,$\blacksquare$,$\bullet$) and at 4.2 K ($\triangle$,$\square$,$\bigcirc$) for TPP in PhR ($\blacktriangle$,$\triangle$), PI ($\blacksquare$,$\square$), and PMMA ($\bullet$,$\bigcirc$).

The changes in relative hole depth, $\Delta A/(\Delta A_0)$, based on its initial value measured just after irradiation at 4.2 K are summarized in Fig. 8 for TPP in PhR, PI, and PMMA. The hole measured at annealed temperature becomes shallow with the increase in the annealed temperature, T, and disappears for T>60 K in PMMA and PI and for T>100 K in PhR. The measurement of hole profile burnt at 4.2 K during step-wise increase in temperature gave the same results. But the hole partly recovers when the sample was cooled and measured again at 4.2 K. This reversible part of hole disappearance would be caused by the hole broadening induced by dephasing due to electron-phonon interaction. Fig. 8 shows the partial hole recovery at 4.2 K after cycle annealing up to 110 K for TPP in PMMA, 120 K for TPP in PhR, and 130 K for TPP in PI. The irreversible disappearance of hole observed in cycle annealing even after cooling to 4.2 K is totally a ground state phenomenon and is thought to depend on the change in microenvironment of the chromophore due to molecular relaxation of matrix polymers. Thus stiff molecular structure of the aromatic polyimide provides better restored hole depth measured at 4.2 K for TPP than for TPP in PMMA. The change in hole recovery at 4.2 K depending on the annealed time at 80 K is shown in Fig. 9 for TPP in PhR. The initial rapid irreversible hole disappearance proceeds in 5 min, and then a very slow process of structural diffusion follows. Inhomogeneous progress of the kinetic process such as this is frequently observed for reactions in amorphous polymer solids, and is sometimes related to the free volume distribution or energy level distribution of reaction sites[6].

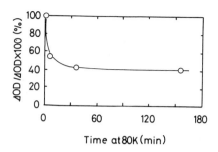

Figure 9.   Dependence of hole depth measured at 4.2 K on the annealed
time at 80 K for the hole of TPP burnt at 4.2 K by He-Ne laser
in PhR.

Recently we found that larger holes could be burnt at 30 K than at
4K for the same irradiation energy when both were observed at 4 K, owing
to the broadening of homogeneous line width[9]. This suggests that the
burning of photochemical hole at 80 K would not be so difficult for the
system where the hole burnt at 4 K can be observed at 80 K.   Thus, we
tried to and could burn and observe the holes at 80 and 100 K for TPP in
phenoxy resin[16]. Fig. 10 shows a photochemical hole of TPP in phenoxy
resin burnt and measured at 80 K by using liquid nitrogen cryostat. The
$\Delta\omega_h$ is about 26 cm^{-1} and $\Delta A/A_0$ is 5.3 % after 5 min burning at 80 K with
30 mW/cm^2 ring dye laser. The burning and observation of hole at 80 K
were recently reported also for sulfonated TPP in poly(vinyl alcohol)[17].
However to our knowledge, Fig. 10 is the first experiment of PHB without
liquid helium as a refrigerant.

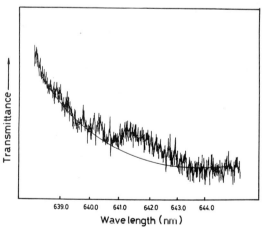

Figure 10.   Photochemical hole of TPP in PhR burnt and observed at 80 K
by using a liquid nitrogen cryostat. Irradiation: ring dye
laser, 30 mW/cm^2, 5 min, 641.3nm.

CONCLUSION

The efficiency of hole formation and the temperature stability of holes burnt at 4.2 K in PHB of quinizarin and TPP were studied in various main-chain aromatic polymer matrices. The hole of TPP in phenoxy resin burnt at 4.2 K was detected even at liquid nitrogen temperature (80 K), and moreover the hole could be burnt and observed even at 80 and 100 K for the same system.

ACKNOWLEDGMENTS

This work was supported in part by Grant-in-Aids for a special research project on "High Efficiency Photochemical Processes" and for scientific research in priority area of "Macromolecular Complexes" from the Ministry of Education, Science, and Culture, Japan.

REFERENCES

1. J. Friedrich and D. Haarer, Angew. Chem. Int. Ed. Eng., _23_, 113 (1984).
2. R.M. Macfarlane and R.M. Shelby, J. Luminescence, _36_, 179 (1987).
3. W.E. Moerner ed., "Persistent Spectral Hole-Burning; Science and Applications,"Springer, Berlin, (1988).
4. A.R. Gutierrz, J. Friedrich, D. Haarer and H. Wolfrum, IBM J.Res. Develop., _26_, 198 (1982).
5. W.E. Moerner, J. Mol. Electronics, _1_, 55 (1985).
6. K. Horie and I. Mita, Adv. Polym. Sci., _88_, 77 (1989).
7. T. Tani, H. Namikawa, K. Arai and A. Makishima, J. Appl. Phys., _58_, 3559 (1985).
8. K. Horie, K. Hirao, K. Kuroki, T. Naito and I. Mita, J. Fac. Eng. Univ. Tokyo, _39_, 51 (1987).
9. K. Horie, T. Mori, T. Naito and I. Mita, Appl. Phys. Lett., _55_, 935 (1988).
10. K. Horie, A. Furusawa, K. Kuroki, I. Mita, S. Kubota, T. Koyama, K. Hanabusa and H. Shirai, unpublished results, (1988).
11. K. Horie, K. Kuroki, I. Mita, H. Ono and S. Okumura, unpublished results, (1988).
12. F. Graf, H.K. Hong, A. Nazzal and D. Haarer, Chem. Phys. Lett., _59_, 217 (1987).
13. S. Volker and R.M. Macfarlane, IBM J. Res. Develop., _23_, 547 (1979).
14. K. Horie, K. Kuroki, T. Naito, I. Mita and K. Hirao, Polymer Preprints Jpn., _36_, 3510 (1987).
15. S. Volker and R.M. Macfarlane, J. Chem. Phys., _73_, 4476 (1980).
16. A. Furusawa, K. Horie, K. Kuroki and I. Mita, Preprints Jap. Soc. Appl. Phys., 881117-01, p.175 (1988).
17. K. Sakoda, K. Kominami and M. Iwamoto, Jap. J. Appl. Phys., _27_, L1304 (1988).

MATERIALS FOR MASTERING AND REPLICATION PROCESSES

IN OPTICAL VIDEO, AUDIO AND DATA DISC PRODUCTION

Peter E.J. Legierse and Jan H.T. Pasman

Optical Disc Mastering (ODM)
Lodewijkstraat 1, Building DBD
5652 AC Eindhoven
The Netherlands

The manufacturing processes and materials for the currently commercially available optical discs are described with emphasis on disc mastering. A distinction is made between optical discs addressing two different market segments:

1. **High End:** characterized by high demands on pit or groove definition, process control and overall quality (LaserVision, CompactDisc-Video and recordable data storage discs).

2. **High Volume:** characterized by large quantities of replica discs, faster turnaround times and larger tolerances in pit shapes and production processes (CompactDisc-Audio, CompactDisc-ROM (Read-Only-Memory)).

The intrinsic system tolerances and complexities of the different optical disc systems are considered to discuss different mastering processes and materials. The photoresist mastering process is described in some detail.

INTRODUCTION

In the early seventies various companies such as Philips and MCA investigated the possibility of recording and replication of video information on optically read-out discs.[1,2,3] In the beginning, the efforts were mainly directed toward recording of the video information onto the first disc shaped carrier (master), the so-called mastering process. The mastering was established partly by using principles, techniques and processes that were used in the manufacture of integrated circuits. However, as a light source an intense focussed monochromatic laser spot had to be used to transfer the high frequency video signal with exposure times of some 70 ns per pit into a light sensitive (photoresist) layer of thickness of about 0.1 μm, only a fraction of the then used IC photoresist thickness (figure 1). Therefore, IC techniques and processes had to be modified to allow for such small dimensions in vertical and lateral directions.

After a decade of optical video disc experience and technology development, the commercially available LaserVision (LV) video disc system and the 120 mm CompactDisc (CD) digital audio disc system were successfully introduced in the first half of the eighties.[4] The CD mastering process was derived from the well established video disc photoresist process,[5] optimized for the specific CD parameters and yielding the high process latitude and reliability that is necessary for very large scale CD mass production.[6]

A third optical disc application is the recordable optical data storage disc which exploits the high information storage density as compared to magnetic disc storage. In this case the actual information is not prerecorded as is the case with video and audio discs, but only the preformat for addressing, tracking and synchronisation is recorded.[7] The actual data are written onto the disc by the end-user into a recordable and possibly erasable material.

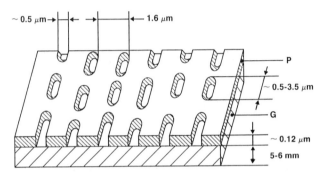

Figure 1. Schematic diagram of a recorded master surface. P is the photosensitive material and G is the glass substrate.

The recently introduced CD-Video disc is based essentially on the combination of digital CD audio and analog LV video. The 120 mm, the so-called videoclip, contains about 20 minutes of CD audio and about 5 minutes of LV video with audio of CD quality. The 200 mm (CDV-EP) and 300 mm (CDV-LP) discs contain video with CD audio.

DISC REPLICATION

At present, commercially available LaserVision video discs are produced by injection molding of polymethylmethacrylate (PMMA) into 1.25 mm thick, 300 mm diameter discs[8] or by photopolymerization of liquid acrylate monomers on PMMA substrates (figure 2).[9] After deposition of a high reflectivity metal (e.g. Al) and a subsequent protective coating, two discs are glued together into a double sided symmetrical sandwich.

Preformatted data discs are manufactured by injection molding of polycarbonate (PC) into 5 1/4" discs or by photopolymerization on glass or thermoset plastics (12" and 5 1/4" formats). The discs are then vacuum or solvent coated with the recordable layer. Recording can take place by melting or ablation of metals, alloys, polymer-dye combinations, by creating phase transitions in the crystalline or amorphous state, or by thermally induced switching of the magnetisation direction in magneto-optic layers.[10]

CD discs are produced by injection or injection-compression molding of polycarbonate (PC) into 1.2 mm thick, 120 mm diameter discs, which after metallizing, coating and labelling form single sided discs. Presently CD discs, being produced by the millions, form a High Volume consumer market. Compared to analog video discs digital CD audio disc production appears to be much more tolerant to variations in production methods, processes and materials. As a consequence, there is a drive towards development of cheaper production methods for the High Volume CD market.

The 120 mm CDV videoclip is basically manufactured in the same way as CD audio discs. The 200 and 300 mm formats are produced similarly to video discs. The mastering and replication processes, however, have to satisfy much tighter tolerances than just the combination of the individual CD and video processes.

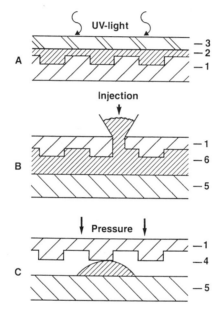

Figure 2. Principles of disc replication methods: (A) photopolymerization, (B) injection moulding, (C) compression moulding. 1 denotes nickel stamper, 2 liquid acrylate monomers, 3 plastic substrate, 4 cake of viscous thermoplast, 5 mirror block, and 6 molten thermoplast.

Figures 3 to 7 show scanning electron micrographs of the characteristic information patterns on replica disc surfaces of different optical disc applications.

SYSTEM TOLERANCES

In the case of digital CD-Audio discs the length of the pits and the spaces between them can only have discrete values as defined by the EFM digital coding scheme.[1,4] These lengths correspond to pit and land durations of only a limited integer number (3,4,5,..,11) of clock pulses of 231 ns, each corresponding to about 0.3 μm length (figure 6). Therefore, read-out must detect whether in between two successive clock

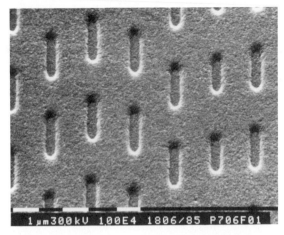

Figure 3. Scanning electron micrograph of a LaserVision replica disc
surface. One horizontal white bar equals 1 μm.

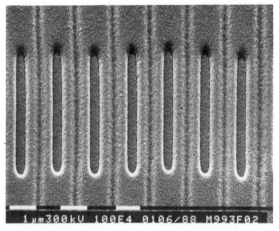

Figure 4. Scanning electron micrograph of a CCS data storage replica disc
surface. One horizontal white bar equals 1 μm.

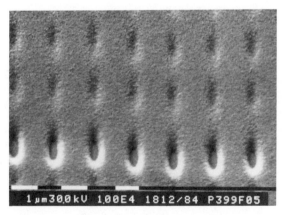

Figure 5. Scanning electron micrograph of a LD1200 data storage replica
disc surface. One horizontal white bar equals 1 μm.

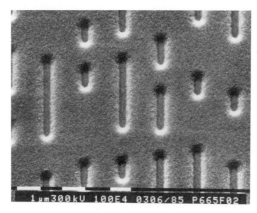

Figure 6. Scanning electron micrograph of a CD-Audio replica disc
surface. One horizontal white bar equals 1 μm.

Figure 7. Scanning electron micrograph of a CD-Video replica disc
surface. One horizontal white bar equals 1 μm.

moments there is a pit/non-pit transition, thus requiring only moderate
accuracy in the pit dimensions in the tangential direction. Once the
read-out signal, as determined by pit width and depth, exceeds some
minimum signal-to-noise ratio perfect digital signal retrieval is
possible. Moreover, the CD system compensates for some extent of over or
under dimensioned pits (asymmetry caused by over or under exposure and
development of the photoresist). The CD code furthermore enables the
detection and correction of errors in the read-out signal caused by local
defects in the mastered pit structures or process failures during
replication.[4] The CD specification allows up to 3% of the data blocks to
contain some faulty information corresponding to a block error rate (BLER)
of 220.

For video formats like the LaserVision and CD-Video standardized
systems and preformatted recordable data storage systems (12" WriteOnce,
5 1/4" WriteOnce and Erasable, Recordable CD, etc.) the tolerances on pit
or groove dimensions are much tighter. For instance, in the video format
where the composite video signal is FM encoded in the pit frequency, the
length of each individual pit is modulated in a continuous analog way by
the (analog or digital CD) audio signals (figures 3 and 7).[11] As a
consequence, the pit length has to be controlled between much narrower
limits to give good audio signal retrieval and to avoid intermodulation

between luminance signal and audio and color signals to which the analog video signal is susceptible. Also, LaserVision being an analog system, each extra dB in the signal-to-noise ratio, gained by better dimensioning of pits or reduction of noise, adds to the quality of the video image. Error correction, moreover, is not possible and each dust particle on the information surface will cause visible defects in the video picture.

For preformatted data storage disc systems not only the tangential and radial dimensions of pits are important but also the depth of the information tracks is exploited. The Composite Continuous Servo (CCS) 5 1/4" format, for example, uses pits of depth of approximately 1/4 of the read-out wavelength for addressing and synchronizing, and for radial tracking grooves are used of approximately 1/8 of the wavelength (figure 4 and 8).[22] In some formats like LMS's 12" LD1200 WriteOnce format the, on the average, 1/8 wavelength deep groove is even modulated sinusoidally in depth to enable continuous synchronization (figure 5).[7] These intermediate depth levels are mastered by varying the exposure intensity in an analog way such that after development a three dimensional structure is formed in the photoresist layer. The photoresist, therefore, must have a well defined and controllable relation between exposure dose (recording intensity) and development rate resulting in the desired depth profile.

On the basis of system tolerance and complexity considerations we can distinguish two market segments for optical discs, namely the High Volume market of CD-Audio characterized by large quantities of replica discs and the High End market of Video, CD-Video and preformatted data discs characterized by high demands on quality, dimensional control and a high degree of complexity of the information structures.

Clearly for the High Volume application one might consider using, beside photoresist mastering, different mastering methods even though the overall performance may be only marginally acceptable. For the High End applications, however, one has to resort to the well proven photoresist mastering process. We will now outline the mastering process and then describe in more detail the characteristic behaviour of different recording materials that can be used for mastering.

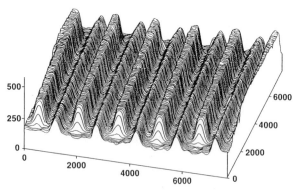

Figure 8.   Scanning tunneling micrograph of a CCS data storage replica disc surface showing sector mark pits between the continuous tracking grooves. Dimensions along the horizontal and vertical axes are given in nm.

MASTERING PROCESS

At present most mastering processes for optical disc systems which
are commercially available,[5-9,13] or that are used in-house by optical
disc manufacturers are based on exposure by a focussed laser beam of light
sensitive organic materials spin coated onto a flat rigid polished master
disc (figure 9). The intensity of the laser beam is modulated in
accordance with the information to be recorded. In case of photoresist
mastering[14] the development rate increases with the absorbed amount of
light and hence, with a scanning recording spot, a latent spatial
information pattern is produced by the local differences in the temporally
varying recording intensity. Subsequent development locally dissolves the
photoresist, resulting in the desired depth profile as shown in figure 1.
The pit depth, as determined by the resist layer thickness, is some
0.12 μm. The length of the pits ranges between 0.5 and 3.5 μm. The
separation between tracks (trackpitch) is of the order of 1.6 μm.

After development, a metal layer (e.g. Ag, Ni, Cr, etc.) is applied
to make the information surface electrically conductive in order to enable[13]
the production of stampers for replication by an electroforming process.
These stampers can be applied in injection molding, compression-injection
molding or photopolymerization processes to replicate the information into
large numbers of discs.

Mastering processes based on photoresist are currently used by all
major optical video, audio and data storage disc manufacturers such as
Philips, PDO, Sony/CBS, Pioneer, 3M Company, Warner, Matsushita and many
others.

In the last few years some companies have come up with alternatives
to photoresist as a light sensitive material in mastering (ODC[15], DISC).
These are ablative materials, which upon exposure directly form a pit
pattern and thus do not require development.
Besides mastering based on photochemical or photothermal principles a
mechanical CD mastering method has recently been announced by Teldec
(Direct Metal Mastering),[16] where the pit pattern is cut into a soft metal
(Cu) by a piezo electromechanical stylus.
The characteristic behaviour of these three alternative mastering
materials will be considered into some detail.

CHARACTERISTICS OF MASTERING MATERIALS

Ablative Polymer/Dye Materials

Ablative direct effect mastering is based on thermal decomposition of
a thin layer of an easily degradable material, such as nitrocellulose,
made absorbent with a dye that locally converts the light energy of the

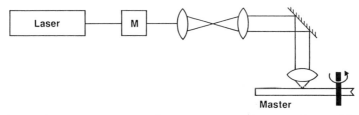

Figure 9. Schematic diagram of master recording. M is an HF optical
modulator.

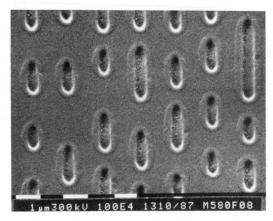

Figure 10. Scanning electron micrograph of a CD-Audio replica disc
surface made with a nitrocellulose mastering process. One
horizontal white bar equals 1 μm.

recording spot into heat.[15] During recording not all exposed material will
decompose into volatile products but some material will melt and form a
typical rim structure around the pit as shown in figure 10. Also, some
material may cluster into small particles that are randomly deposited onto
the information surface giving rise to extra noise (debris formation).

Because of the thermal nature of the pit formation process, the pit
shape is determined by the temperature distribution profile more than by
just the exposure profile. Thermal diffusion may influence the pit shape
in such a way that both depth and width are different at the beginning and
at the end of the pit (torpedo shaped pits), resulting in a nonlinear
relation between recorded pulseform, physical pit shape and read-out
signal. As a thermal diffusion time constant is involved this effect
depends on the linear velocity during the recording (CD: 1.2 - 1.4 m/s,
LV: 8 - 27 m/s).

The direct formation of pits enables the direct read after write
(DRAW) of the recorded information and hence offers direct signal quality
assessment and a feedback possibility during the recording process. By
controlling the recording intensity in this way the pit formation process
is controlled.

Presently a few companies have started using this technique for CD
mastering. However, replica discs appear to be not, or only just, within
the CD disc specifications. Because of the random way in which the rim is
formed (material transport) and the thermal diffusion effects, it is
unlikely however that ablative mastering could be used in the High End
mastering processes with their high demands on definition and control of
the dimensions of the information structures.

Direct Metal Mastering

Recently a new CD mastering method was announced based on the
principle of mechanical mastering as used in the grammophone recording
industry.[16] A diamond stylus is piezoelectrically driven into a soft metal
layer such as Cu resulting in pits with a typically V-shaped cross section
as shown in figure 11. Not only direct read after write is possible but
also electroforming can take place directly after mastering and more
copies can be made from the durable master. It seems unlikely, however,

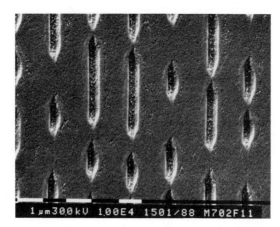

Figure 11. Scanning electron micrograph of a CD-Audio replica disc
surface made with a Direct Metal Mastering process. One
horizontal white bar equals 1 μm.

that this process can be applied in video formats such as CD-V and LV,
because of the higher freqencies involved. Also for preformatting of
complex three dimensional geometrical patterns this process seems
unsuited.

Photoresist Process

     Positive working photoresist as used in mastering consists of an
alkaline soluble polymer (novolak), made insoluble by a light sensitive
component (O-naphthoquinonediazide) acting as a dissolution inhibitor
(Figure 12). Local exposure of the resist photochemically decomposes the
light sensitive component locally into an alkaline soluble reaction
product (3-indenecarboxylic acid)[17] which then makes the exposed area
soluble in an alkaline solution.[17] The photochemical reaction rate
(sensitivity) is highest for exposure with UV light. However, there is
some sensitivity at blue visible wavelengths, thus enabling exposure with
Ar-Ion laser light at 458 nm or He-Cd laser light at 442 nm. Figure 13
shows the absorbance spectra of two different commercially available
photoresist systems. Clearly the sensitivity of photoresist "B" is higher
which would allow the use of a lower intensity laser. However, the

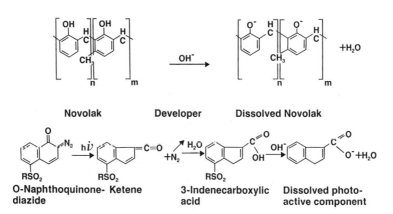

Figure 12.   Exposure and development reactions in a photoresist.

solubility of the unexposed photoresist also has to be considered, because unwanted development between pits will give rise to surface roughness and hence to noise in the read out signal. Therefore, the contrast in development rates of exposed and unexposed resist must also be optimized.

After exposure and development of the resist master, spiral or concentric tracks of depressions are formed in the resist. The individual microscopic dimensions of such depressions (pits or grooves) are determined by a number of parameters. The thickness of the photoresist layer determines the maximum possible pit or groove depth. Because of the low absorbance at the exposure wavelength and the very thin resist layer, the absorption and hence the development rate is, apart from a small standing wave effect, practically constant.

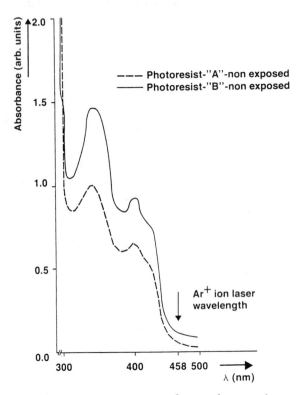

Figure 13. Absorbance spectra of two photoresist systems.

The fraction of the photosensitive component that is decomposed defines the local development rate [18]. This development rate therefore depends in a continuous way on the exposure dose, which is the total amount of light integrated (linearly) over time for a given point. Therefore, individual pit or groove depths between zero and the layer thickness can be formed by varying the locally absorbed light energy (figure 14). With a scanning light spot, the absorbed energy (per unit area) depends on the spatial spot intensity distribution (e.g. spot size, spot intensity), the tangential velocity and the to be recorded temporal signal level (e.g. EFM pulse pattern). The (locally varying) intermediate depth levels characteristic of some preformatted data discs are therefore obtained by varying the spot intensity. The exposure

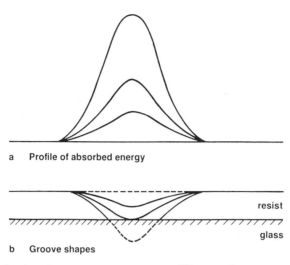

a   Profile of absorbed energy

resist

glass

b   Groove shapes

Figure 14.   Relation between exposure profiles and groove or pit shapes.

profile must be taken as a convolution of the spot intensity distribution with the temporal signal level. Since for LaserVision and CompactDisc the pit length is of the same order as the spot size, a two dimensional exposure profile with smoothed features as compared to the signal pattern results (figure 15).

The overall pit or groove dimensions are determined by the development parameter settings such as developer composition and concentration, development time, temperature and application method. During development the average pit formation process is monitored by diffraction order measurement.[19] A laser beam that is directed through the substrate and photoresist is diffracted by the forming periodic pit or groove pattern which acts as a transmission phase grating (figure 16.A).

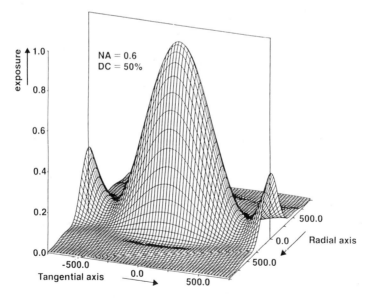

Figure 15.   Calculated exposure profile for LV pits of length 550 nm and tangential period 1100 nm, recorded with NA = 0.6.

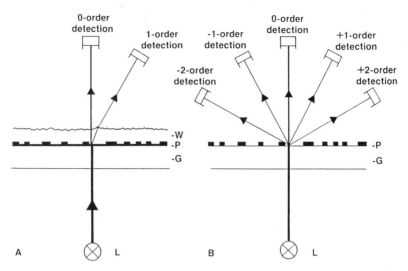

Figure 16.   Schematic diagram of the transmission order measurement set-up
for development control (A) and process analysis (B).   L
denotes the HeNe laser light source, G the glass master
substrate, P the photoresist and W the aqueous developer
solution.

When the first diffraction order intensity, which is a measure of the
developed pit volume, reaches a predefined level the development is
stopped.   In this way small deviations in process parameters are
automatically compensated for.

For the mastering of three dimensional information structures on
preformatted data storage discs and also for the design and control of pit
dimensions on CD and LV discs, it is important to know how these
dimensions depend on photoresist, exposure and development parameters and
how these parameters are interrelated.   Since the process conditions in
mastering (e.g. very short exposure time, focussed spot) and the resulting
submicron resist patterns (e.g. thickness of 0.1 µm) are quite different
from those in conventional IC processes, the investigation into relations
between pit dimensions and photoresist process parameters require special
measurement techniques for the specific mastering situation.   Again,
optical diffraction transmission order measurement can be performed, but
now for the determination of the dimensions of the grooves or pits after
development.[20]

Here, a laser beam is directed through the interface between the
photoresist and air (figure 16.B).   Because the refractive indices of
resist (n_r) and air (n_a) are different, the transmitted optical wavefront
will, just above the resist surface, be modulated in phase ($Phi(x,y)$)
proportional to the difference in optical pathlengths for the parts of the
optical wavefront that traverse the groove and those that traverse the
undeveloped resist:

$$Phi(x,y) = \frac{2 \pi}{\lambda} D(x,y) \left[n_r - n_a \right] \qquad (1)$$

where $D(x,y)$ is the depth profile in the resist.   When this phase
modulation is periodic in one direction, as is the case with a series of

166

identically developed grooves with trackpitch p, the transmitted wave will
diffract into a discrete number of diffraction orders, the direction of
which is given by the well-known grating law.

$$\sin\left[\theta_m\right] = m\,\frac{\lambda}{p} \qquad (2)$$

where m is the order number (.., -2, -1, 0, +1, +2,..). The intensities
of the orders, however, depend on the groove dimensions. Therefore, the
(macroscopic) measurement of the order intensities gives information about
the (microscopic) dimensions of grooves. For the present case of
transmission order measurement the optical contrast between photoresist
and air (= n_r - n_a) is relatively small; as a consequence, the modulation
depth in the phase of the transmitted wave is low. For 150 nm deep
grooves this phase modulation is less than $\lambda/6$. In this case scalar
diffraction theory may be applied to calculate the intensities of the
diffacted orders as a function of the groove geometry.[20] Figures 17 and 18
show results of such calculations. The calculated first order intensity
(normalized with respect to the transmitted intensity) and the ratio
between the second and first order are plotted as a function of the groove
width (here defined as the Full Width at Half the Maximum depth, FWHM) for
gaussian and rectangularly shaped grooves of depth of 100 nm with a
trackpitch of 1.6 μm. For a given groove form (gaussian or rectangular)
the first and second transmitted order intensities both increase to a good
approximation quadratically with the maximum depth in the range of depths
that we are interested in (0-150 nm). Therefore, once the groove form is
taken to be gaussian or rectangular, the ratio between second and first
order intensities depends essentially on the groove width. Consequently,
with some pre-knowledge about the expected groove form, the measurement of
this ratio directly gives the groove width. Next, once the groove width
is known, the groove depth can be derived from the first order intensity.

In general, the groove form will be approximately a gaussian as long
as the grooves are not fully developed all the way through to the glass

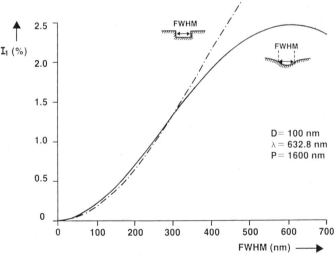

Figure 17.   Theoretical first diffraction order intensities for gaussian
(full curve) and rectangular (broken curve) continuous grooves
of depth D of 100 nm as a function of the groove width (Full
Width at Half Maximum depth, FWHM). The trackpitch P is
1600 nm and the wavelength is 632.8 nm.

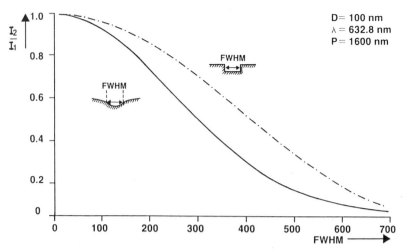

Figure 18. Theoretical curves of the ratio between second and first diffraction order intensities for gaussian (full curve) and rectangular (broken curve) continuous grooves of depth D of 100 nm as a function of the groove width (Full Width at Half Maximum depth, FWHM). The trackpitch P is 1600 nm and the wavelength is 632.8 nm.

substrate surface (as with pregrooves on data discs). When the grooves are developed further the side walls become steeper and the groove approaches more and more a rectangular shape.

It is clear that with the measurement of two independent values I1 and I2/I1 it will always be possible in normal situations to determine the two parameters, i.e. width and depth, that fully specify the shape provided the general groove shape is assumed to be of gaussian or rectangular form or any other form that is fully described by two parameters. For grooves that have only just reached the resist-glass interface, the groove will be more of a trapezoidal form which is defined by three parameters and for such "intermediate" grooves simple transmission order measurements are not sufficient.

Figure 19.A shows an example of an order intensity measurement performed on a master on which grooves were recorded with a spot of size defined by the numerical aperture (NA) of 0.60. As a function of the recording spot intensity the measured first and second order intensities (I1 and I2) have been plotted on a double logarithmic scale.

For low recording intensities, both the first and the second orders increase logarithmically linear with intensity, with a tangent of approximately 4. Since the distance between the first and second order curves is nearly constant, the ratio I2/I1 is almost constant and so the groove width is essentially independent of the recording intensity level. This implies that basically, for a given spot size, the effect of increasing the intensity is only a multiplication of the groove profile in the depth direction in the resist by some factor, without a change in the groove form in the lateral direction.

For higher recording intensities, the order intensity curves bend and tend to saturate, and the ratio I2/I1 decreases. At the intensity where the bending starts, the glass substrate surface will have been reached. The groove depth is now limited by the resist thickness and development

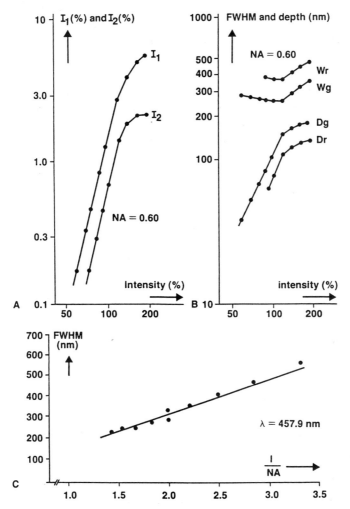

Figure 19. Diffraction measurement analysis: (A) measured first (I1) and
second (I2) diffraction order intensities as a function of the
intensity of the recording spot of NA=0.6, plotted on
logarithmic scales, (B) width (W) and depth (D) calculated
from the data of (A) assuming a gaussian (g) and rectangular
(r) groove shape, (C) theoretically expected groove width FWHM
(full curve) and experimental values (dots) as a function of
1/NA.

can only take place in the lateral direction, thus causing wider grooves.

In figure 19.B the calculated results for groove width W and groove
depth D are shown using the gaussian (subscript g) and (at higher
intensities only) the rectangular (subscript r) approximations. From the
constant tangent (= 2) of the curve (plotted double logarithmically) of
the depth Dg at lower intensities it can be seen that the depth in the
photoresist increases quadratically with the recording intensity, and that
the development rate is therefore a quadratic function of the exposure.
At higher recording intensities the rectangular approximation must be

used. The calculated depth D_r then is seen to saturate at about 150 nm, which was the resist layer thickness, while the width keeps increasing. Again, for fully developed grooves (and also for pits in case of CD and LV) the ratio I2/I1 can be used to determine the final groove (or pit) width.

Next we will consider the effect of the spot size on the groove width. For a given wavelength (λ) and numerical aperture (NA) the spot intensity profile will approximately be an Airy pattern:

$$I(r)= I(0) \left[2 \frac{J_1 \left[2 \pi \dfrac{NA}{\lambda} r \right]}{2 \pi \dfrac{NA}{\lambda} r} \right]^2 \qquad (3)$$

where J1(x) is the Bessel function of the first kind and order one.

For grooves the radial exposure profile can be calculated by integrating the Airy profile in the tangential direction. As we have seen above, this exposure profile must then be squared to obtain the development profile. For grooves that are not too steep, as is the case with grooves in photoresist that do not reach the glass-resist interface, development is only in the vertical direction and hence the depth profile approximately equals the development rate profile. This profile closely resembles a gaussian shape with Full Width at Half Maximum value given by

$$FWHM = 0.36 \frac{\lambda}{NA} \qquad (4)$$

This, therefore, is the theoretically expected FWHM of the groove depth profile. In figure 19.C this theoretical relation is plotted as a function of 1/NA for the case of λ = 457.9 nm (solid line). The plotted dots are experimentally determined values of the FWHM with NA ranging between 0.3 and 0.7 taken from W_g curves such as figure 19.B where the groove depth D_g is 100 nm.

Figure 20. Development time (in s) for three different recording intensities, with the developer concentration as a variable.

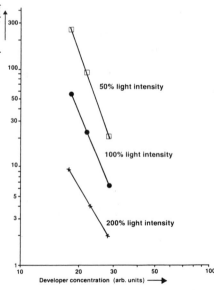

Since, as we have seen, the development rate for given exposure and development conditions increases with the square of the spot intensity, the development time to reach the desired first diffraction order stop criterion will be roughly inversely proportional to the square of the recording intensity. Figure 20 shows experimental results of the effect of halving and doubling the normal recording intensity on the development time for three different developer concentrations.

The developer concentration strongly influences the development time. Figure 21 shows that a variation in the developer concentration of 10% gives a 40 to 50% difference in development time.

In conclusion, as many process parameters can be independently set and controlled, the photoresist mastering process is a very versatile process, which allows the creation of many different three-dimensional structures for present and future optical disc systems.

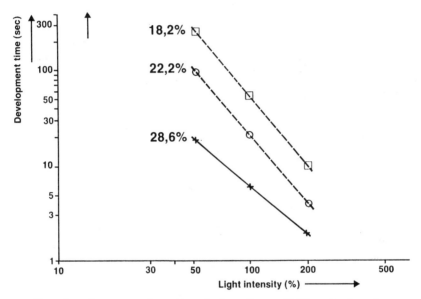

Figure 21. Development time (in s) for three different developer concentrations, with the recording intensity as a variable.

CONCLUSIONS

We have considered the mastering and replication processes for the present optical disc systems. From both a market and a system tolerances standpoints two separate market segments can be distinguished: the High Volume CD market and the High End (CD-)video and data storage markets. Traditionally, photoresist mastering processes have been successfully applied for all optical disc systems (deduced from the fact that more than 90% of the mastering worldwide is done in photoresist). Some companies have started to use other mastering materials and methods for CD production but these discs appear at best to be only marginally within specification. For the more demanding High End mastering, photoresist is still the only practical choice and this is likely to hold for future higher density optical disc systems.

REFERENCES

1. G. Bouwhuis, editor, "Principles of Optical Disc Systems",
    Adam Hilger Ltd, Bristol, 1985.
2. K. Compaan and P. Kramer, The Philips "VLP" system, Philips
    Tech. Review, $\underline{33}$, 178 (1973).
3. K. D. Broadbent, A review of the MCA DiscoVision system, 115th SMPTE
    Tech. Conf. Equip. Exhibit., Los Angeles, Apr. 26, 1974.
4. Compact Disc digital audio, Special issue 6, Philips Tech. Review, $\underline{40}$
    (1982).
5. J. F. Olijhoek, T. H. Peek and C. A. Wesdorp, Mastering technology for
    the Philips optical disc systems, Video Disk Technology Overview,
    Electro '81, New York.
6. W. Verkaik, CompactDisc mastering - An industrial process, Audio
    Engineering Society Premiere Conference, Rye, New York, 1985.
7. J. H. T. Pasman, J. F. Olijhoek and W. Verkaik, Developments in
    optical disc mastering, Third Int. Conf. Opt. Mass Data Storage,
    Los Angeles, SPIE Proc., $\underline{529}$, 62 (1985).
8. J. Hennig, Kunststoffe fuer optische Plattenspeicher, Kunststoffe, $\underline{75}$,
    425 (1985).
9. H. C. Haverkorn van Rijsewijk, P. E. J. Legierse and G. E. Thomas,
    Manufacture of LaserVision video discs by a photopolymerization
    process, Philips Tech. Rev., $\underline{40}$, 287 (1982).
10. A. Huijser in "Principles of Optical Disc Systems", G. Bouwhuis,
    editor, Chap. 6, Adam Hilger Ltd, Bristol, 1985.
11. H. Vaanholt, The coding format for composite PAL video signals and
    stereo sound in the LaserVision optical videodisc system, 4th
    Int. Conf. on Video and Data Recording, Southampton (London:IERE),
    351, 1982.
12. ISO Draft standard for Write-Once Disk, ISO/DP 9171-1(E), 97/23 N 178,
    97/23 N 179.
13. P. E. J. Legierse, Mastering technology and electroforming for optical
    disc systems, Trans. Inst. Metal Finishing, $\underline{65}$, 13 (1987).
14. J. H. T. Pasman, in "Principles of Optical Disc Systems", G. Bouwhuis,
    editor, Chap. 5, Adam Hilger Ltd, Bristol, 1985.
15. Philip de Lancle, DRAW second generation CD mastering, Mix, September
    1988, 182, and
    European Patent EPS 0 051 286, D. L. Atwell, assigned to Discovision
    Associates, 1986.
16. H. Redlich and G. Joschko, CD Direct Metal mastering technology: A
    step toward a more efficient manufacturing process for Compact Discs,
    J. Audio Eng. Soc., $\underline{35}$, 130 (1987).
17. J. Pacansky and J. R. Lyerla, Photochemical decomposition mechanisms
    for AZ-type photoresists, IBM J. Res. Dev., $\underline{23}$, 42 (1979).
18. F. H. Dill, W. P. Hornberger, P. S. Hauge, and J. M. Shaw,
    Characterization of positive photoresist, IEEE Trans. El. Dev., $\underline{22}$,
    445 (1975).
19. J. G. Dil and C. A. Wesdorp, Control of pit geometry on video disks,
    Appl. Opt., $\underline{18}$, 3198 (1979).
20. J. H. T. Pasman, in "Principles of Optical Disc Systems", G. Bouwhuis,
    editor, Chap. 3, Adam Hilger Ltd, Bristol, 1985.

PART III.   POLYMER PHYSICS: RELEVANCE TO OPTICAL RECORDING

MAGNETO-OPTICAL RECORDING MEDIA PROPERTIES AND THEIR RELATIONSHIP TO

COMPOSITION, DEPOSITION TECHNIQUES, AND SUBSTRATE PROPERTIES

T. H. Wallman, M. C. A. Mathur, and M. H. Kryder

Magnetics Technology Center
Carnegie Mellon University
Pittsburgh, PA 15213-3890

A comparison of glass and polymer substrates for rare earth-transition metal magneto-optical recording media has been made. Both rigid (glass and polycarbonate) and flexible (polyester) substrates were investigated. The magnetic, optical and adhesion properties of the deposited films were measured and compared. Film magneto-optical properties and film adhesion were investigated. Films on rigid polycarbonate substrates have been found to have higher coercivities and coercivities more sensitive to the deposition conditions than films deposited on glass substrates under identical conditions. Coercivity variations are thought to be due mainly to substrate surface effects. In flexible tape substrates, film compensation temperatures and compositions are more dependent on the substrate. Magneto-optic films on all substrates had similar Kerr rotation. The film adhesion on glass substrates was found to improve with substrate etching prior to deposition while the adhesion on polycarbonate substrates suffered under substrate etching. The adhesion of films on tape substrates was found to be well above a proposed standard for magnetic tape adhesion.

INTRODUCTION

Magneto-optical (M-O) recording has several attributes which make it a promising technology for high density storage of digital information. Areal densities of optical technology are an order of magnitude higher than those achieved on currently available magnetic recording products. Furthermore, in optical drives the head can be spaced relatively large distances (about 1 mm) from the medium as compared to magnetic recording (less than 0.25 μm). Thus head crashes are not a concern. In addition, optical drives offer the possibility of removable media.

Currently glass is a common substrate for M-O materials and its use is limited to the disk format only. To meet the future needs of inexpensive and durable M-O disks, the use of plastic substrates is desirable. In addition, M-O tape memories would also be desirable since tape provides a much greater recording surface area than disk recording and thus much greater memory capacity per unit volume.[1]

In this report we describe techniques for depositing rare earth-transition metal films onto glass and plastic substrates and compare how the deposition conditions affect the magnetic, optical, and adhesion properties of magneto-optical media deposited onto glass and plastic substrates.

EXPERIMENTAL PROCEDURES

To deposit rare earth-transition metal films onto plastic substrates, it is preferred to use DC magnetron sputtering instead of RF sputtering because DC magnetron sputtering provides relatively high deposition rates with low substrate heating. Accordingly we made use of a Leybold Z-400 sputtering system operating in the DC magnetron mode to deposit the M-O materials. The TbFeCo films were deposited under a range of deposition conditions which are described in the results. To prevent oxidation of the media, the M-O films were sandwiched between two RF sputtered SiO_2 layers of approximately 20 nm thickness.

To eliminate the possibility of damaging or altering the substrate surface with the high temperatures of RF deposition or substrate etching, temperature measurements were made during these processes. The temperatures were measured using Tempilabel temperature sensitive tape labels.[2] The maximum power used in all RF deposition and etching was determined by setting a maximum surface temperature of 100-120°C which is well below the softening temperatures of the substrate materials.

Corning 0211 glass and Makrolon CD 2000 polycarbonate were used for rigid substrates. For tape substrates, polyester of 21.1, 23.4, and 76.2 μm (0.83, 0.92, and 3.0 mil) thickness was used. The 21.1 and 23.4 μm polyester substrates were manufactured with two types of surfaces: a treated surface meant to promote film adhesion and an untreated surface. The substrate samples were approximately 2.5x2.5 cm (1x1 in).

The smoothness of the substrates before and after deposition of the SiO_2 films was characterized by laser spot scanning interferometry (LASSI) and by white light interferometry. The magneto-optical properties of the films were measured using a magneto-optic hysteresis loop tracer developed at Carnegie Mellon University. This apparatus measures the rotation of polarized light (Kerr rotation) reflected off the M-O film as the magnetization of the film is changed under a sweeping applied magnetic field. The measurement results in a M-O hysteresis loop which gives the Kerr rotation (which corresponds to signal level) and the coercivity (the field required to reverse the magnetization of the film). A typical hysteresis loop is shown in Fig. 1.

Measurements of film coercivities as a function of temperature were taken. This is used to determine the compensation temperature of the films which is the temperature at which the coercivity increases to infinity. The compensation temperature is governed by the ratio of rare earth to transition metal in the film so its measurement is also an indirect measurement of film composition.[3,4]

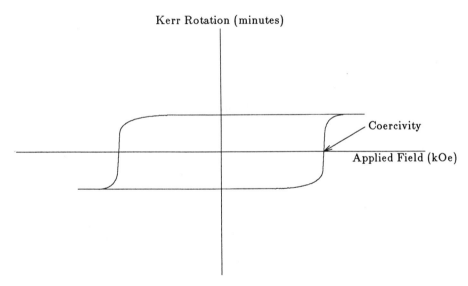

Figure 1. Typical Magneto-Optical Hysteresis Loop.

Film adhesion tests can be performed by several methods.[5,6] Films on
2.5x2.5 cm tape substrates were tested using a recently developed shear
test, which we have shown correlates well with the peel test which has
been proposed as a standard adhesion test for magnetic tape.[7] The peel
test is performed on 38 cm long tape samples that have been scribed across
the width of the tape. The film side of the sample is attached to a
smooth metal plate with double-sided adhesive tape. The metal plate and
the end of the tape sample are attached to the jaws of a universal testing
machine with a jaw separation rate of 25.4 cm/min. With the tape folded
180° on and parallel to the scribe, the adhesion is measured as the force
required to remove any of the film. A comparison of peel test and shear
test results from identical magnetic tape samples is shown in Fig. 2.
This comparison is made to determine the shear test equivalent of the peel
test proposed magnetic tape standard. The M-O films on rigid substrates
were tested using a recently developed tensile test.

RESULTS

To be certain that the RF diode sputter deposition of SiO_2 did not
distort the surface of the plastics, interferometer measurements were made
on the polycarbonate substrates after SiO_2 deposition. White light
interferometry measurements such as the ones in Fig. 3 show no distortion
of the surface. The LASSI measurements in Fig. 4 show that the deposition
of the SiO_2 causes the surface of the polycarbonate to become smoother.

Data on coercivity as a function of temperature, made with the M-O
hysteresis loop tracer, of films on rigid glass and polycarbonate
substrates are shown in Figs. 5 and 6. The films used to collect the data
in Fig. 5 were sputtered under different deposition powers and a constant
chamber argon pressure, and those used to collect the data in Fig. 6. were
deposited under constant power but various pressures. The data show that
the coercivity of films on thick polycarbonate substrates are higher than
those of films deposited under identical conditions onto glass. Also

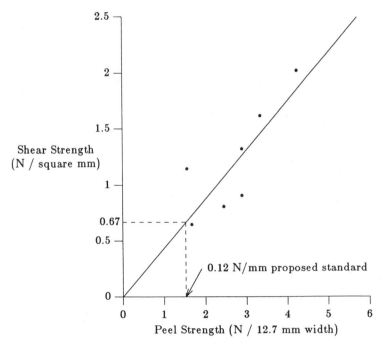

Figure 2. Comparison of Peel and Shear Tests for Adhesion of Magneto-Optical Films on Identical Polyester Substrates.

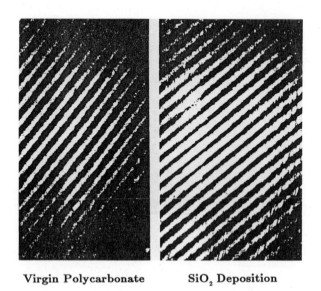

Virgin Polycarbonate **SiO$_2$ Deposition**

Figure 3. Interference Patterns on Polycarbonate Substrate Before and After SiO$_2$ Deposition. Lines are 270 nm apart.

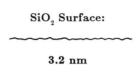

SiO$_2$ Surface:

3.2 nm

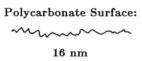

Polycarbonate Surface:

16 nm

Figure 4. Surface Roughness Measured by LASSI (Maximum deviation from center line).

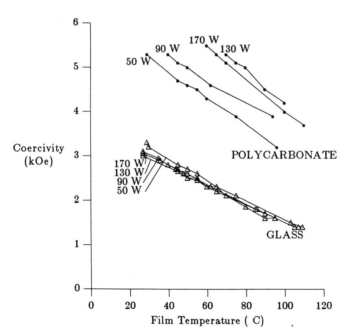

Figure 5. Coercivities of Magneto-Optical Films Deposited on Rigid Substrates at Different Deposition Powers and Constant 5 mT Pressure.

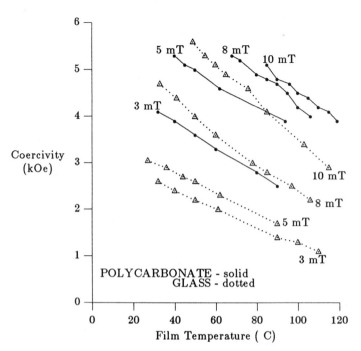

Figure 6.   Coercivities of Magneto-Optical Films Deposited on Rigid
Substrates at Different Deposition Pressures and Constant
130 W Power.

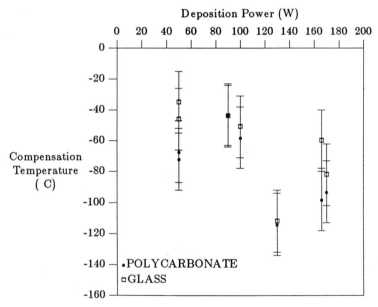

Figure 7.   Compensation Temperatures of Magneto-Optical Films Deposited on
Rigid Substrates at Different Deposition Powers and Constant
5 mT Pressure.

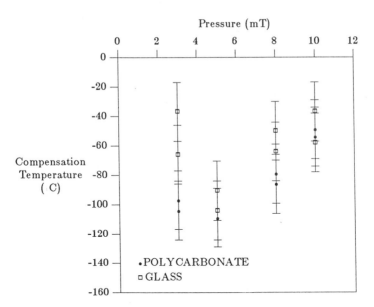

Figure 8. Compensation Temperatures of Magneto-Optical Films Deposited on Rigid Substrates at Different Deposition Pressures and Constant 150 W Power.

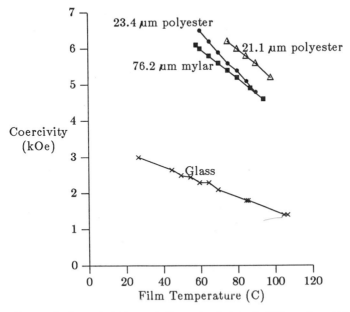

Figure 9. Typical Coercivities of Magneto-Optical Films deposited on Flexible Tape Substrates Compared to Films Deposited on Glass, all Deposited at 5 mT and 80 W.

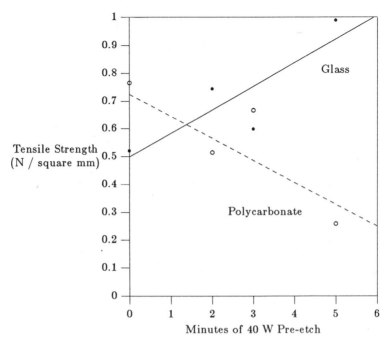

Figure 10. Tensile Adhesion Test Results for Magneto-Optical Films
Deposited on Rigid Polycarbonate Substrates.

there is a stronger dependence of coercivity on deposition power when the
films are on thick polycarbonate than when they are on glass.

The coercivity versus temperature data in Figs. 5 and 6 may be
replotted as H_c^{-1} versus temperature and extrapolated to infinite
coercivity to estimate the compensation temperature. The compensation
temperatures derived by this method are plotted in Figs. 7 and 8.

Although the error bars for the compensation temperatures in Figs. 7 and 8
are relatively large, there appears to be a tendency for the compensation
temperature to be slightly higher on glass than on polycarbonate. A
comparison of the coercivity of films made on flexible tape substrates as
compared to films made on glass is shown in Fig. 9. Again the coercivity
of the films on the polymer substrates is higher than that of films on
glass. A comparison was also made of the magnitude of the Kerr rotation
measured from films deposited on glass and polymer substrates, but the
data showed negligible differences.

The tensile adhesion test results for rigid glass and polycarbonate
substrates are given in Fig. 10. The tension required to pull the film
off the substrate is plotted as a function of the length of time that the
substrate was sputter etched at 40 W before the film deposition. The data
show that the adhesion to glass tends to improve with sputter etching
prior to film deposition, but that adhesion to polycarbonate tends to
degrade when the substrate is sputter etched.

The shear test results for the adhesion of the film to the flexible
tape substrates are plotted in Figs. 11 and 12. The data are once again
plotted for different sputter etching times of the substrate. There is
some indication that a one minute sputter etch at 20 W slightly improves
adhesion on treated polyester and degrades on untreated polyester and
mylar, but the most significant factor is that all the data are well above
the proposed magnetic tape standard.

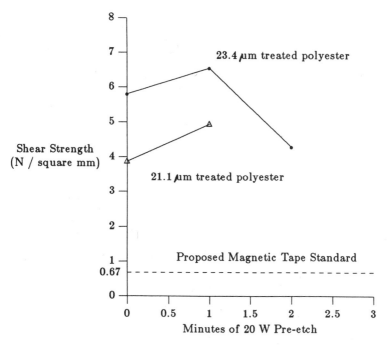

Figure 11. Shear Adhesion Test Results for Magneto-Optical Films
Deposited on Treated Polyester Tape Substrates.

It should be noted that we used a focussed diode laser to write
micrometer sized domains in the films on both the rigid and flexible
polymer substrates. Results were substantially the same as on glass
substrates. The domains were readily written and appeared very similar to
those written on glass.

DISCUSSION AND CONCLUSION

The LASSI measurements of surface roughness indicate that the surface
of the polymer substrate is smoothed by the deposition of SiO_2.
Nevertheless, the data in Figs. 5, 6 and 9 all indicate that films
deposited onto polymer substrates have higher coercivity than films
deposited onto glass. This could be due to the larger surface roughness
of the plastic substrates as compared to glass or to a chemical
interaction between the substrate and the film. The decreased adhesion of
the films caused by sputter etching of the polycarbonate suggests that the
surface properties of the polycarbonate do affect at least the adhesion
properties and it is, therefore, not entirely surprising that they also
affect the coercivity especially considering that the films are less than
200 nm thick.

The fact that the coercivity of films deposited onto polycarbonate
substrates is much more than the coercivity of films deposited on glass,
as shown in Figs. 5 and 6, may be due to the fact that the glass has
better heat conductivity than the polycarbonate and also indicates that
the deposition process alters the interface between the substrate and
film.

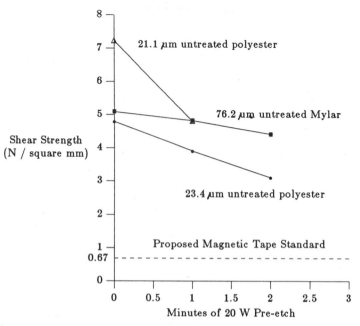

Figure 12.   Shear Adhesion Test Results for Magneto-Optical Films
Deposited on Untreated Polyester Tape Substrates.

Evidence that there are some significant effects of substrate heating
are contained in Figs. 7 and 8 where there appears to be a slightly higher
compensation temperature for films deposited on glass than for films
deposited on polycarbonates.   The very small difference suggests a minor
decrease in the ratio of rare earth to transition metals in the films on
the polycarbonate substrates.   This is consistent with an increased
surface temperature of polycarbonate compared to glass because of the
poorer thermal conductivity of polycarbonate.   Higher substrate
temperature promotes resputtering and rare earths resputter more readily
than transition metals.   Since compensation temperature changes by about
30°C for a 1% change in composition,[4] the data indicate this composition
change is of the order of 1%.

Although there do appear to be some detectable effects caused by
substrate heating, it is very unlikely that the surface of the
polycarbonate actually melts during the processing.   Measurements with
thermally sensitive tape indicated that temperature was always well below
the Vicat softening temperature.   It also seems unlikely that there is a
large chemical change at the interface of the magnetic film and substrate
or there would be a significant change in the Kerr magneto-optic effect,
which was not found.   It, therefore, seems likely that the higher
coercivity of films deposited onto polymers as compared to glass is due to
a slightly larger surface roughness of the polymers.   Adhesion to the
polymers is well above the tape standard, but tends to degrade with much
sputter etching of the polymer substrates, and it is likely that this is a
result of decreased roughness caused by the sputter etching.

In conclusion, although there are some subtle differences between the properties of magneto-optic films deposited on glass and polymer substrates, generally the magnetic, magneto-optical, and adhesion properties are comparable. This plus the fact that domains can be written and read on both rigid polycarbonate and flexible polyester and Mylar substrates suggests that polymer substrates are interesting candidates for both rigid and flexible media. Problems remaining to be solved are the birefringence of the polymer substrates and the fact that films deposited on them corrode more easily than films deposited on glass because the polymer substrates absorb and transmit water vapor and oxygen.

ACKNOWLEDGEMENTS

This work was funded by the IBM Corporation and the U.S. Air Force Rome Air Development Center under PR No. I-7-4074. We are grateful to the Hoescht, Dupont, and Bayer Corporations for their donations of the polymer substrates used in this research. We are deeply appreciative of Clem Kalthoff and Lil Hackett of IBM Tucson for the adhesion measurements.

REFERENCES

1. Peter Vogelgesang and Judith Hartmann, Erasable optical tape feasibility study, Proc. Intl. Soc. Optical Eng., 899, 172-177 (Jan. 1988).

2. Tempil Division, Big Three Industries, Inc., Series 20 Tempilabel Temperature Monitor.

3. Masud Mansuripur and M.F. Ruane, Mean-field analysis of amorphous rare earth-transition metal alloys for thermomagnetic recording, IEEE Trans. Magnetics, MAG-22, 33-43 (Jan. 1986).

4. M.H. Kryder, H.-P. Shieh, and D.K. Hairston, Control of parameters in rare earth-transition metal alloys for magneto-optical recording media, IEEE Trans. Magnetics, MAG-23, 165-167 (Jan. 1987).

5. K.L. Mittal, Ed., "Adhesion Measurement of Thin Films, Thick Films and Bulk Coatings,"STP No. 640, ASTM, Philadelphia (1978).

6. K.L. Mittal, Selected bibliography on adhesion measurement of films and coatings, J. Adhesion Sci. Technol., 1, 247 (1987).

7. ANSI Technical Committee X3B5 of Accredited Standards Committee X3, Proposed American National Standard Magnetic Tape and Cartridge for Information Interchange, Draft 10 (1987).

UNIAXIAL ANISOTROPY STUDIES IN AMORPHOUS Tb-Fe ON POLYIMIDE SUBSTRATES

R. Krishnan[1], M. Porte[1], M. Tessier[1] and J.P. Vitton[2]

1: Laboratoire de Magnetisme, C.N.R.S., F-92195 Meudon,
   France
2: Kodak Pathe, F-71102 Chalon-sur-Saone, France

Amorphous Tb-Fe films have been RF sputter deposited onto water cooled Kapton polyimide substrates at three different Ar pressures and at 80 W RF power. M-H and polar Kerr loops have been studied, Torque measurements have been made in the temperature range from 6 to 300 K. For Ar pressure of 8mTorr, a maximum uniaxial anisotropy K_u value of $2.4.10^6$ erg/cm^3 has been found and is attributed, in part, to stress (compressive) effects. As T is decreased, K increases in contrast to films deposited onto glass substrates. This is explained in terms of the absence of stresses in the sample caused by cooling because the thermal expansion coefficients of the Kapton and the Tb-Fe film are comparable. On the contrary, for a glass substrate which has a low thermal expansion coefficient, additional tensile stresses are induced under such circumstances which lead to a decrease in the value of the stress anistropy term. Cooling in a magnetic field is shown to modify the torque curves.

INTRODUCTION

Amorphous films of rare earth-transition metal alloys are of potential interest for magneto-optical storage media. In addition, they offer very interesting possibilities for the study of certain fundamental physical properties in disordered materials.

Several articles have appeared on the above aspects [1,2]. The uniaxial anisotropy K_u present in these materials, which is of fundamental interest for magneto-optical storage media, has been studied by several authors in order to obtain some insight into the basic mechanisms involved[3]. We have recently reported on the low temperature studies of K_u in Tb-Fe films [4]. Most of the studies reported have been performed on films deposited on glass substrates. On the other hand, amorphous films deposited on flexible polyimide substrates seem to offer an interesting possibility of being suitable for magneto-optical tapes as reported recently [5].

We were interested in studying the anisotropy in such materials because the physical properties of polyimides are very different from those of glass. For instance, polyimides are deformable and their thermal expansion coefficients are comparable to those of metals. So one could expect lesser stresses than are encountered with glass substrates. Consequently, we have prepared some Tb-Fe films on polyimide substrates and have studied their properties as a function of temperature. We describe some of our results here.

EXPERIMENTAL DETAILS

The amorphous films were deposited by RF sputtering. Kapton (Dupont Chemicals) sheets of 50 μm thick were chosen as substrates for our studies. Amorphous Tb_xFe_{1-x} with $18 < x \leq 20$ were deposited onto water cooled Kapton substrates using an RF of 80 W. A composite target was used. Three different argon pressures (P_{Ar}), namely 5, 8 and 10 mTorr, were chosen in order to vary the intrinsic stresses. Since the Tb content of the film is known to increase with increasing P_{Ar}, the composition of the target was modified suitably for each value of P_{Ar} so as to obtain more or less the same Tb content in the film. The samples were about 250 nm thick and were given a protective coating of Al_2O_3 about 12 nm thick before breaking the vacuum.

The amorphous structure was verified by X-ray diffraction, and electron diffraction studies were performed on selected samples about 65 nm thick which were specially deposited on NaCl substrates. The chemical composition of the samples was determined by inductively coupled plasma analysis.

A vibrating sample magnetometer (VSM) was used with an external field H_{max} = 16 kOe to study the perpendicular M-H loop from which both the magnetization (M in $emu.cm^{-3}$) and the coercivity (H_c) were obtained. The polar Kerr magneto-optical effect was also used to obtain the perpendicular loop. Of course the value of Kerr rotation θ_k depends on the thickness of the Al_2O_3 layer and in our cases θ_k was around 0.2^o at the He-Ne laser wavelength of 632 nm.

Torque curves were studied as a function of angle (θ) between the applied field and the film plane. The applied field was varied for each measurement in the range from 5 to 12.5 kOe. Measurements were taken at intervals of 6° through 360°. The angle setting could be adjusted to better than 0.1^o. Considering the torque (L) at $\theta=45^o$ as a function of field H we can write after Miyajima et al. [6]

$$(L/H)^2 = - (M^2V/2K)L + (MV)^2/2$$

where K is the total uniaxial anisotropy and $K = K_u + 2\pi M^2$ where K_u is the intrinsic uniaxial anisotropy and V is the volume of the sample. Plotting $(L/H)^2$ vs L yields M and K through the relation $M = (2B)^{1/2}/V$ and $K = B/(AV)$ where A is the slope of the straight line and B the intercept on $(L/H)^2$ axis. It is relevant here to state that M values from both VSM and torque measurements agreed within experimental error. The main cause of error, about $\pm$ 5%, in these measurements arises from the determination of sample volume. Our K measurements are accurate to about $\pm$ 3%. The data presented here are accurate to about $\pm$ 15% considering all sources of errors. Torque measurements were extended to low temperatures, T < 10 K, only for selected samples.

In some cases, we also studied the effect of cooling in a magnetic field because we observed some magnetic after-effects.

RESULTS AND DISCUSSION

All the samples showed rectangular M-H and θ_k-H loops. Fig. 1 shows θ_k-H loop for the sample II (X = 17). Table I summarizes some magnetic data for the three samples studied. It is seen that the magnetization varies in the range 100-140 emu.cm^{-3}, which agrees with the values published before [4].

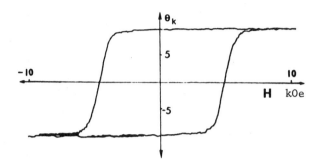

Fig. 1. Polar Kerr (θ_k-H) loop for sample II.

Torque measurements indicated a well defined K_u perpendicular to the film plane. Fig. 2 shows the angular dependence of torque for H=12.5 kOe for the sample II. One notices the presence of rotational hysteresis which is normally found for these samples, particularly when the applied H is smaller than the anisotropy field (2 K_u/M) [4]. Table I shows the value of K_u for the three samples studied. In Tb based films, K_u arises mainly from the large random anisotropy of Tb atoms though stress effects could also play a role. Considering that the Tb content does not vary much in the samples, the high value of $2.4.10^6$ erg. cm^{-3} for sample II is noteworthy. This higher value of K_u could be attributed to larger compressive stress ($\sigma < 0$) induced by the growth process which, in turn, contributes to K_u from magnetoelastic effects. Let us recall that the stress induced anisotropy $K^S = - \frac{3}{2} \sigma\lambda$ where λ is the magnetostriction which is on the order of $+ 2.10^{-4}$ for such amorphous alloys [7]. Hence K^S is positive and adds to the single ion random anisotropy from Tb.

The low temperature measurements gave the following results. In the temperature range 6 to about 150 K the torque curves depended strongly on the magnetic state of the sample. For instance, cooling the sample in the presence of a field totally altered the result. For instance, Figs. 3 and 4 describe the results for sample I which are representative of the results for the other samples as well. Fig. 3 shows the angular dependence of the torque (L) at 99 K after first studying L = f(H) at $\theta = 45°$. The uniaxial anisotropy is suppressed as indicated by L > 0 for $\theta = 45°$ and the change in the symmetry (Fig. 3). The sample was warmed up to 290 K and then cooled in the absence of the field (H = 0) and the L = f(θ) was studied starting with $\theta = 0$ and setting H = 12.5 kOe. The symmetry reappears indicating the presence of K_u. The small peak near $\theta = 90°$ has been observed by us in other samples also, and its origin is not quite clear at present [4]. So these experiments indicate large after-effects at low temperatures and the possibility to strongly modify K_u by cooling in external fields.

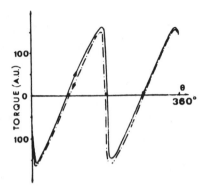

Fig. 2. Torque curve for sample II at 290K at H = 12.5 kOe.

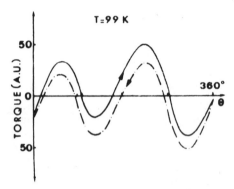

Fig. 3. Torque curve for sample I at 99K. See text for details.

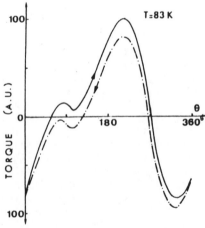

Fig. 4. Torque curve for sample I at 83K, after cooling in zero field.

Table I. Some Properties of the Samples.

No.	Tb at. %	M emu. cm^{-3}	K_u (10^6) erg. cm^{-3}
I	19	100	1.3
II	17	140	2.4
III	18	120	2.0

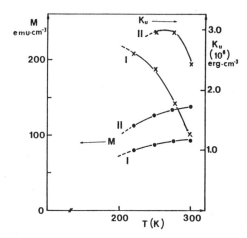

Fig. 5. Temperature dependences of M and K_u for samples I and II.

However, it has been possible to study these properties in a small range of temperature where no such after-effects were seen. Fig. 5 show the results. A sharp increase in K_u is observed while M starts decreasing as one approaches the temperature T_{comp} where the resultant magnetization of the alloy vanishes. For samples prepared on glass substrates K_u, on the contrary, decreased as T was lowered [4]. We are led to conclude that for Kapton samples no additional stresses are created because the thermal expansion coefficient (α) of Kapton is about 2.10^{-5}/K which is close to that of the metal film. On the contrary, for glass α is an order of magnitude smaller which at low T could give rise to tensile stresses ($\sigma > 0$). This then counteracts the initial compressive stress, thereby decreasing the stress anisotropy K^S and the net anisotropy. As λ increases at low T, the contribution from $\sigma\lambda$ plays an important role.

CONCLUSION

We have prepared amorphous Tb-Fe films by RF sputtering on Kapton substrates and have characterised their magnetic properties. The temperature dependence of K_u is different from that observed on samples on glass substrates. Another important observation is the large after-effect of anisotropy. Cooling the sample in magnetic fields modifies K_u. Further experiments are underway to study this aspect in detail.

REFERENCES

1. Y. Mimura, N. Imamura and T. Kobayashi, IEEE Trans. Mag. <u>MAG-12</u>, 779 (1976)
2. M.H. Kryder, J. Appl. Phys. <u>57</u>, 3913 (1985)
3. R.B. Van Dover, M. Hong, E.M. Gyorgy, J.F. Dillon Jr. and S.D. Abbiston, J. Appl. Phys. <u>62</u>, 216 (1987)
4. R. Krishnan, M. Porte, M. Tessier, J.P. Vitton and Y. Le Cars, IEEE Trans. Mag. <u>MAG 24</u>, 1773 (1988)
5. M. Dancygier, IEEE Trans. Mag. <u>MAG-23</u>, 2608 (1987)
6. H. Miyajima, K. Sato and T. Mizoguchi, J. Appl. Phys. <u>47</u>, 4669 (1976)
7. S. Hashimoto, Y. Ochiai, M. Kaneko, K. Watanabe and K. Aso, IEEE Trans. Mag. <u>MAG-23</u>, 2278 (1987)

ORIENTATIONAL AND STRESS BIREFRINGENCE IN OPTICAL DISK SUBSTRATES

M. J. Brekner

Hoechst AG, R+D Informationstechnik Division
6230 Frankfurt am Main 80, F.R. Germany

Starting from a basic analysis of birefringence in glassy polycarbonate, two phenomenologically different sources and thus two types of birefringence, the orientational and the stress birefringence, are described. Both contribute to the overall birefringence of optical disk (OD) substrates. Some evaluations and data for each type of birefringence are presented.

As optical disk substrates are manufactured mainly by injection molding, some related processing parameters, e.g. injection time, cooling time, mold temperature, etc., are related to the substrate birefringence. Finally, various ways to minimize birefringence are discussed and some limiting factors are highlighted.

INTRODUCTION

At present, polycarbonate has no real competitor with respect to the OD-application. Polycarbonate is easily processible by injection molding and exhibits low water absorption, good thermo-mechanical properties and high transparency. However, the limiting property with respect to its use for optical disk substrates is the optical anisotropy. When processed by basically any technique the disk substrates obtained show birefringence patterns which are of concern. It is a general understanding that among various processing techniques injection molding is the most preferred one for optical disk substrates. Reasons for using injection molding are, for example, the rotational symmetry of the birefringence pattern and the enhanced mechanical disk stability. As most of the birefringence referring to a plane perpendicular to the disk plane can be compensated by technical means in the disk drives, this paper will focus only on the in-plane birefringence of OD-substrates.

SPECIFICS OF THE OD-SUBSTRATE BIREFRINGENCE

In order to avoid any additional influence of experimental specifics on retardation, the data presented here refer always to the maximum in-plane birefringence of OD-substrates, i.e. the difference between the

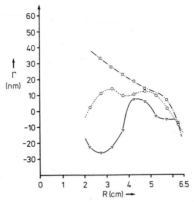

Figure 1. Radial retardation profile (single pass) of injection moulded OD-substrates of various batches (∇ = batch A; O = batch B; □ = batch c) produced by various substrate suppliers.

refractive index in radial direction and that in the tangential direction. The respective retardations refer to single pass measurements with a He-Ne-laser.

Now what are the specifics of birefringence of OD-substrates? First, there is the dependence of the radial birefringence profile of OD-substrates on the injection molding conditions as illustrated in

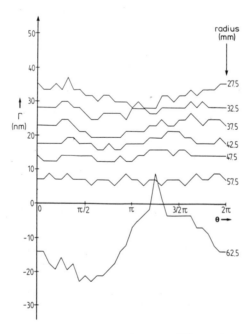

Figure 2. Circumferential retardation profiles at various radii of an OD-substrate.

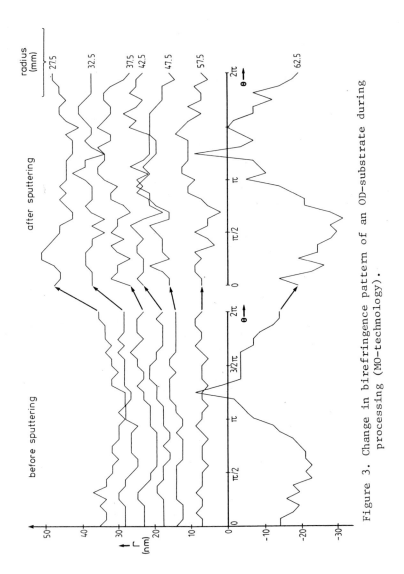

Figure 3. Change in birefringence pattern of an OD-substrate during processing (MO-technology).

Figure 1. The circumferential birefringence profile as exemplified in Figure 2 shows also variations in birefringence. However, this birefringence range is smaller than the radial one.

Another feature of substrate birefringence is its change during disk processing, e.g. during the build-up by sputtering of a magnetooptic layer on the substate surface. Figure 3 shows the influence of an arbitrary sputter process on the substrate birefringence. It has to be pointed out that such change in birefringence depends on both the sputtering conditions and the initial birefringence profile. As the sputter process raises the substrate temperature above room temperature, its impact on birefringence might be compared with an annealing step.

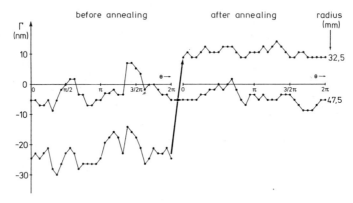

Figure 4. Change in birefringence pattern of an OD-substrate by annealing.

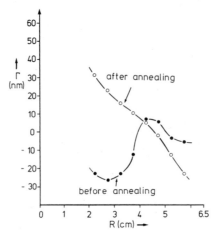

Figure 5. Change in radial retardation profile (single pass) by annealing of an OD-substrate (diameter=130mm) of batch A (see Figure 1).

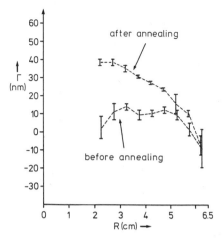

Figure 6. Change in radial retardation profile (single pass) by annealing of an OD-substrate (diameter =130mm) of batch B (see Figure 1).

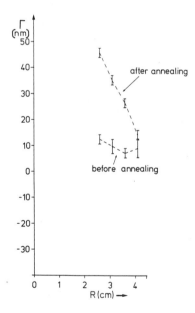

Figure 7. Change in radial retardation profile (single pass) by annealing of an OD-substrate (batch D) with a diameter of 86mm.

197

Figures 4, 5, 6 and 7 show some examples concerning the changes in various birefringence patterns of OD-substrates by annealing the substrates for 5 hours at 124°C. The exemplified substrates were manufactured by different substrate suppliers.

BASICS OF BIREFRINGENCE IN AMORPHOUS POLYMERIC MATERIALS

Before starting to analyse such substrate specifics some basics of birefringence should be considered. Generally, it is assumed that the birefringence of polymeric materials is the result of chain orientation. It can be expressed by:

$$\Delta n = K \, F_0 \, (a_\parallel - a_\perp) \qquad (1)$$

where F_0 is the Herman orientation function, $a_\parallel - a_\perp$ the difference of polarizabilities in Kuhn's model and K a characteristic constant. However, the special case $a_\parallel - a_\perp = 0$, where chain orientation does not yield birefringence, is limited to some very few copolymers and blends and will not be discussed here. Considering the thermodynamic aspects, the birefringence resulting from chain orientation might be called, besides orientational birefringence, entropy birefringence as well.

At temperatures far above the glass transition temperature of the polymer, the chain orientation correlates with deformational stress. Thus a linear relation exists between the applied stress $\Delta\sigma$ and the birefringence :

$$\Delta n = C \cdot \Delta\sigma \qquad (2)$$

with C as the stress-optical coefficient. Such stress-optical coefficients are determined from the melt. For polycarbonate C is in the range of $3.5 - 3.7 \times 10^{-9} \, Pa^{-1}$.

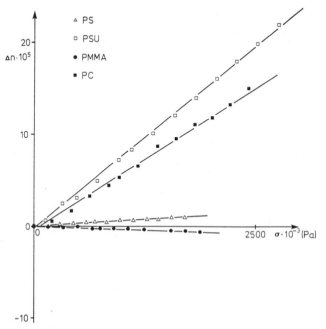

Figure 8.   Stress induced birefringence of glassy polymers
              (PS = polystyrene; PSU = polysulphone;
              PMMA = polymethylmethacrylate; PC = polycarbonate).

198

The phenomenon changes if stress is applied to the glassy polymer. Our stress-optical measurements on polycarbonate at 23°C yield a linear relation similar to Equation (2) with a constant C (index 'g' indicates the glassy state) in the range $6.6 - 7.0 \times 10^{-11}$ Pa^{-1}. The respective data are presented in Figure 8. The difference between C and C_g indicates that in the glassy state the molecular mechanism producing birefringence differs from that in the melt. This difference is very striking in the case of polystyrene where C is negative while C_g is positive. When considering that below glass transition no cooperative flow mechanism exists on molecular level, it is obvious that in this temperature range stress induced birefringence does not originate from chain orientation but from changes in chemical bond angles and bond lengths. These elementary processes are reversible and from a thermodynamic point of view they represent changes in internal energy. Consequently, the stress induced birefringence of glassy polymers might be called, besides stress birefringence, energy birefringence as well.

When comparing the stress birefringence of glassy polycarbonate with that of other glassy polymers, as visualized in Figure 8, it becomes evident that polycarbonate belongs to those polymers which exhibit high stress-optical constants in their glassy state.

COMPENSATION BETWEEN ORIENTATIONAL AND STRESS BIREFRINGENCE IN OPTICAL DISK SUBSTRATES

The fact that melt processing of polymers yields birefringence is a well known phenomenon[3-5]. However, as already mentioned, the birefringence of OD-substrates made of polycarbonate exhibits some interesting features which will be analysed in the following.

Considering first the change in birefringence during annealing, i.e. disk processing, an important question arises concerning the molecular origin

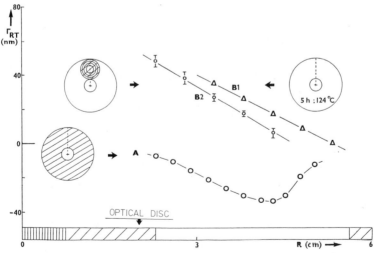

Figure 9.   Stress birefringence
(A = retardation profile of OD-substrate as injection molded; B1 = retardation profile of annealed OD-substrate; B2 = retardation profile of cut out disc; A-B1, A-B2 = stress related changes of retardation).

of such change. Is it molecular orientation or is it some residual stress which changes during annealing?

The experiment presented in Figure 9 demonstrates that the change in birefringence during annealing is due to a reduction of residual stress in the substrate. The radial retardation profile B2 in Figure 9, measured on disk shaped substrate pieces which were cut out of the substrate by avoiding any heat formation, is nearly identical with the retardation profile B1, which corresponds to the annealed substrate. An additional experiment illustrated in Figure 10 shows that the residual stress in the substrate is basically tangential tension. Accordingly, after cutting out of the substrate an arbitrary disk sector, the radial retardation profile of this disk sector corresponds to curve B of Figure 10. When applying to the respective disk sector a momentum M in the disk plane, the profile changes to C indicating that in the tension range of the disk sector the retardation shifts towards the initial birefringence profile A of the disk substrate.

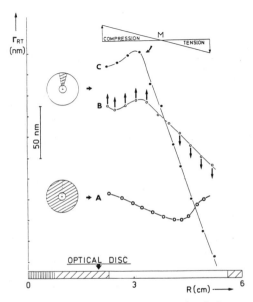

Figure 10.   Tension/compression induced birefringence
(A = substrate retardation; b = retardation of cut out
substrate sector; C = sector retardation with applied
momentum M).

Thus, the difference between the profiles B and A is basically induced by tangential tension which probably results during cooling of the molded disk substrate.

Consequently, when referring to Figure 9, the radial birefringence profile of an injection molded polycarbonate substrate is composed of two contributions: one corresponding to the radial retardation of type B (i.e. B1 and B2), and the other corresponding to the difference between the radial retardation profiles A-B, which represents the stress induced retardation.

The birefringence which gives the profile B is not influenced by any stress because it can not be changed by means of annealing if the annealing temperature does not reach the glass transition temperature. In conclusion, B should correspond to the orientational birefringence. However, this can be demonstrated by means of shrinkage experiments.

On heating an oriented polymer above its glass transition temperature it will shrink in the direction in which it had been oriented. The degree of shrinkage correlates with the degree of orientation. Such experiments were performed on 15 mm discs which were cut out of the substrate at various substrate radii. After annealing at 180°C for 2 hours the shrinkage of these cut out disks was measured in radial and tangential directions, the directions referring to the substrate. Figure 11 presents such data on three arbitrary substrates. The difference between radial and tangential shrinkages is plotted versus the substrate radius. The higher this difference is, the more pronounced is the radial chain orientation. Accordingly, near the substrate center the radial chain orientation is dominant, while at a radius of 5 to 6.5 cm the orientation in tangential direction seems to match that in the radial direction. When comparing this plot with the radial retardation profiles of the respective annealed substrates presented in Figure 12, the similarities are evident. The retardation is very high close to the center of the substrate and approaches zero, i.e. the point of optical isotropy in the substrate plane, at a radius between 5 and 6.5 cm.

These similarities between the asymmetry of shrinkage and the optical anisotropy of annealed substrates suggest strongly that the radial retradation profile of type B in Figure 9 is due to orientational birefringence. Consequently, the high polarizability in radial direction close to the center of the substrate, as reflected by the retardation profile B, correlates with the radial chain orientation in this part of the substrate. This indicates that no stress induced birefringence but only the basic polarizability of the polycarbonate chain ($a_{\parallel} - a_{\perp}$) is contributing to the optical anisotropy of an annealed polycarbonate disk substrate.

In conclusion, there exists much evidence that two types of birefringence, an orientational birefringence and a stress birefringence, are contributing to the substrate birefringence. It is also possible to separate these two contributions by annealing. However, the most important observation with respect to these two birefringence contributions is the fact that they are basically opposed to each other, which means that the stress birefringence induced by tangential tension compensates the orientational birefringence. Such compensation can explain, for example, why the radial birefringence profiles of substrates which are injection molded under various conditions (see Figure 1) exhibit large differences; while the radial birefringence profiles of annealed substrates (see Figure 12) are very similar to each other . It substantiates also that under certain injection molding conditions, substrates can be produced with nearly no in-plane birefringence, although the polycarbonate chain exhibits a high anisotropy of its polarizability. The latter, however, has the disadvantage that such low birefringence levels are not really stable. As shown, for example, in Figure 7, birefringence changes by means of heat impact from a relatively low level towards dominant orientational birefringence.

An explanation for the stress reduction during annealing or processing of the substrates can be drawn from the temperature profile of the loss factor of polycarbonate presented in Figure 13 which shows that indeed there exist local relaxation processes between room temperature and the glass transition temperature of polycarbonate which can be activated by heat impacts.

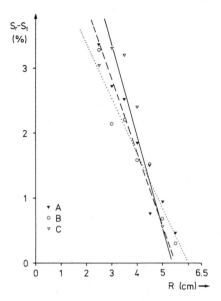

Figure 11.   Shrinkage in OD-substrates at 180°C
(S_r = radial shrinkage; S_t = tangential shrinkage;   A,   B,
C = arbitrary substrates).

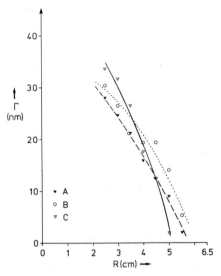

Figure 12.   Radial retardation profile of annealed OD-substrates
(A, B, C=arbitrary substrates).

202

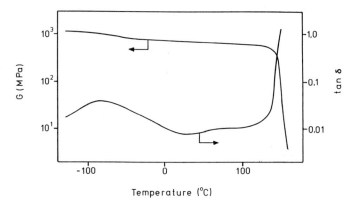

Figure 13.   Temperature dependence of the storage modulus and of the
             loss factor of polycarbonate.

As the reduction of substrate birefringence via stress birefringence is
altered by heat, a really stable reduction of substrate birefringence can
be achieved only by a reduction of the orientational birefringence. Unfortunately, the orientational in-plane birefringence of an injection
molded substrate will hardly be zero because of the tensional flow of the
polycarbonate melt during mold filling [3-5]. Thus, the amount of stress
birefringence is in the same order of magnitude as the orientational
birefringence and has, therefore, to be considered seriously.

DISCUSSION

The basic conclusion from the above analysis is to control the residual
stress of injection molded substrates of polycarbonate. Depending on substrate specifications, it is desirable to have a certain uniform stress
level in the substrate. In order to keep heat induced birefringence
changes small, the stress level should be kept low. If taken alone, from
this criterion one can derive the extreme requirement to have a stress-
free substrate. Such requirement implies, for example, in a first-
approximation isothermal injection molding conditions.

In order to reveal some ways of influencing stress in the substrates,
some calculations were performed with the "MOLDFLOW" program, considering
flow data of an arbitrarily chosen low molecular weight polycarbonate. We
focused basically on the temperature distribution in the injection filled
mold and on ways of attaining isothermal injection molding conditions or
of at least coming close to such extreme conditions. Figure 14 shows how
the injection time influences such temperature distribution. By varying
injection time and nozzle temperature isothermal injection molding can be
attained. Figure 15 illustrates the temperature distribution calculated
for various injection molding conditions. One of them, characterized by a
nozzle temperature of 280°C and an injection time of 0.16 seconds, corresponds to isothermal injection molding. Similar calculations revealed a
very high sensitivity of the temperature distribution in the mold on
changes of the mold shape. Figure 17 presents some examples. The respective mold shapes are illustrated in Figure 16.

Assuming that a large radial temperature gradient induces high residual
stress, one can derive from the presented calculations that low stress
birefringence might be obtained with high injection speed. High melt temperatures produced by internal friction at high injection speed are   en-

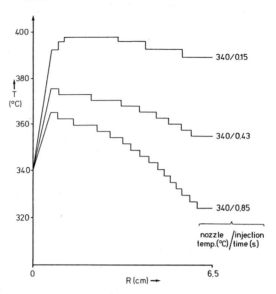

Figure 14. Radial temperature profiles in the mold after mold filling: dependence on injection time (calculations with the MOLDFLOW program for mold shape B of Figure 16).

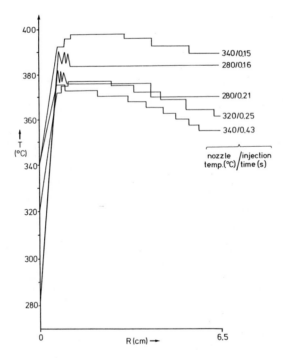

Figure 15. Radial temperature profiles in the mold after mold filling under various filling conditions (calculations with MOLDFLOW program for mold shape B of Figure 16).

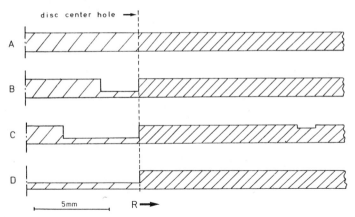

Figure 16.   Considered   mold   shapes   of   central   mold   area   for
MOLDFLOW-calculations.

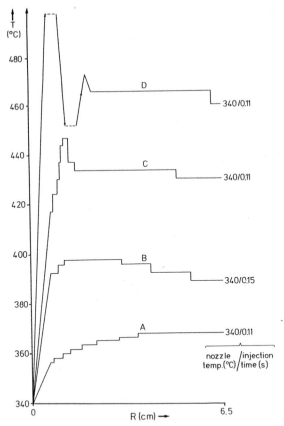

Figure 17.   Dependence of radial temperature profile in the filled mold
on the mold shape: A,   B,   C,   D = different mold shapes
(MOLDFLOW-calculation).

hancing the relaxation processes above the glass transition temperature of polycarbonate and thus reduce also the residual orientational birefringence.

With respect to the compensation of shrinkage during the cooling cycle, it should be pointed out that any such compensation which changes the temperature distribution in the mold renders the production of stress-free substrates.

CONCLUSIONS

The birefringence of polycarbonate OD-substrates is composed of two types of birefringence, the orientational and the stress birefringence. Stress birefringence compensates orientational birefringence and decreases during annealing. In order to avoid heat induced changes of the substrate birefringence, stress has to be controlled carefully via injection molding conditions. Low stress levels should be attained with isothermal injection molding conditions. One way of coming close to such conditions is to choose high injection speeds.

REFERENCES

1. W. Kuhn and F. Grün, Kolloid-Z., 101, 248 (1942).

2. G.H. Werumeus Buning, R. Wimberger-Friedl, H. Janeschitz-Kriegl and T.M. Ford, paper presented at the International Symposium on Optical Memory, Tokyo, Sept. 1987.

3. H. Janeschitz-Kriegl, Polymer melt rheology and flow birefringence, in "Polymers/Properties and Applications", vol. 6, Springer-Verlag, Berlin, 1983.

4. M. Takeshima and N. Funakoshi, J. Appl. Polym. Sci., 32, 3468 (1986).

5. A.I. Isayev and C.A. Hieber, Rheol. Acta, 19, 168 (1980).

BIREFRINGENCE ANALYSIS OF INJECTION-MOLDED POLYCARBONATE(PC) SUBSTRATES

A. Iwasawa and N. Funakoshi

NTT Applied Electronics Laboratories
Tokai, Ibaraki 319-11 Japan

To obtain a polycarbonate (PC) substrate with low optical anisotropy, birefringence and refractive index ellipsoids are analyzed for injection-molded PC substrates annealed under various conditions. The relationship between azimuth and incident angles and retardation shows that, for oblique rays from the fast axis direction, retardation increases directly with incident angle. However, for oblique rays from the slow axis direction, retardation decreases with increasing incident angle up to 15° and then increases. Photo-elastic experiments with PC films show that the fast axis direction has a low refractive index. These results indicate that the PC refractive index ellipsoid is pie-shaped. Optical anisotropy parallel to the disk surface associated with the annealing temperature (T_a) shows two different temperature dependences. In one case, retardation increases with T_a up to $100°C$, and then decreases. In the other case, the opposite occurs. In both cases, if changes in the fast axis direction are considered, the results indicate that annealing causes relaxation of the compressive stress produced thermally below the T_g in the radial direction. The $\Delta n_z(n_x$ or $n_y - n_z)$ of the substrate annealed at T_g varies from 4.5×10^{-4} to 1×10^{-5}. One percent of the disk diameter is constricted. Thus, polycarbonate molecule conformation changes and the residual stress, frozen at T_g during the injection process, relaxes. Therefore, the vertical birefringence($= \Delta n_z$) depends on the residual stress frozen at the T_g during the injection process.

INTRODUCTION

Injection-molded PC substrates are widely used as optical disk substrates. Since PC is highly birefringent[1-3], it has not been considered suitable as a magneto-optic disk substrate. Magneto-optic readout is accomplished by detecting slight rotation in polarization of a linearly polarized beam reflected from the recording film. Thus, the performance of a magneto-optic readout signal is easily degraded by optical retardation. Injection-molding techniques which minimize the in-

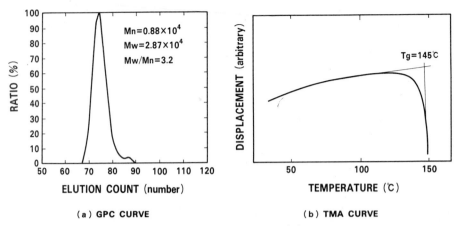

Figure 1. Molecular weight distribution (a) and thermo-mechanical
property (b) of injection-molded PC substrate materials.

plane birefringence ($=(n_x-n_y)$) of PC substrates have been demonstrated.
However, the vertical birefringence causes major retardation in oblique
light rays. Also when the optic axis is tilted relative to the disk
coordinates, the carrier-to-noise ratio decreases due to the imbalance of
the differential detecting system. The retardation magnitude and the
tilting angle to the disk coordinates depend on the injection-molding
conditions and the media fabrication process. To obtain a PC substrate
with low optical anisotropy and to improve performance of the magneto-
optic disk, the refractive index ellipsoid of injection-molded PC
substrates annealed under various conditions is analyzed. This paper
presents some characteristics of the refractive index ellipsoid.
Birefringence based on annealing effects is also discussed.

EXPERIMENTAL

PC Substrate

    Figure 1 shows typical examples of characteristics of injection-
molded PC substrate materials examined. Figure 1(a) shows the molecular
weight distribution of an injection-molded PC substrate. M_n is 0.88×10^4
, M_w is 2.83×10^4 and M_w/M_n is 3.2. Figure 1(b) shows the thermo-
mechanical analysis curve of an injection-molded PC substrate. The glass
transition temperature of the injection-molded PC substrate used is about
145 °C. Annealing was done from 70 °C to 145 °C for 20hr each.

Measurement of Optical Birefringence

    Retardation and the fast axis direction were measured with a
photoelastic modulator (assembled by ORC Manufacturing Co. Ltd., Tokyo
Japan). In this apparatus[4], monochromatic light beams pass in
succession through a polarizer, a photoelastic modulator, a substrate and
an analyzing polarizer, and are then detected by a photodetector, as
shown in Fig. 2. The photoelastic modulator converts the linearly
polarized light into light oscillating between the left and right
ellipticity at its 50kHz resonant frequency. When the modulated light
passes through the substrate, it is divided into two components. The
phase difference between them is caused by birefringence of the substrate
and is detected by the intensity of the modulated light. The sample is
rotated to maximize the intensity. The magnitude of retardation and the

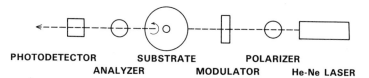

PHOTODETECTOR   SUBSTRATE   POLARIZER
         ANALYZER   MODULATOR   He-Ne LASER

Figure 2. Schematic diagram of the apparatus for the measurement of
optical birefringence.

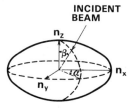

α : AZIMUTH ANGLE
β : INCIDENT ANGLE

Figure 3. Model of refractive index ellipsoid.

fast axis direction can be determined from the intensity of the modulated
light. To analyze the refractive index ellipsoid, the dependences of
retardation on the azimuth (from the radial direction) and incident
angles were also measured.

ANALYSIS OF REFRACTIVE INDEX ELLIPSOID

An index ellipsoid with three principal refractive indices,
n_x, n_y and n_z, is shown in Fig. 3. The cross section of this ellipsoid
perpendicular to the incident beam can be described by

$$(\cos^2 \alpha * \cos^2 \beta / n_x^2 + \sin^2 \alpha * \cos^2 \beta / n_y^2 + \sin^2 \beta / n_z^2)x^2 +$$

$$2\sin \alpha * \cos \alpha * \cos \beta (-1/n_x^2 + 1/n_y^2)xy +$$

$$(\sin^2 \alpha/n_x^2 + \cos^2 \alpha/n_y^2) y^2 = 1 \text{----------- (1)}$$

where β is the incident angle and α is the azimuth angle.

$$x = r\cos \theta \quad , \quad y = r\sin \theta \quad \text{----------- (2)}$$

When formula (2) and the expected values of the three refractive indices
are substituted in formula (1), retardation can be calculated from the
difference between the maximum and minimum values of r. The three
refractive indices should be determined to match the calculated values
to the experimental values.

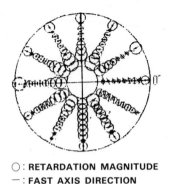

○ : RETARDATION MAGNITUDE

— : FAST AXIS DIRECTION

Figure 4. In-plane retardation and fast axis direction.

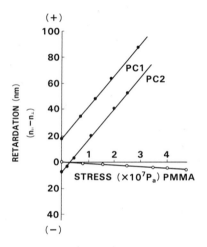

Figure 5. Dependence of retardation on drawing stress for PC and PMMA.

RESULTS AND DISCUSSION

Refractive Index Ellipsoid

In-plane retardation ($(n_x-n_y)d$) and fast axis direction. Examples of the retardation magnitude and the fast axis direction measured with a incident beam normal to the substrate surface are shown in Fig. 4. The diameter of the circles shows the retardation magnitude and the bar direction indicates the fast axis direction. The retardation magnitude and fast axis direction depend on the molding and annealing conditions.

Photo-elastic properties of PC and PMMA. The photo-elastic properties were examined to confirm the relation between the fast axis direction and the drawing direction. The retardation magnitude and the fast axis direction of the PC and PMMA films were measured when the films were drawn at the room temperature. The results are shown in Fig. 5. In the case of PC, retardation increases and the fast axis direction becomes perpendicular to the drawing direction when a drawing stress is applied. The slopes of the straight lines indicate a stress-optical constant of about 6.8×10^{-11} Pa^{-1}. However, in PMMA, the retardation increases with the drawing stress, but the fast axis direction remains parallel to the drawing direction. The stress-optical constant of PMMA is determined to be -0.37×10^{-11} Pa^{-1}. Based on these results, the fast axis direction in this experiment shows a low-refractive index.

210

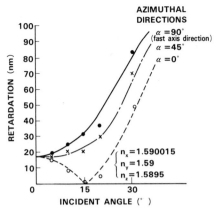

Figure 6. Dependence of retardation on azimuth and incident angles. Lines
show that the calculated results from Equation (1) and (2) match
experimental results.

Table I. Birefringence of Various Substrate Materials.

	$\Delta (n_x - n_y)$	Δn_z
PC	4.0×10^{-6}	5.5×10^{-4}
PMMA	2.0×10^{-6}	4.0×10^{-5}

Azimuth and incident angle dependence on retardation.   The
dependences of retardation on azimuth and incident angles are shown in
Fig. 6.   For the oblique ray from the optic axis direction with a low
refractive index, retardation increases directly with incident angle.
However, for the oblique ray from the optic axis direction with the high
refractive index, retardation decreases with increasing incident angle
up to 15 degrees and then increases.   When n_x=1.590015, n_y=1.59, and
n_z=1.5895 are substituted in formula (1), the estimated curves agree
well with the experimental results.

Model of refractive index ellipsoid.   The results of dependences of
retardation on the azimuth and incident angles and photo-elastic
properties show that the PC refractive index ellipsoid is pie-shaped.
That is, n_z is smaller than n_x and n_y, as shown in Fig. 3.

Some Characteristics of the Refractive Index Ellipsoid

The refractive index ellipsoids of PC substrates injection-molded
under various conditions were examined. Some characteristics of the
refractive index ellipsoid in injection-molded substrates are described
in this section.

Substrate materials.   Table I shows Birefringence for various
substrate materials. The Δn_z (n_x or n_y - n_z)of PC is about 5.5 x 10^{-4}
, larger than Δn_z of PMMA which is about 4 x 10^{-5}.   That is ,the PC
shows large optical anisotropy in the plane vertical to the disk
surface; whereas, PMMA is rather optically isotropic. The difference in
the optical birefringence between PC and PMMA depends mainly on the
anisotropy of polarizability of the phenyl groups all located in the PC
polymer backbone.

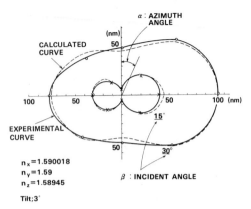

Figure 7. Asymmetry of refractive index ellipsoid tilted to the disk
surface.

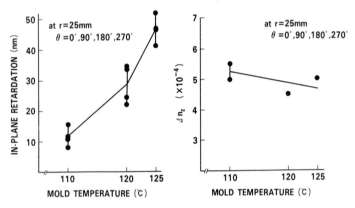

Figure 8. Dependence of in-plane retardation (left) and Δn_z (right) on
mold temperature.

Tilted refractive index ellipsoid. Refractive index ellipsoid
tilted to the disk surface is shown in Fig. 7. The lack of symmetry can
be seen from this pattern. It shows that the optic axis is tilted
slightly relative to the disk surface of the PC substrate. If a
refractive index ellipsoid is tilted relative to the disk surface, the
carrier-to-noise ratio decreases. Therefore, a substrate manufacturing
process which produces a non-tilted n_z axis should be investigated.

Molding temperature and birefringence[5]. The dependences of in-
plane retardation ($(n_x-n_y)d$) and Δn_z on the mold temperature are shown
in Fig. 8 as one example. In-plane retardation increases with mold
temperature, but Δn_z decreases. These results seem to depend on the
cooling process at temperatures between the mold temperature and the
glass transition temperature. They also seem to indicate an annealing
effect. This will be described in the following paragraph.

Annealing Effect

Annealing is one method of reducing noise in a magneto-optic
disk. Therefore, the dependence of birefringence on annealing temperature
(T_a) was studied.

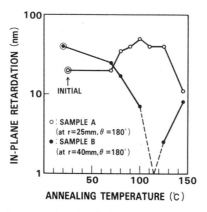

Figure 9. Dependence of retardation on annealing temperature.

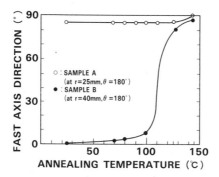

Figure 10. Dependence of fast axis direction on annealing temperature.

Dependence of retardation on annealing temperature. The dependence of optical anisotropy parallel to the disk surface on annealing temperature is shown in Fig. 9. Two different temperature dependences can be seen in this figure. In one, the ratardation increases with annealing temperature up to 100 °C and then decreases. In the other, the opposite occurs. The fast axis direction of one sample is the radial direction and that of the other sample is tangential direction.

Fast axis changes caused by annealing. The dependence of the fast axis direction (radial direction:0) on annealing temperature is shown in Fig. 10. The symbols are the same as in Fig. 9. The fast axis direction of the substrate, shown by open circles, does not change, but remains in the circumferential direction. On the other hand, the fast axis direction, shown by closed circles, changes from the radial to the circumferential direction. Considering the drastic change in the fast axis direction, it is clear that annealing causes relaxation of compressive residue stress along the radial direction.

In-plane retardation. Kanai suggested at the Polymer Symposium[6] in Japan that the in-plane retardation is determined by the relation

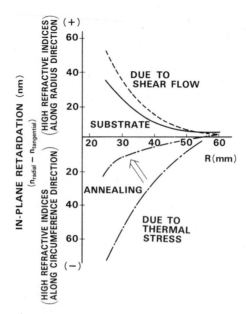

Figure 11. Change in in-plane retardation by annealing.

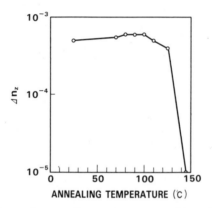

Figure 12. Dependence of Δn_z on annealing temperature.

between the flow-induced chain orientation and the thermal stress, so that a stress-balanced substrate shows slight in-plane retardation. Retardation increases and the fast axis direction changes caused by annealing can be explained as follows. When the substrate is annealed, the compressive stress in the radial direction relaxes, but the flow-induced chain orientation is not relaxed by annealing below T_g. As a result, in-plane retardation becomes mainly dependent on flow-induced chain orientation. This is why in-plane retardation of an annealed PC substrate becomes larger than that of an injection molded substrate. (See Figure 11)

Optical anisotropy in the plane vertical to the disk surface. The dependence of Δn_z on annealing temperature is shown in Fig. 12.

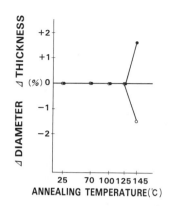

Figure 13. Changes in thickness and diameter by annealing.

Changes in Δn_z caused by annealing below T_g are very small. But when the substrate is annealed at T_g, Δn_z varies drastically from 4.5×10^{-4} to 1×10^{-5}. Thickness and diameter changes of the injection-molded PC substrate, caused by annealing, are shown in Fig. 13. Thickness and diameter changes do not occur until 125 °C. Annealing at the glass transition temperature constricts the disk diameter by 1.5% and increases the thickness by 1.5%. Thus, it seems that polycarbonate molecule conformation changes and all the flow-induced chain orientation relaxes. In other words, Δn_z depends mainly on the flow-induced chain orientation frozen at T_g.

CONCLUSION

1 The PC refractive index ellipsoid model is pie-shaped. That is, n_z is the smallest of the three principal refractive indices.

2 Optical anisotropy in the disk surface is determined by the relation between compressive stress and flow-induced chain orientation in the radial direction. Optical anisotropy perpendicular to the disk surface depends on the flow-induced chain orientation frozen at the glass-transition temperature.

REFERENCES

1. M. Takeshima and N. Funakoshi, J. Appl. Polym. Sci., 32, 3457 (1986)
2. G.H. Werumeus Buning, R. Wimberger Friedl, H.Janeschitz-Kriegl, and T.M. Ford, paper presented at the International Symposium on Optical Memory, Tokyo, Sept. 1987
3. A. Yoshizawa and N. Matsubayashi, in " Proceedings of SPIE " held in San Diego, CA, 1986.
4. F.A. Modine, R.W. Major, and E. Sonder, Appl. Optics 14(3), 757 (1975)
5. A.I. Isayev and C.A. Hieber, Rheol. Acta 19, 168 (1980)
6. T. Kanai, Y. Uryu, and K. Shimizu, paper presented at the Symposium of the Society of Polymer Science, Japan, Tokyo, 1987.

POLYMERS AS MID-INFRARED HOLOGRAPHIC RECORDING MEDIA

S. Calixto

Centro de Investigaciones en Optica
Apartado Postal 948, Leon, GTO
C.P. 37000, Mexico

In this paper it is shown that thin polymethyl-methacrylate films and polycarbonate blocks can be used to record interference patterns having a spatial frequency of a few lines/mm. Light source used in the experiments was a CO_2 laser which emits radiation with a wavelength of 10.6 μm. Diffraction efficiency behavior for red and infrared light is presented.

INTRODUCTION

Information storage can be accomplished with several methods. Among them are optical methods such as photography, holography and optical digital. In photography, an image of the object, the information, is formed by means of a lens over the photosensitive medium. Holography was presented in 1948 as an alternative to photography to store information. Photosensitive materials used to record an interference pattern (the hologram), which originated from the interference of the reference and object beams, should fulfill certain requirements. For example, they should present high resolution and sensitivity. The desirable development and fixing processes should be dry and preferably should be performed in a short time. Insensitivity to solvents and radiation after processing, availability in various thicknesses and sizes, erasibility, infinitely cyclable, and inexpensive are other good features. At the beginning of holography, photographic film was used to record the hologram. It seemed that the experience obtained for over 100 years with photographic emulsions was enough to obtain good holographic film. However, soon it was realized that more research was needed to find better holographic films.

When lasers emitting visible radiation are used to record holograms a variety of photosensitive materials[1] can be used to record the interference pattern. Among these materials are photographic films, dichromated gelatin, ferroelectric crystals, photochromic materials, thermoplastics, photopolymers[2], and others. Spatial modulation induced by visible light on the recording material can be present as a modification of the surface of the material or as the modification of the properties throughout the thickness of the material. In this last case some materials present a spatial modification of the refractive index and others a spatial change of their absorption.

Although light sources emitting radiation in the visible region
of the spectrum can be used when holograms are recorded,but there exists
the [3,4]possibility to use lasers which emit radiation in the ultraviolet
(UV) or the infrared (IR) region. To accomplish this,
materials sensitive to these nonvisible radiations have to be
developed. Films to be used when UV radiation is used to record the
hologram could work on the basis of ejection of photoelectrons because
each photon could have enough energy to cause this phenomenon. For
example,light having a wavelength of [5,6,7]0.3 μm has an energy of about 4
eV. However,when IR radiation is used to record the hologram,
ejection of photoelectrons can not be accomplished because IR photons
have low energy (about 0.1 eV at 10.6 μm wavelength). So another mechanism
should be operative to record the hologram. Recently it has been
shown [5] that thin gelatin films can record IR interference patterns.
However, these films do not show stability with changes in relative
humidity. Among the thermal qualities that holographic IR recording
media should have are their low thermal conductivity and low
diffusivity; most of the polymers possess these desirable qualities.
For example for polymethylmethacrylate (PMMA) thermal conductivity is
about 0.00048 cal/s cm^2 ℃ and diffusivity is about 0.0007 cm ˜s. In
the present paper are described the results of introductory
experiments performed with two polymeric materials as IR recording
media: polymethylmethacrylate (PMMA) films and polycarbonate blocks.
Part I of the paper describes the experiments to characterize the
thin PMMA films through the recording of interference gratings. Also
in this section is presented an application of these thin films in
the recording of some diffractive elements like interference gratings
and zone plates. Part II presents the results of recording
interference gratings when polycarbonate (PC) blocks were used. With
this material transient and permanent interference gratings were
recorded. Measurements of diffraction efficiency, as a figure of merit,
are presented.

EXPERIMENTAL I. THIN PMMA FILMS

To characterize the medium ,as an infrared recording material for
holography, the recording of interference patterns having different
spatial frequencies was done. IR source used in the experiment was a
frequency stabilized CO_2 laser emitting light in the mid-IR region of
the spectrum (wavelength 10.6 μm). An interference optical
configuration which gives a spatial distribution of heat, consisting
of a sinusoidal thermal pattern, was implemented, Fig. 1. The spatial
frequency of the interference pattern was varied by adjusting the
angle between interference beams. To judge the quality of the recorded
interference pattern (an interference grating) a He-Ne laser beam
(wavelength 0.6328 μm) was sent to recording area and first order
diffraction efficiency was monitored during exposure time. To
fabricate thin PMMA recording films the following method was applied.
Two flat glasses (flatnes about λ/10, size 6cm x 6 cm) were chosen,
and one of them had a hole of about 4 cm diameter. They were placed in
close contact [8] and then the set was laid over a leveled table. A
mixture of PMMA and solvent was poured in the hole, and after
evaporation of the solvent, a thin film (10 μm-100 μm thickness) was
found. Film thickness was varied by changing the amount of mixture.
PMMA was used as received without further purification. The
absorption of IR radiation by these films was determined by a
spectrometer. Fig. 2 shows transmittance behavior as a function of
wavelength of radiation, with different film thicknesses. To sustain
the films during exposure time, without a substrate that could absorb
IR radiation and degrade the information to be recorded, a metal

O-ring was glued to the film and after the glue drying period the set film plus ring were detached from the glass. With this method an IR recording film without substrate was obtained. Different films with different thicknesses were placed in the recording configuration and exposures ranging from about 50 msec to about 350 msec were given. Combined recording beam's power was about 28 W/cm^2.

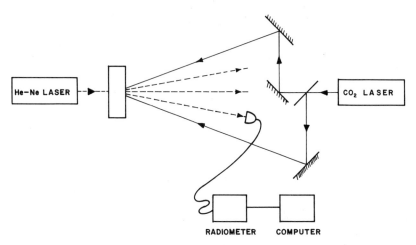

Figure 1.   Optical recording configuration used to characterize polymers as IR recording media.

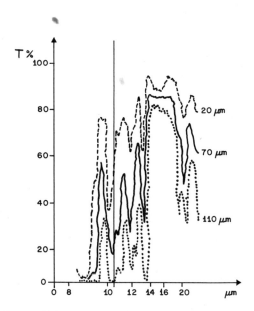

Figure 2.   Transmittance as a function of IR light wavelength Film thickness is the parameter.

TABLE I. Diffraction efficiency values attained when intereference
gratings were recorded on PMMA films having a thickness of
about 20 μm. Pattern spatial frequency was about
5 lines/mm, and IR power density was 28 W/cm^2.

t_{exp} (msec)	Diff. Eff.(%) $\lambda = 0.632$ μm transmission	ALUMINIZED	
		Diff. Eff.(%) $\lambda =0.632$ μm reflection	Diff. Eff.(%) $\lambda =10.6$ μm reflection
160	2.5	< 1	very low
180	25	12.6	0.2
200	26	11.7	0.2
220	20	< 1	very low
240	7.8	< 1	very low

Typical results for a 20 μm thick film are shown in Table I.
Diffraction efficiency measurements, in transmission mode, were made
just after the recording step. This same film was aluminized in a
vacuum chamber and then diffraction efficiency measurements in
reflection mode were made. It should be noted that transmission
diffraction efficiencies show a maximum of about 26%; in reflection
mode diffraction efficiency is about 12% for red light. When IR
radiation was sent to the aluminized gratings about 0.2% diffraction
efficiency was measured. To investigate the reason for this last
result an interference microscope was used to find the behavior of
surface modulation induced in the plastic. It was possible to notice
that the height of the modulation was about 0.3 μm. This small value
confirms the low diffraction efficiency attained with IR radiation.
Experiements were performed on thicker films and results resembled
those shown in table I.

Response of thin PMMA films to higher spatial frequencies (10
lines/mm and 15 lines/mm) was investigated by changing interbeam angle
and measuring diffraction efficiency. Typical results are shown in
Table II. It should be noted that diffraction efficiency values are
lower than those presented in Table I. It is suspected that
modulations comparable to those shown by the 5 lines/mm gratings can
not be achieved because heat transfer between recording lines is
present at the recording step.

TABLE II. Diffraction efficiency values attained for different
gratings showing spatial frequencies of 10 lines/mm
and 15 lines/mm. They were recorded on a PMMA film
having a thickness of about 20 μm, IR power density
was 28 W/cm^2.

t_{exp} (msec)	10 lines/mm Diff. Eff.(%)	15 lines/mm Diff. Eff.(%)
160	1.5	0.8
170	3.5	0.7
180	5.5	0.5
190	4.5	0.4
200	2.5	0.3

The recording of other diffractive structures on PMMA films was made with the aid of standing wave patterns. One such configuration to record interference gratings can be seen in Fig. 3. Light from the CO_2 laser was allowed to impinge normally over a copper mirror creating a standing wave pattern. PMMA thin films were placed in the light trajectory. When the faces of the film were perpendicular to the light trajectory, modulation planes were parallel to the surfaces. However if the faces presented a tilt, a relief structure, a diffraction grating, was present on both faces of the film after the recording step. One example of such grating is shown in Figure 4. Exposure time was about 50 msec. The number of lines per millimeter in the recorded structure can be changed by just tilting the film.

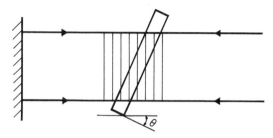

Figure 3.   Optical   recording configuration used to create a   standing wave pattern.

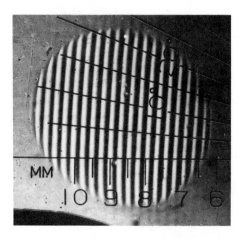

Figure   4.   Interference   grating   recorded   with   the   configuration depicted in Figure 3.

Another recording configuration based on standing wave patterns is shown in Figure 5. A lens focuses the beam at the center of curvature of the curved mirror. The standing wave pattern now consists of spherical wavefronts. If the recording film is placed perpendicular to the optical axis a zone plate will be recorded (Figure 6); if the film is not perpendicular an off axis zone plate will be present. Future studies comprising PMMA films will include the behavior of diffraction efficiencies as a function of exposure time and recording beams power. Modulation depth of the layer and IR reconstruction of these diffractive elements will be studied.

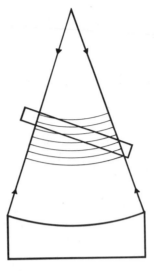

Figure 5.   Recording configuration used to create a circular standing
wave pattern.

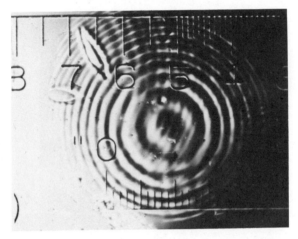

Figure 6.   Zone plate recorded with the configuration depicted in
Figure 5.

EXPERIMENTAL II. POLYCARBONATE BLOCKS.

A polycarbonate[9] (PC) was the second polymeric medium used to
record IR patterns. Blocks of about 5 mm thickness were chosen. No
transmittance of mid-IR radiation by this material was noticed. This
medium presented different behavior depending on the length of
exposure time; the recording beam's power was held constant at 28 W/cm^2.
When recording times were of the order of 20 msec to 1 sec,
transient interference effects were present. However, when exposures
times in the range of 2-3 sec were used, a permanent grating was
recorded.

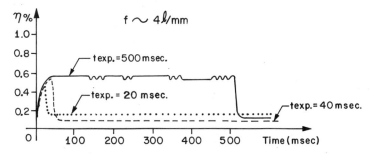

Figure 7. Diffraction efficiency behavior of transient interference
gratings having a spatial frequency of about 4 lines/mm. A
polycarbonate (Lexan) was the recording material

Transient gratings having a spatial frequency of about 4 lines/mm
were recorded in PC using the optical configuration depicted in figure
1. Typical diffraction efficiency behavior, for three exposure times,
can be seen in Figure 7. Modulation of the polymer was enough to
obtain a visible diffracted light pulse of about 0.5% diffraction
efficiency. As can be seen in the figure, diffraction efficiency
diminished exponentially with time immediately after the shutter is
closed. Transient interference gratings having about 10 lines/mm were
recorded but, unfortunately, diffraction efficiency values were lower
than 0.1%.

Fast recording and recovering of the normal thermal conditions of
PC allow its usage as an optical switch. Figure 8 shows diffraction
efficiency behavior as a function of time when such pulses were given.
This method presents a means to detect information carried by IR
wavelengths and detected by visible radiation. The recovering of the
normal thermal condition was not present anymore if light pulses
impinged on the plate for more than about 15 seconds.

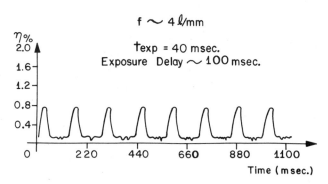

Figure 8. Diffraction efficiency behavior when a series of recording
pulses, 40 msec , were used to write the information.

223

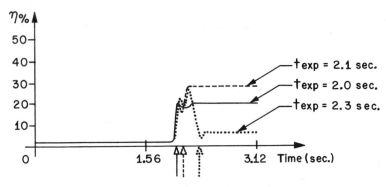

Figure 9. Diffraction efficiency behavior when permanent gratings were recorded in Lexan. Different exposure times were used.

Figure 10. Three fields of view of an interference microscope when interference gratings were investigated. Exposure times were a) 2.0 sec, b) 2.1 sec and c) 2.3 sec.

Permanent gratings were recorded when exposures times on the order of 2-3 seconds were used. Results in Figure 9 show a typical diffraction efficiency behavior, for red light, when 3 exposures were given; it can be noticed that the maximum diffraction efficiencies attain a value of about 30%. To find out why diffraction efficiency diminished when overexposures times were used, gratings surface modulation was investigated with an interference microscope. Figure 10 shows three fields of view of three different gratings recorded with three exposures times, 2.0 sec, 2.1 sec, and 2.3 sec. In the first photograph it should be noticed that the grating profile is like a square. In the second photograph a sinusoidal profile is shown. With respect to the modulation depth, it can be inferred by seeing the photograph that an over exposure will cause poor modulation depth.

Interference gratings having 9 lines/mm were recorded in PC. It was found that diffraction efficiency values lower than those shown in Figure 10 were present.

Future studies with PC blocks will include behavior when higher spatial frequencies and recording powers are present in. the interference pattern. Recording of interference patterns other than interference gratings will be performed. Also reconstruction of diffractive elements with IR radiation will be performed. Finally in a second part of the project other polymers will be tested.

SUMMARY

It has been shown that PMMA, in the form of thin films, and PC blocks can be used to record interference patterns when a mid-IR light source is used. Course diffraction gratings and zone plates can be recorded. Diffraction efficiencies, for red light, can reach about 26%. For the moment several hypotheses have been raised about the recording mechanism; however, more studies should be done to test them fully.

ACKNOWLEDGEMENTS

Support of this work by the CONACYT (Mexican National Council of Science and Technology) and the ACS-Petroleum research Fund (U.S.A.) is gratefully acknowledged.

I would like to thank Carmen Menchaca for his help in some experiments. Thanks are also given to Jose Castro, Jorge Vargas and the technical staff of the workshops (CIO). Word processing by Francisco Cuevas is acknowledged.

REFERENCES

1. H. M. Smith, Editor, "Holographic Recording Materials," Springer Verlag, New York, 1977.
2. S. Calixto, Appl. Opt. 26(18), 3904 (1987). See references therein.
3. T.P. Sosnowsky and H. Kogelnik, Appl. Opt. 9, 2186 (1970).
4. K.J. Ilcisin and R. Fedosejevs, Appl. Opt. 26, 396 (1987).
5. S. Calixto Appl. Opt. 27, 1977 (1988).
6. M. Rioux, M. Blanchard, M. Cormier, R. Beaulieu and D. Belanger, Appl. Opt. 16, 1876 (1977).
7. E.M. Barkhudarov, V.R. Berezouski, G.V. Gelashvili, M.I. Taktakishvili, T. Ya Chelidze and V.V. Chichinadze, Sov. Tech. Phys. Lett. 2, 425 (1977).
8. Polymethylmethacrylate, BDH Chemicals
9. The polycarbonate used is obtained under the trademark Lexan, a General Electric Product.

FACTORS AFFECTING POLYIMIDE LIGHTGUIDE QUALITY

C. Feger (1), R. Reuter (2), and H. Franke (2)

(1) IBM T. J. Watson Research Center, Yorktown Heights, NY 10598, U. S. A.

(2) Universität Osnabrück, Fachbereich Physik, Osnabrück, F. R. G.

Low loss lightguides were fabricated from three commercial polyimides of which one contains one, the others two hexafluoroisopropylidene (6F) groups. The latter utilizes the all para and the all meta isomer of the same diamine, respectively. As the number of 6F groups increases the optical losses of the corresponding lightguides decreases. In thick lightguides of the two 6F groups containing polyimides loss values below 0.1 dB/cm can be realized using optimized conditions. Two mechanisms — ordering with or without charge transfer complex formation and voids or pinholes — are found to be responsible for optical losses. The second type of losses can be reduced by cure optimization. Where ordering is possible annealing leads to increased optical losses. Geometrical restraint of the ordering as in sufficiently thin films, however, leads to loss reduction for otherwise identical conditions. Losses observed in the bulk are always higher than in the top and bottom layers of the polyimide films.

INTRODUCTION

Lightguiding in high temperature stable polymers is necessary in many opto-electronic applications. Classic polymeric waveguide materials such as poly(methylmethacrylate) (PMMA) and polycarbonates exhibit good optical properties; however, they do not possess high thermal stability. This is a severe limitation especially in areas where high process temperatures (up to 400 °C) are encountered as in the electronics and aerospace industry. On the other hand, the existing high temperature stable organic materials, e. g., polyimides, exhibit usually high optical losses [1]. In the last few years, however, new optically clear polyimides have been synthesized [2]. In the following systematic changes in the structure of commercial polyimides have been used to elucidate the mechanisms leading to optical losses in planar optical waveguides made from such materials.

EXPERIMENTAL

The materials investigated were HFDA-ODA synthesized from hexafluoro isopropylidene - 2, 2′ - bis (phthalic anhydride) (HFDA) and ODA (Figure 1., obtained from DuPont), HFDA-HFDAM-44 and HFDA-HFDAM-33 (SIXEF-44® and SIXEF-33® both from American Hoechst) synthesized from the same dianhydride, HFDA, and hexafluoroisopropylidene - 2, 2'- bis(4-aminobenzene) (HFDAM-4) and the all meta isomer of HFDAM-4, hexafluoroisopropylidene - 2, 2'- bis(3-aminobenzene) (HFDAM-3), respectively. HFDA-ODA was obtained in the polyamic acid form in NMP and HFDA-HFDAM-44 and HFDA-HFDAM-33 as powder. HFDA-HFDAM-44 was partially (75%) and HFDA-HFDAM-33 was fully imidized.

Planar waveguides of HFDA-ODA were produced by spincasting films on glass substrates. Films of thickness 5 μm and 1.5 μm were obtained. Thick films (thickness > 50 μm) for dynamic mechanical testing and absorption spectra measurements were prepared by doctorblading. The same procedure was applied to obtain planar waveguides of HFDA-HFDAM. Curing was performed by heating in an oven.

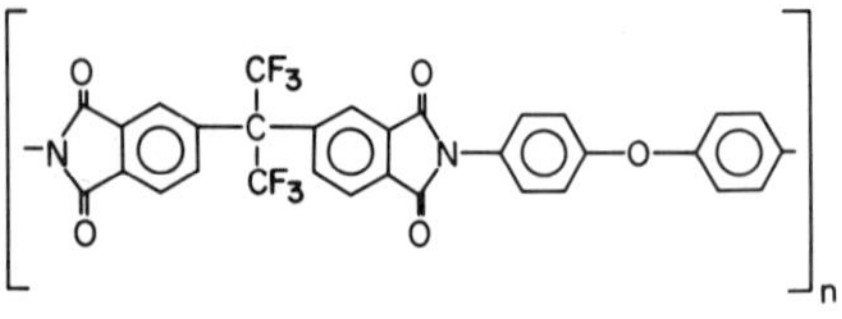

HFDA-ODA

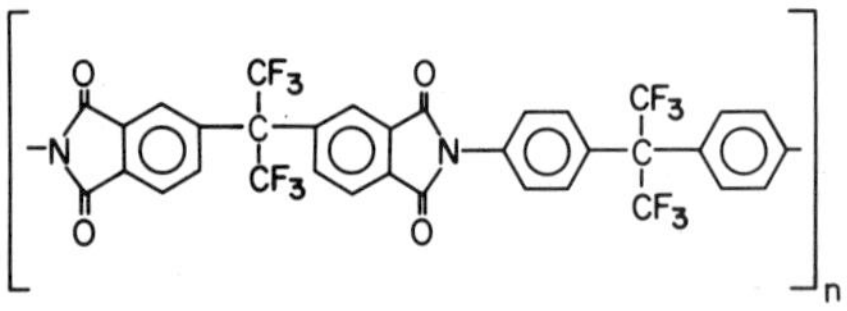

HFDA-HFDAM-44

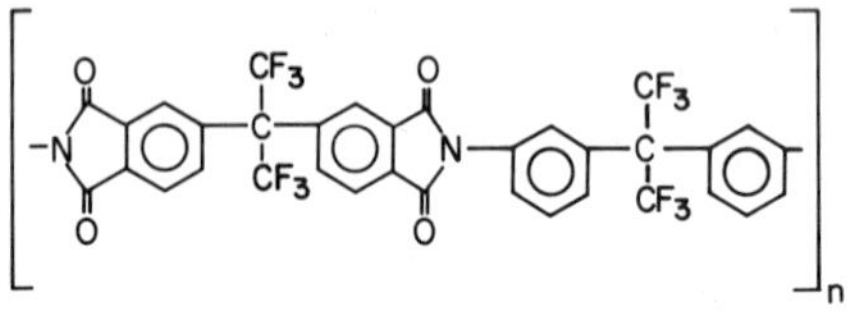

HFDA-HFDAM-33

Figure 1. Molecular structure of polyimide a) HFDA-ODA; b) HFDA-HFDAM-44; and c) HFDA-HFDAM-33 (taken from ref. 9 by permission).

Dynamic mechanical thermal analysis (DMTA) was conducted with a Polymer Labs DMTA at a frequency of 1 Hz, small deformations, and between 30 and 400 °C. Free standing films predried at 80 °C have been used. Moduli are calculated for sample dimensions before cure.

UV/visible absorption spectra were recorded with a CARY 17D spectrometer. The waveguide measurements, mode-line spectroscopy, and waveguide losses were recorded using a 633 nm HeNe laser. These experiments have been described in detail elsewhere [3].

RESULTS AND DISCUSSION

In previous studies [3] it was found that the polyimide cure can be optimized to reduce waveguide losses. Thus, the cure of the polyimides was studied at various heating rates. Figure 2.a shows the DMTA spectra for HFDA-ODA at a heating rate of 5 °C/min. The cure is characterized by the glass transition of the complexed polyamic acid and the decomplexation of the polyamic acid/NMP complex [4], partial imidization at 162 °C, and the glass transition of the partially imidized sample at 295 °C. Imidization in this system is complex and might occur in two steps [5]. It is finished only after reaching about 300 °C. At temperatures about 365 °C a small modulus increase is observed possibly due to radical formation and subsequent reaction [6]. The fully cured material (Figure 2.b) shows a sharp modulus drop at the glass transition temperature (310 °C). This indicates a fairly amorphous material. The flat elastic modulus plateau above 310 °C., however, indicates that physical or chemical crosslinks are present.

As reported previously [3, 4] cured, unannealed polyimide films show usually lower degrees of ordering and give lower loss lightguides than slowly cured or annealed films.

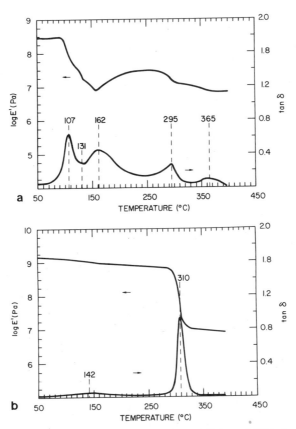

Figure 2. DMTA spectra of a) the cure of HFDA-ODA at 5 °C/min and of b) fully cured HFDA-ODA (taken from ref. 9 by permission).

229

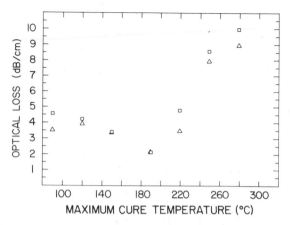

Figure 3. Total optical loss of fast cured HFDA-ODA film for parallel (□) and perpendicular (△) polarized light vs maximum cure temperature (taken from ref. 9 by permission).

Therefore, HFDA-ODA films were ramped fast to a given temperature (up to 310°C) and subsequently cooled. The total optical loss of such a fast cured film (thickness > 3.5 μm.) for both parallel (TE) and perpendicular (TM) polarized light is shown in Figure 3 plotted against the maximum cure temperature. The optical loss shown is the average over the first three modes. Films cured to moderate temperatures (i. e. partially imidized films) exhibit relatively low losses. This indicates a low amount of scattering centers which influence TE and TM polarizations in the same way. As imidization goes to completion, refractive index and optical losses increase. Annealing aggravates the problem. Curing and annealing lead to a shift in the UV spectrum away from the blue spectral range as also

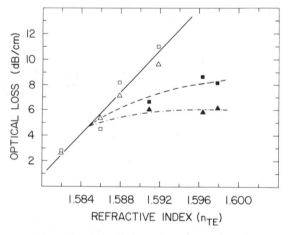

Figure 4. Optical loss values for thick (open symbols) and thin (filled symbols) planar waveguides of HFDA-ODA as a function of the refractive index of parallel polarized light, n_{TE} (taken from ref. 9 by permission).

evidenced by the increased yellow coloration. The latter is thought to originate from charge transfer complex formation [7].

The observed losses are larger than the ones calculated from absorption. As no evidence for Rayleigh- or Mie-scattering was found it is concluded that the size of the loss causing scattering centers is of the order of the wavelength of the guided light.

It was further observed that the highest loss occurs in the bulk and the lowest in the interface layers of the polyimide waveguides. Thus thinner films should exhibit lower losses than thicker ones. This was found to be true in films with a thickness after curing of less than 1.5 μm. The results are illustrated in Figure 4 which shows the optical loss values of the zero order modes for a thin HFDA-ODA film on glass as a function of the refractive index of parallel polarized light, n_{TE}. The latter is used as a measure of the extent of cure (fully cured HFDA-ODA reaches a value of n_{TE} = 1.600). For comparison the behavior of a 3.5 μm film is also shown. The losses in the thin film reach plateau values of 8 and 6 dB/cm, respectively. This indicates that the growth of the scattering centers is inhibited by the layers close to the substrate and the surface. Thus, the interface regions should be more amorphous than the bulk regions in polyimide films. This is in agreement with differences in the behavior of the bulk and the surface layer which has been reported by Diener and Susko [8]. A further corroboration is the observation that in these films the highest losses are measured in the 0th order mode [9] contrary to the behavior of other step index guides [10].

Waveguides made from uncured HFDA-HFDAM-44 by deposition from NMP solution and subsequent drying for 1 h at 90 °C show losses of less than 1 dB/cm. Values as low as 0.1 dB/cm can be realized. These losses are likely to be caused by impurities and do not constitute an inherent material property. Curing and annealing of the polyimide to temperatures of up to 200 °C introduces additional losses of only about 0.5 dB/cm as shown in Figure 5. The fact that no changes are observed in the the absorption spectra before and after annealing is further indication that very few structural changes occur upon annealing. The optical losses obtained after annealing of HFDA-HFDAM-33 films at 200 °C show no significant differences from equally treated HFDA-HFDAM-44 films.

To obtain low loss values in HFDA-HFDAM guides, curing conditions were crucial. A heating rate below 20 °C/min. followed by annealing at 200 °C leads to values of about 0.5 dB/cm; whereas a heating rate of 100 °C/min and annealing leads to saturation values around 3 dB/cm. Increasing the annealing temperature to 300 °C leads to increased loss values leveling below 3 dB/cm for the slowly cured guides and at about 6 dB/cm for the rapidly cured guides (Figure 5). These changes are observable in the corresponding absorption spectra.

Table I. Refractive indices of the investigated polyimides after drying and full cure.

Material	Cure	n_{TE}	n_{TM}	$\Delta n \times 10^3$
HFDA-ODA	dried	1.5705	1.5616	8.9
	300 C	1.5949	1.5870	7.9
HFDA-HFDAM-44	dried	1.5523	1.5493	3.0
	300 C	1.5405	1.5275	13.0
HFDA-HFDAM-33	dried	1.5543	1.5474	6.9
	300 C	1.5466	1.5432	3.4

Analysis of the refractive indices reveals information concerning anisotropy. Highly anisotropic polyimides such as PMDA-ODA reach values around 1.7, while the fluorinated polyimides are all below 1.6. With increasing amount of CF_3-groups per repeat unit, the index of refraction decreases reaching for parallel polarization 1.595 for the fully cured HFDA-ODA and 1.54 and 1.547, respectively, for the two HFDA-HFDAM materials (Table I). The birefringence of HFDA-ODA remains almost unchanged by heat treatment. As mentioned before, moderate loss waveguides with this material were obtainable only when the polyimide was only partly cured or when very thin films were used. It is assumed that in both cases isotropic amorphous materials are being produced. However, as curing continues the polymer undergoes ordering. In the thin films ordering seems restrained by the limited dimensions.

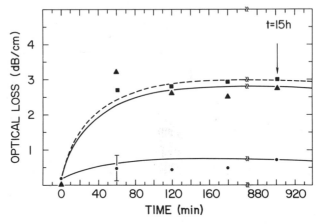

Figure 5. Optical loss for zero order modes of HFDA-HFDAM as a function of anneal time at 200 and 300 °C. HFDA-HFDAM-44 at 200 °C (●) and 300 °C (■); HFDA-HFDAM-33 at 300 °C (▲) (taken from ref. 9 by permission).

CONCLUSIONS

The introduction of the CF_3 groups into the polyimide repeat unit improves the optical transparency of polyimides remarkably. The optical losses drop from very high values in the case of PMDA-ODA to moderate losses in HFDA-ODA to low losses in HFDA-HFDAM. There are two origins for optical losses: Ordering processes can lead to refractive index fluctuations producing scattering centers, and also can favor the formation of charge transfer complexes [7] thus leading to increased absorption. Secondly, the evaporation of complexed or trapped casting solvents as well as water produced during imidization can cause voids or pinholes. Losses produced by the latter mechanism may be reduced by optimizing the curing procedure. However, where ordering processes interfere such as in HFDA-ODA, the processing window remains small. Annealing allows ordering to occur and thus charge transfer complexation and refractive index fluctuations are introduced.

Optical losses can be avoided in HFDA-HFDAM films if optimized curing procedures are followed. Annealing to 300 °C increases the loss values slightly indicating that ordering processes in these polyimides are mostly suppressed.

ACKNOWLEDGMENTS

We are indebted to American Höchst, USA, and Fa. Höchst, W. Germany, for providing samples of HFDA-HFDAM. R. Reuter and H. Franke thank the Stiftung Volkswagenwerk for financial support.

REFERENCES

1. T. P. Russell, H. Gugger, and J. D. Swalen, J. Polymer Sci., Polym. Phys. Ed. 21 , 1745 (1983).

2. A. K. St. Clair, T. L. St. Clair, and W. S. Slemp, in "Recent Advances In Polyimide Science and Technology," W. D. Weber, M. R. Gupta, Editors, p. 16, SPE Mid-Hudson Section ,Poughkeepsie, NY, 1987.

3. H. Franke, H. Knabke, and R. Reuter, SPIE Proceedings, 682 , 191 (1986).

4. C. Feger, Soc. Plast. Eng. Tech. Pap., XXXIII , 967 (1987).

5. R. Mathisen, E. Pyun, and C. S. P. Sung, Soc. Plast. Eng. Tech. Pap. XXXIII , 1103 (1987).

6. J. Shouli, Y. Ligang, Z. Zikang, Z. Zhiming, and Z. Qiyi, Conf. Rec. 1985 Intl. Conf. Prop. Appl. Dielectr. Mat., 2 , 583 (1985).

7. T. A. Gordina, B. V. Kotov, O. V. Kolninov, and A. N. Pravednikov, Vysokomol. Soyedin., B, 15 , 378 (1973).

8. C. E. Diener and J. R. Susko, in "Polyimides: Synthesis, Characterization, and Applications," K. L. Mittal, Editor, Vol.1, p. 353, Plenum Press, New York, 1984.

9. R. Reuter, H. Franke, and C. Feger, Appl. Optics, 27 , 4565 (1988).

10. R. Schriever, H. Franke, H. G. Festl, and E. Kraetzig, Polymer, 26 , 1423 (1985).

PART IV.   BULK/SURFACE CHEMICAL CONSIDERATIONS IN
MAGNETIC RECORDING

CHEMISTRY OF AN EPOXY-PHENOLIC MAGNETIC DISK COATING

J. M. Burns, R. B. Prime, E. M. Barrall, M.E. Oxsen, and S.J. Wright

IBM General Products Division
5600 Cottle Road
San Jose, CA 95193

An understanding of the chemistry of a magnetic memory disk coating system, and an ability to manipulate this chemistry through formulation and processing, is essential in the attainment of optimal performance properties. In this paper we discuss the chemical and physical phenomena associated with the cure of an epoxy-phenolic binder system filled with γ-Fe_2O_3. The magnetic coating formulation consists of a binder system containing a DGE-BPA epoxy, an allyl capped resole phenolic, poly(vinyl methyl ether), and a dispersing agent, plus a filler system consisting of the magnetic particle and alumina. The chemical and physical phenomena occurring during cure have been characterized by a number of analytical techniques, including thermal analysis (DMA, DSC, and TGA) and spectroscopic analysis (NMR, FTIR, and TGA/MS). Using these techniques, the cure of this epoxy-phenolic coating is shown to involve initial crosslinking via non-oxidative reactions, followed by further crosslinking via oxidative reactions. Decomposition of key binder components accompanies the crosslinking to produce a high T_g, microporous coating.

INTRODUCTION

This paper describes the chemistry associated with the cure of particulate magnetic media used in direct access storage device (DASD) technology.[1,2] The production of a final disk product involves several process steps, including coating preparation, spin coating and cure. Control of the chemistry at these early process steps ensures proper downstream processing and attainment of the required coating properties. The resulting coating is a highly crosslinked, microporous, infinite network exhibiting good magnetic properties and good durability.

DASD Technology

Magnetic recording technology includes the magnetic storage medium (the disk), the read and write heads, the recording channel electronics, the data encoding and clocking logic, and the technology that controls the head-to-disk spacing. The key performance parameter is areal density, which is a measurement of information stored per area of disk.

Increased areal density has been achieved through thinner particulate coatings having improved magnetic properties, reduced spacing between the head and the disk, and other advances in head design and materials. The continuous drive towards thinner coatings and smaller head-to-disk spacings has required significant improvements in coating technology, both from a processing and from a functional standpoint.

The disk consists of a rigid aluminum substrate with a thin magnetic layer which may be either a metallic film or a magnetic particle-polymer binder coating. This paper focuses on particulate magnetic coatings which consist of a magnetic particle (typically acicular γ-iron oxide) and a load bearing particle (typically aluminum oxide) imbedded in a highly crosslinked polymer network. When properly processed this coating bonds to the substrate and forms a durable, microporous coating with the desired magnetic properties. Microporosity in the magnetic coating enhances the retention of a lubricant which, in turn, enhances the mechanical durability of the coating and the head-disk combination.

The production of a particulate magnetic disk consists of several process steps. The first step is preparation of the coating ink, which involves mixing and milling to attain the necessary particle dispersion and ink rheology to give the correct spin coating characteristics and coating thickness. The coating is then subjected to several characterization measurements to assure that it is ready for application. These include measurement of solvent and resin concentrations, pigment volume concentration (PVC),[3] dispersion quality and cure behavior.[3] The coating ink is then spin coated onto an aluminum substrate. This step, which includes magnetic particle orientation and solvent evaporation, results in a coating with a uniform submicron thickness and a small amount of residual solvent. Cure, which is discussed in detail in a later section, is accomplished by heating the disks in air for 1-3 hours at temperatures in excess of 200°C. Following cure the disks are buffed or polished to provide the smooth surface necessary to fly a head, and washed to remove loose debris from the buff step. Next a lubricant is applied. Following testing of magnetic and durability properies, the disks are ready to be assembled with the rest of the components into a DASD product.

BACKGROUND

It is useful to review the composition and preparation of typical disk coatings. Coatings have been characterized by dynamic mechanical analysis (DMA) and thermogravimetric analysis (TGA), which are described in the EXPERIMENTAL section, as well as by measurement of disk performance.

The Coating Ink

Coatings used for the manufacture of rigid disk magnetic media were first described by Johnson, Flores and Vogel.[4] The coatings discussed in this paper consist of an acicular magnetic oxide, e.g. γ-Fe_2O_3 with an appropriate dispersing agent, a small amount of Al_2O_3, solvents and polymer binder system. The binder for such disk coatings is formed from a mixture of epoxy and phenolic resole resins. Poly (vinyl methyl ether), PVME, is an important additive which aids in film formation during spin coating and also functions as a porosity forming agent. A mixed solvent system is used to impart the right characteristics to the coating ink during spin coating.

Figures 1 and 2 illustrate how coating properties depend on composition. Figure 1 illustrates crosslinking of the binder to form a rigid matrix of high T_g. It can be seen that both oxygen and the iron oxide surface play major roles in attaining the high glass transition temperatures needed for buffing and durability. Cure was found to follow a square-root dependence on oxygen partial pressure;[5] Al_2O_3 has no catalytic effect on cure. The porosity forming role of PVME is illustrated in Figure 2 in terms of the amount of lubricant able to be retained in the coating, as well as an oxidative weight loss (OWL)

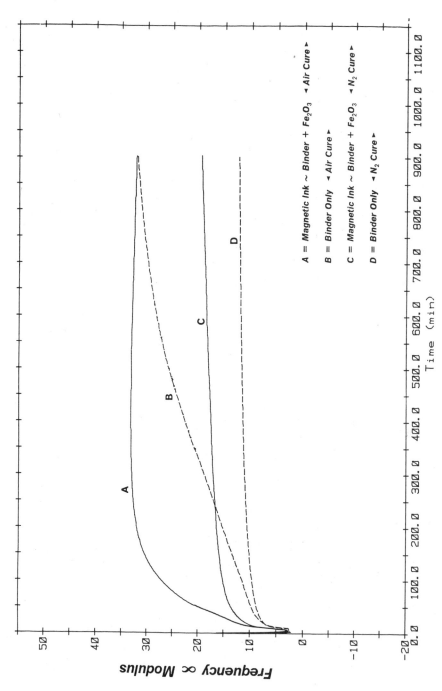

Figure 1. Monitor of crosslinking via Dynamic Mechanical Analysis. Modulus versus time at isothermal cure temperature ~230°C.

which is interpreted as a measure of the porosity forming capability of the ink. A small increase in the glass transition temperature is also observed with increasing PVME content of the coating.

The Cure Process

Cure involves the simultaneous evaporation of residual solvents, reaction of binder resins to form a crosslinked network, and phase separation and degradation of PVME to create microporosity. The cure behavior of a similar magnetic recording ink has been studied in detail by TGA and DMA.[5] During cure the PVME phase separates, as evidenced by a secondary damping peak near its T_g, as seen in Figure 3 (214°C/2h). This

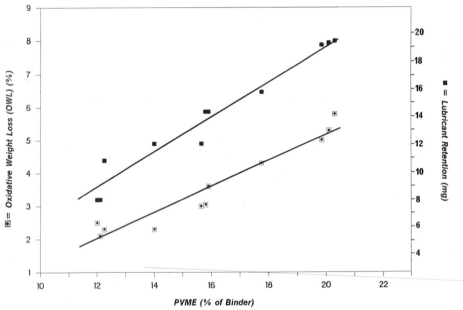

Figure 2. Cause and effect relationship between PVME concentration, OWL, and lubricant retention in the magnetic coating.

behavior is similar to that observed for rubber-modified epoxies.[6] Above 200°C oxidative reactions promoted by iron oxide make a major contribution to the crosslink density. Characteristic parameters for this coating are $T_{g\infty,N_2} \approx 95°C$ and $T_{g\infty,air} \approx 270°C$, the glass transition temperatures for coating cured to completion in nitrogen and air, respectively. Note that prolonged curing in air (>10 hours at 230°C) will eventually result in a decrease in T_g. Isothermal TGA in air and nitrogen showed evidence for oxygen uptake followed by an oxidative weight loss under nominal curing conditions. The curing of this coating thus involves phase separation of the PVME, chemical and oxidative crosslinking, and chemical and oxidative weight loss. Iron oxide, oxygen content of the purge gas, and flow rate of the purge gas were observed to affect cure behavior.

From a practical standpoint, cure of magnetic disks may be viewed as a time/temperature baking process which imparts specific properties to the coating. Curing on a manufacturing scale, which involves many disks in a single oven, is a quasi-isothermal process. Disks at room temperature enter an oven preheated to >200°C, where the disk temperature ramps exponentially to a steady-state temperature. We have

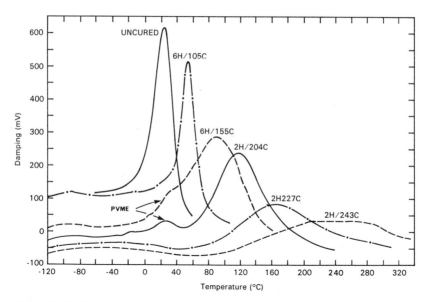

Figure 3. DMA characterization of the variably air cured magnetic coating.

found that cure under these conditions is best described by an equivalent isothermal time, EIt, which uniquely combines the time/temperature profile (a process parameter) with the activation energy for cure (a material parameter) via the Arrhenius equation.[7]. EIt, which is directly related to cure (e.g. degree of cure or T_g) may be calculated for any t/T profile-material combination. Figure 4 shows T_g against $EIt_{230°C}$ (reference temperature of 230°C) for coating cured at a variety of isothermal and quasi-isothermal conditions. Note that EIt may be calculated in real time and thus can be used for cure monitoring and process control.

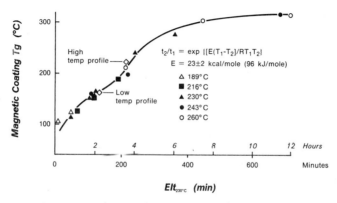

Figure 4. T_g versus Equivalent Isothermal Time.

241

Optimum cure can be described in terms of the cure process itself and response of the disk coating to further downline processing. Minimum cure is determined by the sol fraction, where uncrosslinked resin is a potential head/disk contaminant. As can be seen in Figure 5, the point of minimum cure may be precisely defined. Due to the oxidative nature of cure, the glass transition temperature continuously increases over the practical range of cure. The buff process places an upper limit on cure, where increasing cure causes a decrease in abrasion resistance as shown in Figure 6, and an increased tendency toward scratching of the disk surface.

EXPERIMENTAL

Materials

The *phenolic resin* examined in this study is predominantly a mixture of mono-, di, and tri-methylol allyl phenyl ethers and their oligomers and by-products. A variable amount of residual inorganic base (0.2-2% by weight), remaining from the synthesis, is also present.[8]

The *epoxy resin* examined in this study is a diglycidyl ether of bisphenol A with an epoxide equivalent of ~950.

The *poly(vinyl methyl ether)* has a molecular weight of ~50,000.

The *coating ink* is a typical magnetic formulation and consists of a DGE-BPA epoxy resin, a partially capped phenolic resole, PVME, magnetic iron oxide, alumina, dispersing agents and solvents.

Characterization Techniques

Dynamic Mechanical Analyses (DMA) were performed on a Dupont Model 982 Dynamic Mechanical Analyzer in the horizontal clamping mode. The DMA was programmed with either a Dupont 1090 or 9900 Thermal Analyzer. A 6.5 mm spacing

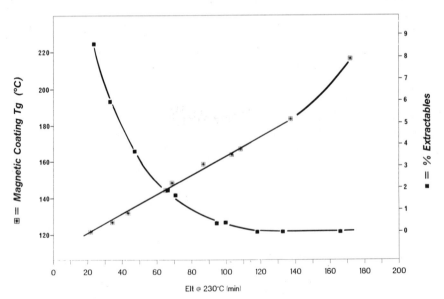

Figure 5. Percent extractables and T_g as criteria for minimum and maximum cure.

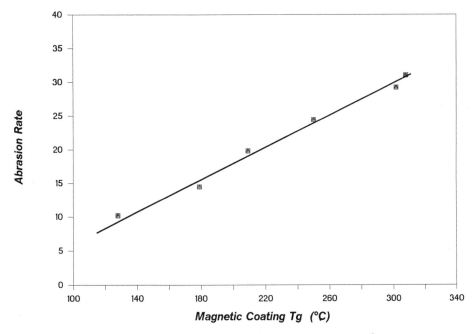

Figure 6. Abrasion rate at buff process versus extent of cure.

between the passive and driven arms, and a 0.1 mm oscillation amplitude, and a 400 cc/m N_2 purge were fixed in these studies. Samples were cut to 16-20 mm lengths before being mounted in the DMA. The DMA glass transition temperature, T_g, is taken as the maximum in the damping peak associated with the α-transition.

For a typical DMA analysis, a sample of the magnetic coating or its components is coated onto stainless steel wire mesh ribbons (84 X 84 wires/inch, wire diameter = 0.0035 inch, purchased from Tetko, Inc., and allowed to dry in a hood for at least 24 hours). Three types of ovens were utilized. The first was the DMA itself. Another was a well controlled laboratory oven. Coated mesh ribbons (3-8 inches in length) cured in this oven were suspended from an oven rack by clips (No. 20 binder clips) so that the mesh ribbons were in the center of the oven. The third was a typical manufacturing oven, where mesh samples were cured along with actual coated disks by mounting ~2 inch long samples in a special disk fixture.[3]

Thermogravimetric analysis (TGA) was performed on a Perkin-Elmer TGS-2 Thermogravimetric Analyzer programmed with an Omnitherm 35053 Three Module Controller/Data System. Air and nitrogen purge gases were utilized depending on the application, as described below. A constant purge rate of 100 ± 5 cc/min. was used; at this flow rate a five minute purge at the beginning of an experiment, or when switching gases, was shown to be sufficient to fill the balance and furnace tube with the respective atmosphere. Special aluminum sample pans (P/N X9901; Ominitherm Corp.) identical in size to the platinum pans supplied by Perkin-Elmer, were used so that they could be discarded after each run.

The *TGA/Mass spectrometer (TGA/MS)* used in these studies consists of a Perkin-Elmer TGS-2 thermogravimetric analyzer and a Sciex TAGA 6000 MS/MS tandem triple quadrapole mass spectrometer.[9] Typical experimental parameters used for TGA/MS consisted of a sample weight of 1-2 mg, a purge gas flow of 100 cc/min, and a heating rate of 10°C/min. The atmospheric pressure chemical ionization (APCI) ion source of the Sciex system permits the TGA to operate under a variety of atmospheric

243

environments. In this study air and nitrogen were used. The systems are physically coupled by a heated, all glass transfer line. The relatively high gas velocities plus the ~200°C interface temperatures, including the TGA furnace tube, help to avoid condensation of less volatile materials on the interface walls. Chemicals entering the ion source from the transfer line are converted to their quasi-molecular or molecular ions, which are then focused into the high vacuum analyzer portion for standard MS or MS/MS analysis.

Nuclear Magnetic Resonance (NMR) characterization was carried out on an IBM Instruments, Inc. NR/200 Model AF spectrometer. High resolution ^{1}H and ^{13}C solution spectra were obtained using a Bruker BB probe, while the ^{13}C CP/MAS spectra were obtained using a Doty Scientific, Inc. BB solids probe.

Fourier Transform Infrared Spectroscopy (FTIR) was carried out on an IBM Instruments Model 85 FTIR spectrometer equipped with a disk checker with a 65° reflection angle and p-polarized light source. Infrared reflectance spectra of thin film samples on a polished aluminum substrate were collected at 4 cm^{-1} resolution with 64 scans.

CHEMISTRY ASSOCIATED WITH CURE

The chemistry which occurs during cure is responsible for two processes which are vital to the production of a high performance magnetic disk coating. The first, *the formation of a binder network*, is produced by the chemical reactions which create crosslinks between the binder resins. This process determines the important mechanical properties of the "glue" which holds the magnetic oxide particles in place and adhered to the aluminum substrate. The second process which arises due to the chemistry occurring during cure is *the formation of microporosity* in the cured binder network. The development of microsporosity in the cured binder network occurs due to the oxidative degradation of specific binder components and is essential to the retention of lubricant on the disk. Although the processes of *network formation* and *microporosity formation* overlap to a significant extent, it simplifies matters to discuss them separately. An initial discussion of the chemistry which occurs during network formation lays the foundation for the subsequent discussion of microporosity formation.

Network Formation

The chemistry which produces crosslinks which harden the coating can be separated into two reaction categories: those which can occur in an oxygen-free environment, which we will designate as *non-oxidative*; and those which occur in the presence of O$_2$, which we will designate as *oxidative crosslinking*. The non-oxidative crosslinking reactions primarily occur during the early stages of cure and are characterized via substitution and condensation reactions of the functional groups initially present on the binder resins. The oxidative crosslinking reactions occur during the later two-thirds of the cure and are characterized via the generation, degradation and combination of radicals.

Non-oxidative Crosslinking. One approach to discussing the types of reactions which characterize the non-oxidative crosslinking in this coating is to discuss each reactive component of the formulation separately. After describing the fate of each reactant, a mental picture of total cure becomes easier to assimilate.

The epoxy resin may be involved in at least three crosslinking reactions during the anaerobic cure. Two of these, with the phenolic resin, are the substitution reactions: (A) phenol hydroxyl on the oxirane ring and (B) hydroxymethyl on oxirane, while the third reaction (C) is a substitution reaction of the epoxy resin main chain hydroxyl on an oxirane group (see Figure 7).

244

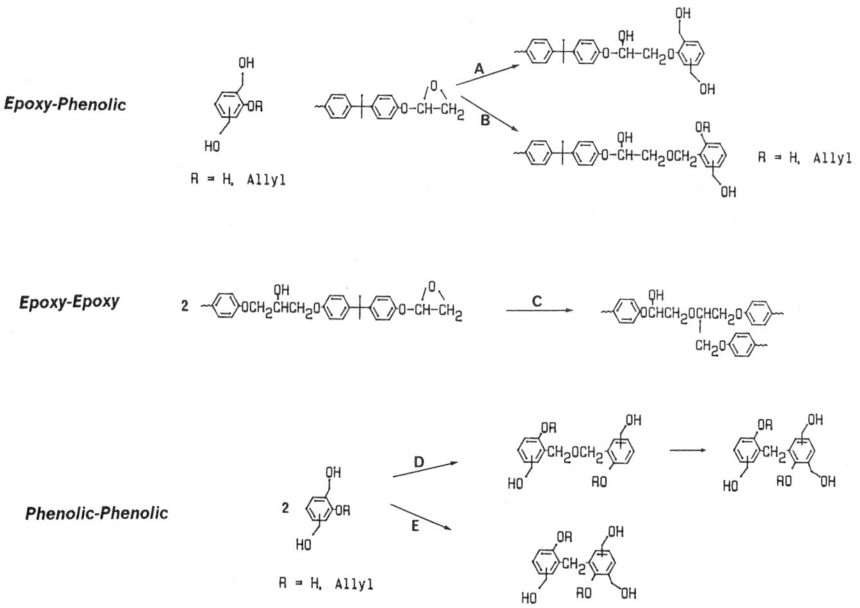

Figure 7. Non-oxidative crosslinking reactions of the magnetic coating binder resins.

The phenolic resole is involved in the substitution reactions mentioned above plus it can form a crosslink with a second phenolic molecule via a condensation reaction of two hydroxymethyl groups (D) or a substitution reaction on the ring (E) (see Figure 7). Another reaction of the phenolic resole is the loss of the allyl group to produce free phenol functionality. Athough this reaction does not produce a crosslink, as will be discussed later, the creation of the phenol functionality may influence the oxidative cure reactions. The precedents for these non-oxidative crosslinking reactions of the epoxy resin and the phenolic resole are documented in the literature.[10]

The PVME resin does not have any functional groups which are active during non-oxidative crosslinking, and it forms a separate phase from the crosslinked matrix. The eventual fate of the PVME during cure will be discussed in more detail in the next two sections.

The magnetic Fe_2O_3 role in the non-oxidative crosslinking is somewhat complex. Some chemistry definitely occurs at the iron oxide surface as demonstrated by the partial conversion of the Fe_2O_3 to Fe_3O_4 upon cure in a nitrogen atmosphere.[11] This reaction is well documented in the literature and has been used to manipulate the coercivity of the magnetic oxide.[12] The fact that the oxide is reduced indicates that the organic resins must be oxidized. This is a portent of subsequent oxidation reactions involving the binder resins and the iron oxide, in which the Fe_2O_3 acts as an oxidation catalyst which is cyclically reduced by the binder and then oxidized by O_2. This subject is discussed in more detail in the next section.

Some of the parameters which affect the type and extent of non-oxidative crosslinking are variance in resin stoichiometry and lot-to-lot variance in the starting materials. The second phase of cure, the crosslinking due to oxidation, is also sensitive to time/temperature variance. Resin ratio variance can typically be controlled and rarely is a contributor to cure variance. Resin ratios can also be purposely varied, allowing some

245

manipulation of the ultimate magnetic coating properties. The most significant contributor to non-oxidative crosslinking variance is the lot-to-lot variance of the starting materials. The most variable of the starting materials is the phenolic resole which is subject to the typical synthetic variables affecting the degree of allyl capping of the phenol, the concentration of methylol groups, and variable base concentration due to work-up variance.

Before we consider the reactions occurring during the oxidative portion of the cure, it is instructive to consider what "starting materials" have been produced by the non-oxidative crosslinking. After the non-oxidative cure we have a crosslinked epoxy-phenolic matrix, containing a separated, unreacted phase consisting of PVME and some epoxy resin (see the microporosity formation discussion below), surrounding the magnetic iron oxide particles. Due to the lack of a catalyst for the non-oxidative reactions, the T_g of the coating at this stage is relatively low, but will increase significantly during the next stage of oxidative cure. The types of functionality present in this cured coating include the *glyceryl ether* groups in the DGE-BPA main chain, the *ether* groups of the PVME main chain, the *uncapped phenol* functionality of the phenolic resole, and the *ferric oxide* surface of the magnetic particle. In the next section we will discuss the effects of the oxidative stage of cure on these and other functionality present after the non-oxidative cure.

<u>Oxidative Crosslinking</u>. Since the reactions occurring between the DGE-BPA and phenolic resins, during the non-oxidative cure, have been well documented in the literature, we have concentrated our efforts to understand the chemistry which occurs during the oxidative stage of cure. Further, it is the oxidative crosslinking which contributes most significantly to the ultimate properties of the disk coating. Again, we will discuss the oxidative crosslinking by first examining each coating component individually. Although the rates of oxidation of the binder resins in the absence of the catalytic effect of the iron oxide are much slower, similar levels of oxidation can be estimated via monitoring an extended cure of the resin via DMA. When the T_g of the oxidized resin has attained a value similar to that observed for the filled coating, spectroscopic analyses can be used to characterize the oxidation chemistry (Refer to Figure 1).

The epoxy resin is quite susceptible to oxidative crosslinking as evidenced by a comparison of the neat epoxy after exposure to N_2 and air cure cycles. The nitrogen cured epoxy resin attains a maximum DMA T_g of ~130°C, while the air cured resin can reach T_g values in excess of 270°C (see Figure 8). In addition to crosslinking, the epoxy resin undergoes oxidative degradation as seen from TGA studies.

The ^{13}C CPMAS NMR analyses of variably cured epoxy resin support a mechanism of oxidative crosslinking involving a degradation of the main chain glyceryl ether group and substitution on the aromatic rings. This is shown, in Figure 9, by the gradual disappearance of the glyceryl ether carbons ($\delta \sim 70$ ppm) and the broadening of the aromatic region due to the generation of additional chemical shifts via substitution on the rings. TGA showed oxidative degradation in air between 230-290°C to be accompanied by 20-21% weight loss with an activation energy of 96 kJ/mole (23 kcal/mole). By comparison, weight loss in N_2 was <1% under similar conditions. The major volatile products identified by TGA/MS were isopropanol and acetone. Based on the NMR and TGA/MS evidence, the products of oxidation of the epoxy resin are consistent with those shown in Figure 10.

The phenolic resole attains the same degree of cure whether cured in N_2 or air. When cured in the presence of iron oxide in air the phenolic resin does not undergo significant oxidative degradation. Although the phenolic resin does not appear to participate in the

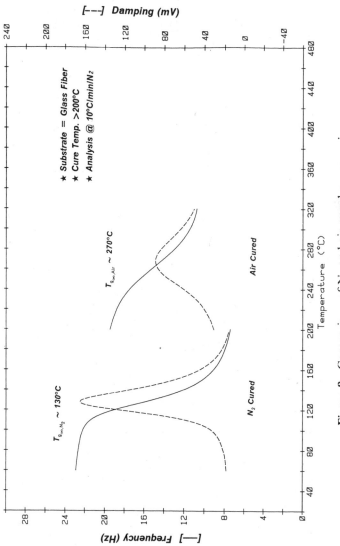

Figure 8. Comparison of N_2 and air cured epoxy resin.

247

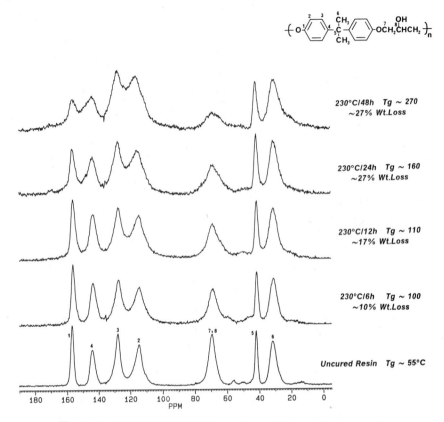

Figure 9. ^{13}C CPMAS NMR monitor of air cured epoxy resin.

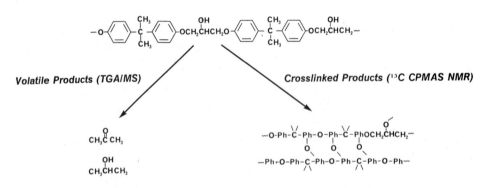

Volatile Products (TGA/MS)

Crosslinked Products (^{13}C CPMAS NMR)

Some aromatic fragments

Figure 10. Products from the oxidation of the epoxy resin during air cure, as measured via TGA/MS and ^{13}C CPMAS NMR.

formation of an oxidatively crosslinked site, it does play a role in the oxidation mechanism. One of the steps in the synthesis of the resin is an allylation of the phenolic hydroxyl. The capping of the phenolic oxygen is not complete and, depending on the lot, the resin consists of ~35-50% free phenol functionality.[13] This phenol functionality is thought to behave as an antioxidant during the oxidation reactions of the binder, thus accounting for the observation that magnetic coating formulations employing phenolic resin lots having higher concentrations of phenol functionality produce air cured coatings with lower T_g than those phenolic resins having lower concentrations of phenol functionality. Further circumstantial evidence supporting the antioxidant behavior of the phenolic resin is the observation that as the resin ages it becomes less effective in inhibiting the oxidative cure due to ambient oxidation of the phenol functionality.

The PVME resin appears to be the most susceptible of the binder resins to the effects of oxidation and undergoes both oxidative crosslinking and degradation during the oxidation portion of the cure. During non-oxidative curing, PVME has no sites that will initiate crosslinking reactions. However, crosslinking of the PVME to itself and the epoxy-phenolic matrix can be achieved at sites which are labile during the oxidative cure step. The oxidation chemistry of PVME has been studied via the thermal analytical techniques of TGA and DMA and the spectroscopic techniques of ^{13}C CPMAS NMR, TGA/MS, and FTIR.[14]

Thermogravimetric studies employing isothermal heating of PVME in air and N_2 show a large weight loss due to degradation in air versus negligible weight loss in N_2. Dynamic mechanical measurements under identical conditions to those used for TGA show significant crosslinking occurring in air, at ~20% weight loss. Based on the thermal analytical data, it is apparent that, under oxidative conditions, PVME experiences a combination of degradation and crosslinking. The mechanism of the oxidation of PVME is clarified by the spectroscopic studies described below.

A ^{13}C CPMAS NMR monitor of an air-cured PVME coating (shown in Figure 11) is characterized by a simultaneous disappearance of the methine and methyl carbons and the appearance of olefinic and carbonyl functionality. A monitor of the changes in

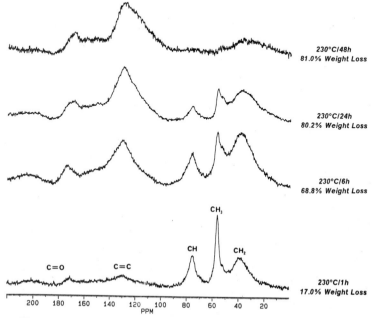

Figure 11. ^{13}C CPMAS NMR monitor of air cured PVME resin.

PVME during air cure via FTIR correlates well with the NMR data, showing a disappearance of the ether bond and aliphatic backbone accompanied by the appearance of carbonyl, hydroxyl, and olefinic functionality. TGA/MS shows methanol to be the major decomposition product in both air and nitrogen. Methyl ether and vinyl methyl ether were also abundant air and nitrogen decomposition products. In addition, methyl acetate and acetic acid were major products of oxidative degradation *only*. The activation energy, determined via TGA, for the oxidative weight loss is 50-60 kJ/mole for the first 35% loss in weight, followed by a rapid rise thereafter, reaching a maximum of 380 kJ/mole at 75% loss in weight.[14] Based on the NMR, FTIR, and TGA/MS data the products of the oxidation of PVME are consistent with those shown in Figure 12.

The Fe$_2$O$_3$ acts as an oxidation catalyst during the cure of the binder. The catalyst activity is clearly delineated by a comparison of the T$_g$ of the cured binder formulated with or without the Fe$_2$O$_3$ filler (see Figure 1). Under a N$_2$ atmosphere, the unfilled binder attains the same T$_g$ as the filled coating. When cured under air, the iron oxide filled binder has a T$_g$ which is more than 100°C higher than the air cured unfilled resin system. Clearly, the oxide acts as a cure catalyst in the presence of O$_2$. Further, its role is that of a localized catalyst where the greater the surface area of the oxide, the greater the degree of oxidation. This is borne out via experiments which show faster rates of oxidation with increasing concentration of oxide at a constant dispersion quality or with increasing dipsersion quality at constant concentration. Further evidence supporting the role of iron oxide as an oxidation catalyst is the reduction of the Fe$_2$O$_3$ to Fe$_3$O$_4$ during an inert atmosphere cure as seen via conversion of the brown colored Fe$_2$O$_3$ to the black colored Fe$_3$O$_4$ and an increased coercivity The oxide does not appear to crosslink with the binder based on an experiment which shows the T$_g$ of the cured matrix to be the same before and after dissolution of the Fe$_2$O$_3$ filler with concentrated HCl. The role of iron oxide as an oxidation catalyst is well documented in the literature[15] with a number of observations which are consistent with the activity described above.

Although the oxidation chemistry can be characterized for the individual resins, it is important to demonstrate that these same oxidation mechanisms are operable when the resins are cured as part of the magnetic coating formulation. As shown in Figure 13, ^{13}C CPMAS NMR, FTIR, and TGA/MS data from analyses of the curing of the total formulation are consistent with the observations from the individual resins. The behavior observed for the epoxy resin and PVME resin cured individually is mirrored in the NMR data obtained for the air cured, fully formulated coating (after HCl dissolution of the

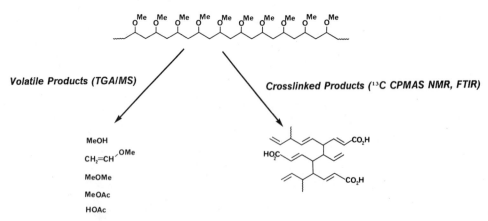

Figure 12. Products from the oxidation of the PVME resin during air cure, as measured via TGA/MS, ^{13}C CPMAS NMR, and FTIR.

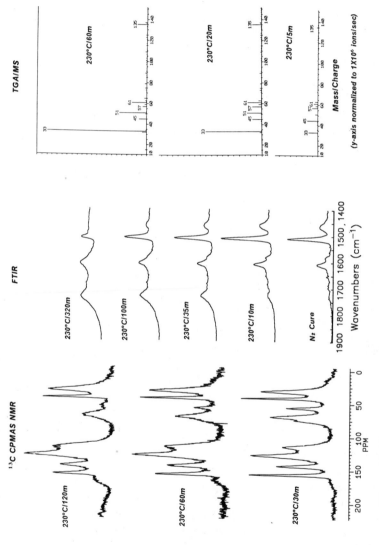

Figure 13. ^{13}C CPMAS NMR, FTIR, and TGA/MS monitor of the air cured magnetic coating.

Fe_2O_3). The PVME is shown to disappear due to degradation, there is a broadening of the aromatic region indicating substitution, and the appearance of carbonyl functionality. TGA/MS showed that methanol was the major volatile product during oxidative weight loss (OWL), as expected from the decomposition of PVME. Other less abundant volatile products were identical to those previously observed for the oxidative degradation of the individual PVME and epoxy resins. As shown in Figure 2, the amount of oxidative weight loss correlates very strongly with the amount of PVME in the coating.

Formation of Microporosity

The effects of oxidation are not only important for adequate development of crosslinking of the binder, they are also required for the generation of microporosity which functions as a reservoir for the disk lubricant. The mechanism for the formation of microporosity is the result of a combination of physical and chemical processes. The physical process involves the phase separation of PVME during the early stages of cure and establishes the morphology of the porosity which is generated upon the oxidative degradation of the phase separated PVME.

First Stage of Porosity Development. During the early portion of the cure, the epoxy and phenolic reactive functionalities begin to crosslink and form the cured binder matrix. The PVME, which has no reactive functionality, can not crosslink into the developing matrix and is forced to form a separate phase which consists primarily of PVME and has a much lower T_g than the crosslinked epoxy-phenolic matrix. The separated phase is clearly in evidence upon dynamic mechanical analysis (DMA) of a N_2 cured or arrested air cured coating (see Figure 3 for an example). Upon extracting the coating with $CHCl_3$, the T_g due to the separated phase disappears accompanied by an increase in T_g of the cured matrix. ^{13}C NMR analysis of the extract shows it to be primarily unaltered PVME plus a small amount of unreacted epoxy resin. The presence of the epoxy as part of the separated phase blend helps account for the T_g difference between that of the separated phase and that expected for pure PVME. Blending the higher T_g epoxy resin with the PVME shifts the T_g to higher values. The DMA T_g's observed before and after extraction are indicative of the majority of the PVME phase separating during cure with a minor portion of it blending with the cured matrix and acting as a plasticizer. It is worth noting that not all of the phase separated PVME can be extracted from an arrested, air cured coating, providing further evidence that some of the PVME has oxidatively crosslinked into the matrix.

The extent and morphology of the PVME phase separation has been found to be primarily dependent upon an incompatibility between PVME and the phenolic resin which is attenuated by a variable interaction between the iron oxide surface and the phenolic resin. A consideration of the following experimental observations helps confirm the statements above.

- Epoxy/Phenolic/PVME resin mixtures, cured in the absence of iron oxide, show phase separation of the PVME. The T_g of the separated phase is dependent upon phenolic lot-to-lot variance. Those lots with a higher concentration of free phenolic functionality and residual base contamination produce a lower separated phase T_g.

- Addition of Fe_2O_3 to the above resin mixtures eliminates the phase separation, as measured by DMA. In addition, the T_g of the cured matrix decreases by 40-50°C. Again, the shift in T_g is dependent upon the lot of phenolic resin. Those lots with the greater concentration of free phenolic functionality and residual base contamination cause the greater T_g shifts.

- NaOH doping of a phenolic resin lot produces a magnetic coating with a lower separated phase T_g.

- Addition of ferric ion to the epoxy/phenolic/PVME/Fe_2O_3 mixtures above is accompanied by a return of the phase separation of PVME upon curing.

- The phenolic resin complexes strongly with ferric ion. It is well known that phenol functionality complexes with ferric ion.[16]

- In the absence of Fe_2O_3, and at a constant PVME concentration, the extent of PVME phase separation increases with increasing phenolic/epoxy ratio.

- PVME does not adsorb appreciably on Fe_2O_3. [17]

In concert, these observations suggest that an interaction between the oxide and the phenolic resole influence the PVME phase separation. As mentioned previously, the phenolic resole is only partially (50-65%) capped with allyl groups with the degree of capping varying from lot to lot. In addition, the greater the concentration of uncapped phenol functionality, the greater is the amount of a residual base contamination introduced during synthesis. These difficult-to-control synthetic parameters produce a lot-to-lot variance of free phenol/phenolate concentration. Based on the observations above, it would appear that the phenolic resin can interact or adsorb onto the iron oxide surface via the phenolate functionality. This interaction can be "broken" by addition of ferric ion which has a greater affinity for the phenolate functionality. The phase separation behavior of PVME is primarily influenced by the presence or absence of the phenolic resin from the cured matrix. The more the phenolic resin is adsorbed onto the surface of the iron oxide, the less is its concentration in the matrix. A comparison of calculated[18] solubility parameters (δ) for the three resins shows the epoxy resin ($\delta \sim 10$) and PVME ($\delta \sim 9$) to be similar, while the phenolic resin ($\delta \sim 12$) is more polar. The large difference in polarity between the PVME and the phenolic resin causes them to be incompatible, immiscible mixtures, hence the tendency for PVME to phase separate to a greater extent in resin mixtures having a higher phenolic concentration. Those phenolic resin lots containing a higher concentration of phenolate functionality adsorb onto the oxide to a greater extent, lowering the concentration of phenolic resin in the binder solution. The lower the concentration of phenolic resin in the developing matrix, the more compatible the PVME is with the matrix and the less the tendency to phase separate. The above observations are also consistent with the changes in T_g of the separated phase. As the concentration of matrix phenolic decreases (due to phenolate adsorption on the oxide), less of the epoxy crosslinks into the matrix and more is available to blend with the separating phase of PVME, shifting its T_g to higher values. The scenario described above can obviously be attenuated via the incorporation of a surface active agent into the magnetic coating formulation. The surfactant will compete for iron oxide surface thereby affecting the amount of phenolic resole which can adsorb on the oxide. The net effect of adding the surfactant is to lessen the phase separation morphology variances due to phenolic lot-to-lot variance. This attenuation *is* observed in actual practice.

Not only is the extent of phase separation affected by the relative concentration of phenolic, but also the morphology of the phase separated domains. The phase separation of the PVME occurs upon the nucleation of the incompatible PVME into domains. The rate of formation and coalescence of the domains is dependent on the degree of incompatibility. The lower the compatibility, the faster the rate of nucleation and coalescence. The mobility of the separated domains is dependent on the rheology of the reacting coating system and quenched at the gel point of the curing matrix.[6] Since the rheological and gelation behaviors are very similar for each magnetic coating batch, incompatibility is believed to be the primary driving force influencing the variance in the morphology. Therefore, in a coating in which the PVME is more compatible, the rate of formation and coalescence is slower and the resulting domains are smaller. In a coating where PVME is less compatible, the rates are faster and the resulting domains are larger. In the absence of iron oxide, variable phase separation behavior, dependent on resole lot-to-lot variance, is still observed, but now the morphological development *is* dominated by resin mixture viscosity and gel time considerations. Those resin mixtures formulated with the phenolic resin lots having a greater concentration of phenol/phenolate functionality show significantly smaller phase separated domains, due to faster curing, than resin mixes containing phenolic resin lots having lower concentrations of phenol/phenolate functionality.

As will be seen in the next section, the variance in domain size of the separated phase has a controlling effect on the amount of porosity produced in the second stage.

Second Stage of Porosity Development. The second stage of porosity formation is characterized by the oxidative degradation of the phase separated PVME. Prior to the onset of oxidative degradation, the morphology of the separated phase has been locked in at the gel point. Depending on the size of the phase separated PVME domains, the surface area of PVME exposed to oxidation will vary. The smaller the domains, the greater the area (assuming the total volume of the separated phase is equivalent). At an equivalent rate of O_2 diffusion into the coating, the smaller domains will be oxidatively degraded faster than larger domains. Therefore, in a given period of time, e.g. during a standardized cure schedule, the smaller domains will degrade and volatilize faster than larger domains. Based on this, and the scenario presented in stage one, a variable rate of porosity development can be summarized in the flow chart below.

Greater Phenolate Functionality	*Lesser Phenolate Functionality*
▼ ▼ ▼	▼ ▼ ▼
Greater Phenolic Adsorption on Fe₂O₃	*Lesser Phenolic Adsorption on Fe₂O₃*
▼ ▼ ▼	▼ ▼ ▼
PVME More Compatible w/ Matrix	*PVME Less Compatible w/ Matrix*
▼ ▼ ▼	▼ ▼ ▼
Slower Domain Growth Rate	*Faster Domain Growth Rate*
▼ ▼ ▼	▼ ▼ ▼
Smaller Domains	*Larger Domains*
▼ ▼ ▼	▼ ▼ ▼
More PVME Surface Area	*Less PVME Surface Area*
▼ ▼ ▼	▼ ▼ ▼
Faster Rate of Oxidative Degradation	*Slower Rate of Oxidative Degradation*
▼ ▼ ▼	▼ ▼ ▼
More Porosity	*Less Porosity*

The variable rates of oxidative degradation are experimentally shown via oxidative weight loss (OWL) measurements using thermogravimetric analysis (TGA). An OWL measurement consists of monitoring weight loss during an isothermal air cure following an initial N_2 cure which eliminates solvent and sets up the non-oxidized, cured matrix. As shown in Figure 14, the higher the concentration of phenolate functionality in the phenolic resole used in a formulation, the faster the rate of weight loss and the greater the weight loss over a given cure time.

The experimental evidence which supports that PVME is the primary contributor to OWL and the ultimate degree of of porosity includes:

- Coatings formulated without PVME show very low OWL values and retain considerably less lubricant. For such coatings, lubricant resides predominantly on the surface and not within the coating.

- OWL values are directly proportional to the amount of PVME in the coating (Figure 2).

- A major portion of the starting PVME can be extracted from a N_2 cured coating (which before extraction retains low lubricant levels) producing a coating which retains levels of lubricant similar to air cured coatings.

- Air cured coatings formulated with radiolabeled PVME show little of the PVME remaining after cure.

- A TGA/MS monitor of the air cure of the coating identifies molecular fragments due to PVME as the primary contributor to the degradation volatiles during cure (Figure 13).

SUMMARY

The main components in the particulate magnetic coating described in this paper are epoxy and phenolic resins, poly(vinyl methyl ether), and iron oxide. Each plays a key

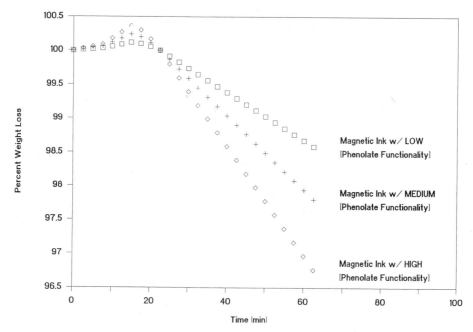

Figure 14. TGA Oxidative weight loss (OWL) versus the molar concentration of the phenolate functionality in the phenolic resin at an isothermal cure temperature of ~250°C.

role in the cure process. Cure begins with non-oxidative reactions leading to a lightly crosslinked coating in which PVME has phase separated. This early portion of cure occurs as the disks are heating up to the steady-state cure temperature. As the temperature exceeds ~200°C, oxidative reactions begin to take place. These include further crosslinking of the coating matrix and decomposition of the PVME into volatile products which leaves a microporous structure in the coating. Formulation, selection of raw materials and cure parameters all play a significant role in producing a high T_g, microporous coating with very low levels of uncrosslinked, soluble resins.

ACKNOWLEDGEMENTS

There are many who contributed in some way to the collection of the data presented in this paper. We would like to mention some particularly key individuals and apologize for any glaring omissions. Special thanks to Meta Karmin, Connie Moy, Hoa-Binh Tu, Bob Long, Fayelle Whelihan, Al Noda and David McClurg for their excellent sample preparations and thermal analytical support. Thanks to Yu-Sze Yen, James Wong, and George Shrout for their FTIR analyses and interpretations. Thanks to Richard Larrabee for his work with radiolabeled PVME. Thanks to the late Gordon Mol for his early TGA studies on the interactions between the binder resins and iron oxide. Thanks to Charles Hignite, Al Ward, Richard Balanson, and Roy Hannan for their managerial support. Thanks to Michael Metts for his assistance in preparing the figures for this paper.

REFERENCES

1. J. M. Harker, D. W. Brede, R. E. Pattison, G. R. Santana, and L. G. Taft, IBM J. Res. Develop., 25(5), 677 (1981).

2. R. B. Mulvany and L. H. Thompson, IBM J. Res. Develop., 25(5), 711 (1981).

3. R. B. Prime, J. M. Burns, M. L. Karmin, C. H. Moy and H.-B. Tu, J. Coat. Tech., 60 (761), 55 (1988).

4. D. D. Johnson, R. Flores and M. J. Vogel, U.S. Patent 3,058,844, 1962.

5. R. B. Prime, J. Thermal Anal., 31, 1091 (1986).

6. L. T. Manzione, J. K. Gillham and C. A. McPherson, J. Appl. Polym. Sci., 26, 889 (1981).

7. R. B. Prime, Proceedings of the 14th NATAS Conference, 137 (1985).

8. J.M. Burns and M.L. Karmin, Proceedings of the 13th NATAS Conference, 369 (1984).

9. R.B. Prime and B. Shushan, Anal. Chem., in press.

10. L. Shechter and J. Wynstra, Ind. Eng. Chem., 48, 86 (1956).

11. The Fe_2O_3 changes from brown to black, accompanied by an increase in coercivity.

12. V.M. DePalma, M.F. Doerner and A.W. Ward, IEEE Trans. Mag., Mag-18, 1083 (1982).

13. The variance in degree of allyl capping was determined via ^{13}C NMR using an inverse gated heteronuclear decoupling method.

14. R.B. Prime, E.F. Whelihan and J.M. Burns, Soc. Plast. Eng. Tech. Papers, 34, 1268 (1988).

15. D.H. Soloman and D.G. Hawthorne, "Chemistry of Pigments and Fillers", Wiley-Interscience, New York, 1983.

16. R.M. Roberts, J.C. Gilbert, L.B. Rodewald, and A.S. Wingrove, "An Introduction to Modern Experimental Organic Chemistry", Holt, Rinehart and Winston, New York, 1969.

17. This was shown via the absence of significant adsorption of radiolabeled PVME from a mixture of Fe_2O_3 and PVME in the magnetic coating formulation solvents.

18. T.C. Patton, "Paint Flow and Pigment Dispersion", Wiley-Interscience, New York, 1979.

256

BIOMIMETIC ROUTES TO MAGNETIC IRON OXIDE-POLYMER COMPOSITES

Paul Calvert[*] and Andy Broad

School of Chemistry, University of Sussex, Brighton, UK
[*]Now at Dept. of Materials Science and Engineering,
University of Arizona, Tucson, AZ 85721

Composites of iron oxides and iron in polymers have been prepared by the precipitation of particles from solutions of iron compounds in solid polymer. The morphology and structure of the oxide particles is discussed. Super-paramagnetic films are frequently produced by hydrolysis of iron(III) chloride in polymethylmethacrylate. These films are believed to contain magnetite. Some reduction of iron(III) to iron(II) apparently occurs. A number of routes have been studied to reduce iron compounds dissolved in polymers to iron metal. The most successful method was sodium naphthalide reduction of iron(III) chloride in polyvinylidenefluoride. A magnetic film was formed, but lost magnetism on exposure to air, presumably due to oxidation of the iron.

INTRODUCTION

Mineralized biological tissues include mammalian bones and teeth, mollusc shells and a host of other less familiar tissues in plants, animals and bacteria. Many are characterized by sophisticated microstructures containing organized arrays of very fine inorganic particles packed to high densities. They are also characterized by the fact that the particles are formed in-situ in an organic matrix, as opposed to conventional composites where preformed particles are dispersed into a polymer. We have been exploring a number of synthetic routes to growing composites in situ with a view to exploring the advantages offered and the constraints imposed. This has recently been reviewed[1].

A number of biological materials are reinforced by iron oxides. These include mag-netotactic bacteria[2], and magnetite or hematite in teeth of some limpets[3]. We have been exploring the possibility of making magnetic films by the growth of particles of magnetic iron oxides or iron in synthetic polymers. The possible advantages include small particle size, no agglomeration and greater uniformity than dispersed particle systems. The main focus of our studies has been the hydrolysis of ferric chloride molecularly dispersed as a solute in poly-methylmethacrylate (PMMA). We have also been studying routes to the reduction of polymeric solutions of iron compounds to form dispersions of metallic iron.

The production of magnetic recording media presently involves the synthesis of single domain particles with sufficient coercivity that they will resist reversal of magnetization under conditions of use but can be reversed by the writing or erase heads. The particles are then dispersed into a polymer binder which is cured. The most widely used iron oxide for recording is γ-Fe_2O_3, maghemite. Others are magnetite, cobalt-doped oxides and barium ferrite. Luborsky[4] has reviewed the production of elongated magnetic particles.

PRECIPITATION OF IRON OXIDES BY HYDROLYSIS OF FERRIC CHLORIDE

Aqueous Formation of Iron Oxides

The aqueous and solid state chemistry of ferric oxides and hydroxides is extremely complex, and the literature on the subject is comprehensive. Only work of interest in the possible growth of particles in polymers will be discussed, as there are already some excellent reviews available.[5,6] The morphology and composition of the oxides/oxyhydrides formed on the hydrolysis of iron(III) salts in aqueous solution is dependent on the temperature,[6-11] pH[6-9,11-14] Fe^{3+} concentration,[6,9,11-13,14,15] the nature of the anion and its concentration,[6,-11,12,15,19,19-23] and the length of time the solution is left (ageing) before the products are removed.[8,14,20,21,24,25]

The solid state transformations of iron oxides are generally better understood. The exact nature of the species which constitute the gel phase in the hydrolysed material is still unclear. The process of hydrolysis of $Fe(NO_3)_3$ solutions with NH_4OH has been investigated by Van de Geissen[26] and he has suggested that the gel may be considered to be $FeOOH.nH_2O$ or $Fe_2O_3.-nH_2O$. Towe and Bradley[27] state that the gel is not oxide particles with adsorbed water but that the water is integral to the structure. Other workers have discovered a species structurally similar to hematite which has been called protoferrihydrite.[10]

In acid media the precipitate formed from iron hydroxide gels converts to α-FeOOH or Fe_2O_3 at room temperature.[19] It is also reported that in the presence of chloride ions in acidic media, ß-FeOOH forms[15-20] and that α-FeOOH sometimes as a mixture with γ-FeOOH forms in the presence of the hypochlorite ion. Takada[11] investigated the hydrolysis of various iron (III) salts at temperatures between 50 and 90°C with both sodium and lithium hydroxides as well as in acidic media. The formation of the variety of products formed is explained in terms of the hydrolysis of iron (III) polynuclear complexes. When LiOH in alkaline solution was the precipitant, hydrolysis products were identified as the ferromagnetic $LiFe_5O_8$. This ferrite was identified even at the lower temperatures used (50°C) and its formation seems to be independent of the anion present. Okamoto et al.[28] have found that even some nearly completely amorphous ferric precipitates, formed under certain conditions, are strongly ferromagnetic.

Quirk et al have studied the mechanism of formation and the polymerization of spherical ferric hydroxy polycations 1.5-3 nm in diameter.[16-19] It has been shown that the initial formation and structure of these polycations is independent of the anion present, but the subsequent ageing processes were modified by the anion.

Matijevic and Scheiner[12] have studied the products of hydrolysis of ferric solutions containing chloride, nitrate or perchlorate anions in acidic solution at elevated temperatures. Particles thus generated in solutions containing Cl^- consisted of either ß FeOOH or α Fe_2O_3 depending on the concentrations of both the chloride and ferric ions. The other anions produced particles consisting of entirely α Fe_2O_3. The conditions used dramatically affect the particle shape as well as the composition, and particles of cubic, ellipsoidal, pyramidal, rodlike and spherical morphology were reproducibly obtained as the solutions were varied. This work especially shows how a small change in the reaction conditions can radically alter the products.

Experimental

Films of iron (III) chloride in poly(methylmethacrylate) were prepared by codissolving anhydrous iron chloride (Aldrich) and polymer (polymethylmethacrylate, Aldrich, high molecular weight) in acetone (dried by distillation from $CaSO_4$. Solutions were prepared containing varying amounts of iron in 10wt% solution of polymer in acetone.

Films were cast by pipetting 2ml of solution onto glass microscope slides, and drying overnight in a circulating air oven at 70°C. Freshly cast films were yellow; dried films containing more than 25wt% $FeCl_3$ were dark brown and could easily be removed from the slides. Films were treated with either 0.880 ammonia, 2M aqueous sodium hydroxide solution, 2M aqueous lithium hydroxide solution or deionized water for twelve hours, and thoroughly washed with deionized water.

Electron Microscopy. The morphology of the films was investigated using a JEOL C35 scanning electron microscope, and the nature of the oxide and particle size was established using transmission electron microscopy.

Samples for SEM were fractured in liquid nitrogen and mounted edge up in epoxy resin on sample stubs. Samples for examination in the TEM were prepared by dissolving the polymer in hot dimethyl sulfoxide(DMSO) and centrifuging the resultant solution. The pelleted oxide was then twice resuspended in dry acetone and recentrifuged to ensure a polymer free sample. The oxide was resuspended in alcohol and dropped onto 300 mesh carbon coated copper grids.

X-Ray Diffraction. The nature of the precipitated oxide was established by X-ray diffraction. The oxides were investigated both in situ and removed from the polymer. The polymethylmethacrylate was removed by dissolving the composites in hot DMSO and pelleting the oxides by centrifugation. The oxides were washed with acetone and recentrifuged twice to obtain a polymer-free sample. The oxides were also isolated by burning off the polymer at 350°C for one hour.

Burned out samples were also fired at 700°C for two days in order to obtain sharper x-ray diffractograms.

Rate of Chloride Loss. 0.08mm thick films of $FeCl_3$ in PMMA (15wt%) were analyzed for chloride content with respect to time when treated with base at room temperature.

Films were prepared as previously described. Each individual film was treated with 30ml base solution or deionized water in a 50ml beaker for a specified time. The films were removed from the solution and rinsed with deionized water over the respective reaction vessels, and the chloride content of both the films and the solutions were then determined by titration with aqueous silver nitrate and potassium chromate (Mohr's titration). Films were dissolved in 5ml acetone prior to titration.

Untreated films were dissolved in acetone and the chloride content was titrated to determine the 100% chloride content. The base was first neutralized with dilute nitric acid using phenolphthalein indicator. Only the 0.08mm films were studied for the base treatment of films.

Iron Analysis. Base solutions and water from the treatment of films were analyzed for iron content in order to determine the amount of iron leached from the films during the precipitation process. Iron was determined spectrophotometrically as the Fe^{2+} complex with 1,10-phenanthroline by monitoring the optical density at 515nm and comparison with a standard curve.[29]

A standard solution was made up by dissolving 0.713 g ferric ammonium sulphate dodecahydrate in 1 liter of distilled water containing 3ml concentrated sulfuric acid. Hydroxylamine hydrochloride solution (10% solution, pH 4.5), sodium citrate solution (250 g per litre) and 1,10-phenanthroline (0.3% monohydrate in water) were also prepared.

5ml of the standard solution was placed in a beaker and one drop of bromophenol blue indicator added. Sodium citrate was added until the indicator changed color and the volume of citrate added was noted. 5ml of standard solution was placed in a 100ml volumetric flask and the same volume of citrate added as was noted previously. To this was added 1ml hydroxylamine hydrochloride solution and 3 ml 1,10-phenanthroline solution. The mixture was allowed to stand for one hour after dilution to 100 ml, and the optical density (OD) measured at 515 nm.

As 1,10-phenanthroline only complexes with iron(II), this method was used to measure the ratio of Fe^{2+} to Fe^{3+} in dried films prior to hydrolysis. Following the same procedure without the addition of the hydroxylamine hydrochloride reducing agent, the OD was measured at 515nm. Another sample containing the same volume of solution to be determined is then analyzed with the reducing agent added. The former measurement gives the amount of Fe^{2+} in the sample, and the latter the total iron content.

Hysteresis Loop Measurements. The samples were measured using the conventional AC method by placing them in the center of a Helmholtz coil driven by a 50Hz mains current. An EMF proportional to the rate of change of magnetization was picked up by a backed off coil placed in the center of the Helmholtz pair. This EMF was integrated to give a voltage proportional to the magnetization and was plotted on the vertical axis of an oscilloscope against a voltage proportional to the coil current and hence the applied magnetic field. Measurements were made with fields up to 200 oersteds (16 kiloamps per meter).

The saturation magnetization of samples was measured using a 4 inch electromagnet containing a pair of backed-off pick-up coils connected to a fluxmeter. Fields of up to 8500 oersteds (680 kiloamps meter) were amply sufficient to saturate the samples. The equipment was calibrated using a small cylinder of 99.998% pure nickel.

Results

Ferromagnetic films containing iron oxides precipitated in PMMA were produced by base treatment of dried films of PMMA doped with iron (III) chloride, as reported previously.[30] Preliminary screening for ferromagnetism was done with a permanent bar magnet. As shown in Table I, the appearance of ferromagnetism depends on the iron loading and the precipitating base. Scanning electron microscopy of fracture surfaces of the films show that the particles are of the order of 1 μm in diameter and are distributed evenly through the films (fig. 1). Films which are not oven dried and contain residual acetone have fine structure in the polymer resembling a honeycomb of holes with oxide particles both in the holes and in the surrounding polymer. The micron size particles appear as spherical agglomerates. Films containing excess residual acetone form larger holes. Air drying of films always leads to some residual acetone, the amount being related to the loading of $FeCl_3$. (Table II). Films with this honeycomb structure consistently show a surface layer of the order of 10 μm thickness with no holes or particles apparent (fig 2).

Transmission electron microscope studies of oxide particle size are not consistent with the data obtained from the SEM studies. Particles are considerably smaller than the 1μm particles seen in the fracture surfaces of the films, typically 20nm in diameter. Particles appear to be small crystals in the case of ammonia treatment, the particles having straight edges. LiOH and NaOH treatments give very small particles in the range 15-20nm. This indicates that the particles seen in the SEM are agglomerations of the much smaller particles formed by the precipitation process. TEM pictures of freshly prepared samples show extremely small particle sizes, with larger particles formed by agglomerations of these minute particles (fig 3). If the films are aged, the particle size appears to increase slightly, with the larger agglomerations looking more like discrete particles (fig 4). Agglomerates of the small particles are also clearly seen. SEM pictures show the formation of agglomerated particles in the "holes" of undried aqueous base treated films (fig 5). Careful washing of the polymer with hot DMF reveals a particle size range of between 100 nm and 5 μm (fig 6).

Studies of the rate of chloride loss from the films show that 90% of the chloride is removed into solution after the first two hours of treatment.

Iron loss studies reveal that less than 1% of the iron contained in the films is lost into solution on treatment with aqueous LiOH and NaOH, around 8-10% by ammonia treatment, whereas between 85-95% of the iron is removed by water treatment of dried films (Table III). It is important to note that less than 1% of iron is removed into solution by water treatment of films containing residual acetone. Water treated films which have lost nearly all of their iron to solution show no internal structure in the SEM. Phenanthroline determination of the iron(II)

Table I. Appearance of Ferromagnetism in PMMA/$FeCl_3$ films

Wt% $FeCl_3$	Ammonia Treatment	NaOH Treatment	LiOH Treatment
5%	X	X	X
5% dried	W	W	W
10%	VW	VW	X
10% dried	M	M	M
20%	VW	VW	M
20% dried	M	M	M
30%	W	W	M
30% dried	S	S	S
50%	S	M	M
50% dried	VS	S	S

X: Not magnetic, VW: Very Weak, W: Weak, M: Medium, S:Strong

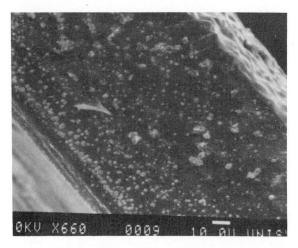

Figure 1 Fracture surface of FeCl$_3$/PMMA film, oven dried and base treated, showing oxide particles.

Figure 2 As figure 1, but film air dried, not oven dried before base treatment.

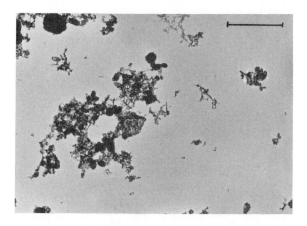

Figure 3 Transmission electron micrograph of particles from base-treated FeCl$_3$/PMMA film by polymer dissolution and ultrasonic dispersion. Freshly treated film. Bar=100nm

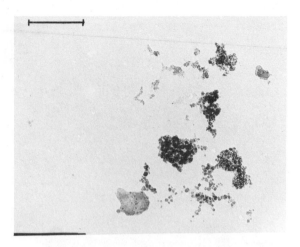

Figure 4 As figure 3, but film aged for 2 days, showing larger particles. Bar=100nm

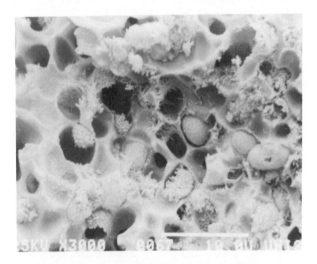

Figure 5 Fracture surface of an air-dried, base treated and aged film shows particle substructure.

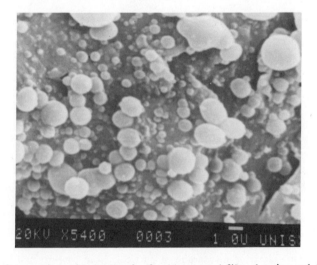

Figure 6 Careful washing with DMF of a freshly treated film showing gel-like particles.

Table II. Ratio of Acetone to $FeCl_3$ in air-dried PMMA films.

wt% $FeCl_3$	Mole ratio Acetone/$FeCl_3$
10	1.1
20	1.4
30	1.6
50	2.5

content of PMMA/$FeCl_3$ films dried at 70°C indicates that up to 50% of the iron has been reduced to iron(II) before base treatment.

Base treated films were found to be insoluble in usually good solvents for PMMA, suggesting that during treatment cross-linking of the polymer occurs. Dried films prior to base treatment readily redissolve in acetone revealing that the base is an integral part of the cross-linking mechanism. FTIR studies of PMMA films cross-linked in this way show strong absorption bands at 1640 cm^{-1} not usually associated with PMMA. These data are consistent with the formation of ionized carboxyl groups in the polymer, possibly formed by hydrolysis of the methacrylate ester and elimination of methanol.

X-ray diffraction of the oxide containing polymer films prepared by the hydrolysis of 50 wt% $FeCl_3$/PMMA films gave very weak diffuse diffraction patterns which proved impossible to assign unambiguously. Freshly prepared samples show no X-ray diffraction peaks whatsoever. Removal of the polymer by dissolution in hot DMF proved to be possible for LiOH and NaOH treated films; the ammonia treated films are completely insoluble, unless dissolved immediately on removal from the treatment solution. X-ray studies of the washed out powders did not give any more useful information, so the polymer was removed by burning out at 400°C for two hours. Patterns of the burned out powders were still fairly diffuse, and even firing at 700°C for two days did not produce very sharp diffraction patterns. LiOH treated films gave no observable peaks either in the polymer or after the polymer had been washed out. After burnout at 400°C diffuse peaks appear which are consistent with Fe_3O_4. Firing at 700°C for two days gives sharper peaks characteristic of γFe_2O_3. NaOH treated films give diffuse peaks when still in the polymer and when washed out. These are not easy to characterize unambiguously. Burned out films show peaks characteristic of γFe_2O_3. The oxide washed out of ammonia treated films show diffuse peaks from (γ-FeOOH) and the burned out films show a mixture of γ- and α-Fe_2O_3.

Magnetic measurements show that the films contain particles which are superparamagnetic, that is they have zero coercivity and an S-shaped hysteresis loop (figure 7). This is consistent with the small particle size seen in the TEM; superparamagnetic behavior typically being shown by particles of the order of 10 to 20nm in diameter if spherical. The saturation magnetization varied between samples, the ammonia samples having approximately twice the value of the LiOH and NaOH treated films. The magnetization of an ammonia treated sample was 43.4 emu/g. The published value for γFe_2O_3 is 74 emu/g, implying that only 58% of the oxide in the film is γFe_2O_3, the rest presumably being α-Fe_2O_3 or FeOOH (Table IV).

Table III. Iron Loss from 50wt% $FeCl_3$/PMMA films during precipitation.

Precipitant	% Iron in solution
0.880 Ammonia	10%
2M NaOH	<1%
2M LiOH	<1%
Water	90%

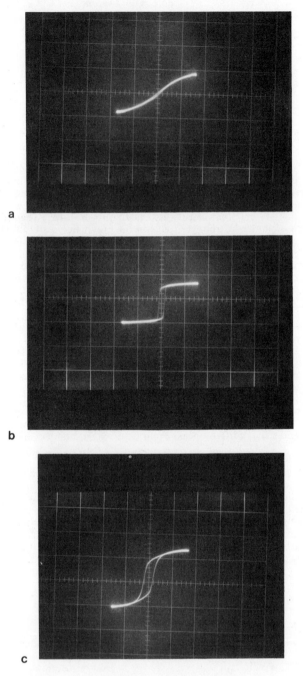

a

b

c

Figure 7 Magnetic hysteresis in (a) particles from an oven-dried and base treated film, (b) iron powder and (c) nickel wire under similar conditions.

Table IV. Magnetization of samples and wt% -Fe$_2$O$_3$.

Precipitant	Magnetization emu/g	Wt% Fe$_2$O$_3$
Ammonia	43.4	58
LiOH	16.4	22
NaOH	20.8	28

Discussion

The morphology and composition of the final product of FeCl$_3$ hydrolysis will depend on the local pH, diffusion kinetics and reaction rates within the polymer. Oven dried films differ from films containing residual acetone. In dry films, the iron center is complexed with the ester carbonyl of the methacrylate and acts to cross-link and embrittle the films. The FeCl$_3$-acetone complex in the undried films acts as a plasticizer as the films are very soft. The FeCl$_3$ acetone interaction is stronger than the ester carbonyl-iron interaction, as the films have to be heated to drive off the residual acetone. On treatment with water, the iron chloride in the undried films is already solvated with a strong ligand, and the water does not remove any of the chloride into solution. Conversely, the iron chloride in dried films is only weakly bound by the polymer side chain, and the iron chloride is rapidly lost to solution on water treatment, with little hydrolysis occurring either in the film or solution. Treatment with base is a different case, the OH$^-$ not only being a strong enough nucleophile to displace the complexing ligands but to react with the iron center displacing the chloride.

It was shown that the polymer becomes cross-linked and insoluble in dried films. This may be hydrolysis of the methacrylate ester to form iron carboxylates.

The honeycomb structure which forms in undried films may reflect shrinkage of the initial hydrated gel particles to denser oxide. However, films treated with 0.880 ammonia show no honeycomb structure whether oven dried or not, whereas aqueous ammonia gives results identical to the other aqueous systems. The holes may also be caused by HCl released during hydrolysis, which would be an insoluble gas in the polymer. Films exposed to moist air form very acid surface droplets, within the polymer there would be no free water to solubilize the HCl. The plasticity of the undried, acetone containing, films allows bubbles to form.

The processes governing agglomerated particle size in this system will be the mobility of the hydrolysed species in the polymer and the mechanism of agglomerate formation. It would be expected that the formation of agglomerates is more restricted in the dried films, and that the resultant particles will be much smaller.

The initial precipitation event is the formation of a gel particle of about 1μm. This gel is initially amorphous or contains very small crystals which slowly ripen during aging.

PRECIPITATION OF IRON IN FILMS BY REDUCTION

Introduction

From the point of view of magnetic properties it could be desirable to form films containing metallic iron particles in polymer. Metal particles have been produced in aqueous solution by the reduction of metal salts with borohydrides.[31,32] Several workers have probably produced composites of metal particles in polymers by decomposition of metal carbonyls in situ but these particles rapidly convert to oxides on exposure to air.[33-36] Polymer solutions have also been used to stabilize dispersions of fine metal particles.[1,37-40] We expected that treatment of a film containing a dissolved iron compound, with a reducing agent, would lead to an iron-polymer composite. Four methods have been studied: reduction with aqueous sodium borohydride, reduction with organic solutions of sodium naphthalide, reduction with hydrogen gas at high temperature and codeposition of metal vapor and monomer.

265

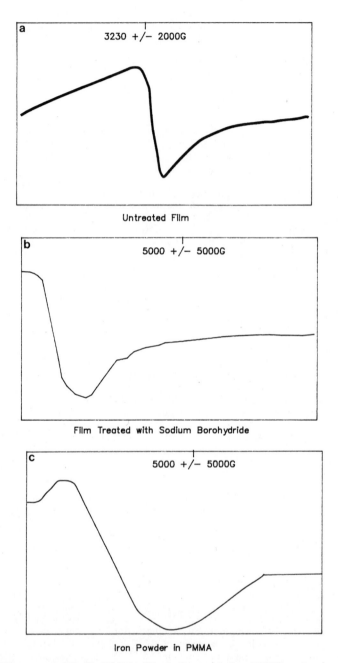

Figure 8 ESR spectra of FeCl$_3$/PMMA film before and after NaBH$_4$ reduction, and iron powder in PMMA for comparison.

Sodium Borohydride Reduction

Sodium borohydride is a commonly used reagent in organic chemistry for reducing aldehydes and ketones. It is a milder reducing agent than lithium aluminum hydride, and in contrast it is insoluble in ether and soluble in methanol, ethanol and isopropanol. When dissolved in water and other hydroxylic solvents, $NaBH_4$ reduces aldehydes and ketones rapidly at 25 °C, but is essentially inert to other functional groups. Sodium borohydride will also reduce solutions of some metal salts.

Solutions were prepared containing 10 wt% polymer and the required amount of iron compound to give films containing from 5 to 50wt% iron compound to polymer after drying.

Films of PMMA were prepared from solution in acetone and polyvinylidenefluoride (PVF_2, Kynar, Pennwalt Co.) was cast from 2-butanone (MEK). Films were cast by pipetting one ml of this solution onto clean microscope slides, peeling the film from the slide and allowing the films to dry thoroughly. Films were then placed in beakers containing 30 ml deionized water and 0.2g sodium borohydride. Iron(III)acetylacetonate ($Fe(acac)_3$) was not reduced in either PVF_2 or PMMA, although when powdered it reacted with aqueous sodium borohydride to give a black ferromagnetic precipitate which liberated hydrogen on the addition of acid (i.e. metallic iron).

$FeCl_3$ was quickly reduced in both PVF_2 and PMMA to give black ferromagnetic films with some of the black particulate material free in solution. When treated with acid, the films lost their black color and hydrogen was evolved. Films were also analyzed by ESR spectroscopy. Untreated films and reduced films were compared with the spectra of iron powder in PMMA. The untreated film gives a sharp peak corresponding to iron chloride while the reduced film shows a broad asymmetric peak similar to the spectrum of iron powder in PMMA (Figure 8).

$FeCl_3$ in PMMA was not reduced by sodium borohydride in methanol, ethanol or iso-propanol. The films rapidly disintegrated in the reducing solution releasing the iron salt into solution. Sodium trimethoxy borohydride is not a powerful enough reducing agent to precip-itate iron from iron(III) chloride solutions, but appeared to reduce the solution to iron(II).

Rate of Chloride Loss. Chloride loss from 15wt% $FeCl_3$ films was studied for film thicknesses of 0.08 mm and 0.38 mm for the $NaBH_4$ treated films, and results show fast initial rates of reaction followed by slower chloride removal. For the 0.08 mm thick films, more than 70% of the chloride was lost from films during the first 30 minutes of treatment, at which time only about half of this amount had been removed from the thicker (0.38 mm) films. It took 23 hours to remove 90% of the chloride from the 0.38 mm films, which took only 2 hours for the 0.08 mm films. (fig.9).

Morphology. Fig.10 shows a typical micrograph of the edge and surface of an untreated PMMA film containing 30 wt% $FeCl_3$. Fig.11 shows the surface and edge of a reduced film revealing the internal structure of the polymer after treatment. It was expected that the particles would be distributed evenly through the polymer, but it is now clear that the particles are only present at the surface of the films (fig.12). Higher magnification of the surface shows the particles to be about 1 μm in diameter (fig.13). X-ray analysis (EDAX) of the area shown in fig.12 indicated that both iron and chloride are present, whereas analysis of the particles gave a spectra of iron with very little chlorine present.

All PMMA films containing $FeCl_3$ (5,10,15 and 30 wt%) gave similar results, particles being small and at the surface of the polymer. If the films were left in the reducing solution overnight, particles with a platey structure were observed (figs.14), apparently as a result of slow conversion to a hydrated oxide. Six PMMA films containing 15wt% $FeCl_3$ were prepared from the same acetone solution, and were reduced using the 0.2g $NaBH_4$ in 30ml water for different times. The first four samples reduced for 30 seconds, 2 minutes, 2 and 4 hours, respectively, all showed solid particles; whereas films left in the solution for 12 and 24 hours showed the modified structure. After twenty-four hours the films lost their ferromagnetism.

It was found that, by adding 5 ml aqueous potassium dichromate to the reaction vessel after the initial reduction to iron had occurred (4 hours) and leaving the films in this solution for 24 hours, the formation of oxides is greatly inhibited. Films treated in this way were still ferromagnetic after three months standing in air. It is proposed that this is due to the adsorp-tion of chromate ions onto the surface of the iron particles. EDAX shows that chromium is present in the particles. The particles may also be stabilized by placing the reduced films in a saturated vapor of a solvent for the polymer until the polymer softens slightly.

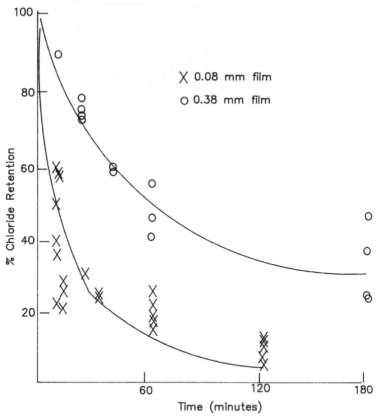

Figure 9 Loss of chloride from films of $FeCl_3$/PMMA, of two thicknesses, on treatment with $NaBH_4$ solution. Titration of chloride released into solution.

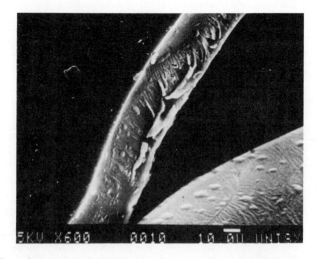

Figure 10 Scanning electron micrograph of $FeCl_3$ (30wt%)/PMMA before treatment.

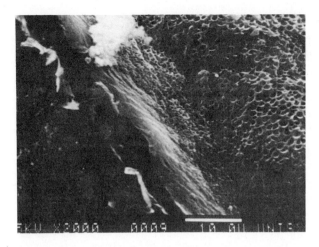

Figure 11 Scanning electron micrograph of the surface and edge of PMMA/FeCl$_3$ film treated with NaBH$_4$

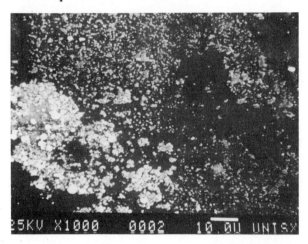

Figure 12 Scanning electron micrograph of the surface of PMMA/FeCl$_3$ film treated with NaBH$_4$

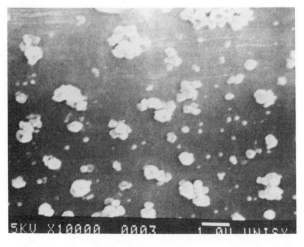

Figure 13 Scanning electron micrograph of iron particles on surface of film after reduction.

Treatment of Films with Sodium Naphthalide

Sodium naphthalide as reducing agent was used in the expectation that, since it is more organophilic than $NaBH_4$, it would lead to precipitation throughout the films rather than just on the surface. The reducing power of sodium naphthalide is greater than that of borohydride, and the solvents it may be used with are more limited; tetrahydrofuran(THF) or ethyleneglycol dimethylether being the solvents generally used. The only systems found so far which do not either reduce or dissolve the polymer is $Fe(acac)_3$ in PVF_2 reduced with NaC_8H_{10} in THF. PMMA and polystyrene were soluble in all the reaction solvents which we tried. $FeCl_3$ is lost rapidly to solution from PVF_2 films.

In order to minimize impurities in the polymer films which could lead to deactivation of the reducing agent, all polymer samples were cast as films in a dry nitrogen atmosphere. The films were placed in Schlenk tubes and kept evacuated for 20 hours to remove any solvent residues. The reducing agent was prepared under nitrogen and transferred to the Schlenk tubes via a 5cc syringe.

The reductions of 5wt% $FeCl_3$ and $Fe(acac)_3$ in PVF_2 gave positive results. Both iron compounds were discolored by the reducing system in PVF_2, the resultant films being dark brown/black. Only the $FeCl_3$ film was shown to be ferromagnetic. On exposure to air a rapid lightening in color to a red/brown was observed for both films. This has been interpreted as oxidation of iron particles in the films by atmospheric oxygen.

In reduced films containing 50 wt% $FeCl_3$, magnetic properties were retained for several hours in air. This retention in high loading films could be explained in terms of particle size: larger iron particles will only oxidize on the surface initially, the oxide layer preventing further reaction. Small particles will oxidize faster due to surface area factors, and may oxidize completely almost immediately. If particle size is related to film loading this could give us useful information about possible control of particle growth.

Discussion

The factors which affect the morphology and distribution of *in situ* precipitates in polymers are complex. The precipitate may form at the surface of the polymer, be distributed through the whole of the film, or occur in solution. Contributing factors will include the mobility of the metal salt in the polymer, the solubility and diffusion of the reducing/precipitating species in the polymer, the extraction of the metal salt from the polymer and the rate of the precipitation reaction itself.

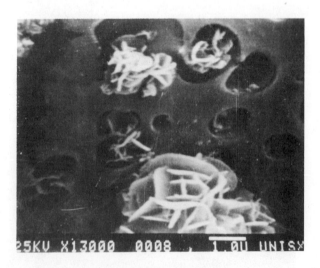

Figure 14 Scanning electron micrograph of iron particles on film surface after being left overnight.

When single phase films of PMMA containing iron chloride are placed in sodium hydroxide solution, there is no apparent loss of $FeCl_3$ to the solution. Conversely, when the same experiment is done with iron(III) chloride in PVF_2, the $FeCl_3$ is removed into the water very quickly.

The evidence from the SEM studies shows the oxide particles in the base treated films to be distributed through the polymer matrix. This indicates that the precipitating species (OH^-) must diffuse into the polymer, and the Cl^- must therefore diffuse out.

The difference between the base treated and the borohydride treated films is that in the former films the particles are found to be distributed through the polymer and in the latter the reduced particles are only found at the surface, whereas oxide particles form internally to the film. This may be explained in terms of the solubilities and mobilities of the various species; the permeability of borohydride into $FeCl_3$-PMMA is much less than that of hydroxide.

HYDROGEN REDUCTION OF FILMS

Films of 30 wt% $FeCl_3$ in polyethersulfone (ICI Victrex) were placed in a furnace at 300°C for 4 hours under flowing hydrogen. The films were strongly magnetic and metallic in appearance but some decomposition of the polymer had occurred. Hydrogen reduction at 200°C gave no apparent change in the films.

METAL VAPOR DEPOSITION

Methylmethacrylate monomer was exposed to iron vapor in vacuum. A red film initially forms on the vessel walls indicating organo-iron complexes. The film later became brittle and appeared metallic but turns brown on exposure to air.

CONCLUSIONS

Iron(III) chloride films in polymethylmethacrylate hydrolysed to ferromagnetic films. There is evidence for some reduction of iron(III) to iron(II), possibly due to interaction with acetone or polymer. The magnetic precipitate is either γ Fe_2O_3 or magnetite and is super-paramagnetic with a primary particle of about 200nm.

Treatment of $FeCl_3$ films with $NaBH_4$ in water gives ferromagnetic composite materials containing iron. The iron particles are located at the surface of the polymer films and the optimum reduction time for the films studied is four hours. If films are left in solution or removed and remain in air, oxidation of the iron results with corresponding loss of ferromagnetism.

Iron(III)acetylacetonate is not reduced by sodium borohydride in either PVF_2 or PMMA although it is reduced in free solution. Reduction of $FeCl_3$ in PVF_2 with sodium borohydride also results in ferromagnetic composites with the iron particles at the surface.

Base treatment of PMMA/$FeCl_3$ films results in polymer/iron oxide films, as does the hydrolysis of Fe(acac)$_3$ in PVF_2. The latter compound is not hydrolysed in single phase films of PMMA even though it is reduced in free solution. The nature of the oxides or oxyhydrides formed by this process is not known, although it is believed to be either amorphous hydrated iron oxide or goethite. The oxide particles are distributed throughout the polymer, as seen in the SEM studies. Treatment of base hydrolysed films with iron sulphate solution yielded ferromagnetic films believed to contain magnetite.

Investigations of fracture surfaces of the $NaBH_4$ reduced composites revealed no particles in the body of the film, only at the surface. This indicates that either the particles are too small to be seen at the magnifications used (up to x10,000), that the particles are present but obscured by a coating of polymer, or that no particles form in the center of the film. If no particles form there, it may be assumed that the reducing agent is not penetrating the polymer film. Therefore the iron chloride is mobile within the polymer films and diffuses to the surface where reduction takes place. As the reducing agent in solution is basic, and films of $FeCl_3$ hydrolysed with base show oxide particles throughout the polymer, some of the $FeCl_3$ may be hydrolysed in the body of the reduced films, but the oxide particles are too small to observe with the techniques used to date.

The rapid oxidation of the particles produced by borohydride reduction has been identified as a problem with the system, although treatment with potassium dichromate and softening of the polymer immediately after reduction have proved partly successful in prevention of this occurrence.

ACKNOWLEDGEMENTS

We would like to thank the MIT-Industry Ceramics Processing Consortium for support of this work.

REFERENCES

1) P.D. Calvert and S. Mann, J.Mater.Sci. 23, 3801 (1988)
2) R.B.Frankel and R.P.Blakemore, J.Magnetism Magnetic Mater. 15-18, 1562 (1980)
3) H.A.Lowenstam and J.L.Kirschvink in "Magnetite Biomineralization and Magnetoreception in Organisms" eds. J.L.Kirschvink, D.S.Jones and B.J.MacFadden, Plenum, New York, 1986
4) F.E. Luborsky, J.Appl.Phys. 32(Supp.3), 171s (1961)
5) N.C. Datta, J.Sci.Ind.Res. 40, 571 (1981)
6) J.D. Bernal, D.R. Dasgupta and A.L. Mackay, Clay.Min.Bull. 4 15, (1959)
7) A.N. Christensen, Acta.Chim.Scand. 22, 1487 (1968)
8) R.C. Mackenzie and R. Meldau, Miner.Mag. 32, 153 (1959)
9) S.V.S. Prasad and V. Sitakara Rao, J.Mater.Sci. 19, 3266 (1984)
10) S.P. Saraswat, A.C. Vajpei, V.K. Garg, V.K. Sharma and N. Prakash, J.Colloid Interf.Sci. 73, 373 (1980)
11) M. Kiyama and T. Takada, Bull.Inst.Res.,Kyoto Univ, 58, 193 (1980)
12) E. Matijevic and P. Scheiner, J.Colloid Interf.Sci. 63, 509 (1978)
13) R.J. Knight and R.N. Sylva, J.Inorg.Nucl.Chem. 36, 591 (1974)
14) W. Feitknect and W. Michaelis, Helv.Chim.Acta. 45, 212 (1962)
15) R.H.H. Wolf, M. Wrischer and J. Sipalo-Zuljevic, Koll. Zeit. Zeit. Poly. 215, 56 (1967)
16) P.J. Murphy, A.M. Posner and J.P. Quirk, J.Colloid Interf.Sci. 56, 29 (1976)
17) P.J. Murphy, A.M. Posner and J.P. Quirk, J.Colloid Interf.Sci. 56, 312 (1976)
18) P.J. Murphy, A.M. Posner and J.P. Quirk, J.Colloid Interf.Sci. 56, 284 (1976)
19) P.J. Murphy, A.M. Posner and J.P. Quirk, J.Colloid Interf.Sci. 56, 270 (1976)
20) Von U. Schwertmann, Z.Anorg.Allg.Chem. 293, 337 (1955)
21) M. Ozaki, S. Kratohvil and E. Matijevic, J.Colloid Interf.Sci. 102, 146 (1984)
22) J. Dousma, D.Den Ottelander and P.L. De Bruyn, J. Inorg.Nucl.Chem. 41, 1569 (1979)
23) R.S. Sapiesko, R.M. Patel and E. Matijevic, J.Phys.Chem. 81, 1061 (1977)
24) E. Matijevic, Pure.Appl.Chem. 50, 1193 (1978)
25) P.H. Hsu, Clay.Clay.Min. 21, 267 (1973)
26) A.A. Van der Giessen, Philips.Res.Rep.Supp. 12, " Chemical and Physical Properties of Iron(III) Oxide Hydrate"(1968)
27) K.M. Towe and W.F. Bradley, J.Colloid Interf.Sci. 24, 384 (1967)
28) S. Okamoto, H. Sekiwaza and S.I. Okamoto, in "Proc. Seventh. Symp. Reactivity Solids", p.341, John Wiley, New York, 1972
29) A.I.Vogel "Textbook of Quantitative Inorganic Analysis" 4th ed. 1978, Longman, London. p.742
30) C.A. Sobon, H.K. Bowen, A. Broad and P.D. Calvert, J.Mater. Sci. Lett. 6, 901 (1987)
31) A.L. Oppegard, F.J. Darnell, and H.C. Miller, J.Appl.Phys 32, 184 (1961)
32) J. van Wonterghem, S. Morup, C.J.W. Koch, S.W. Charles, and S. Wells, Nature 322, 622 (1986)
33) D. Shuttleworth, J.Phys.Chem. 84, 1629, (1980)
34) F. Galembeck, C.C. Ghizoni, C.A. Ribeiro, H. Vargas and L.C.M. Miranda, J.Appl.Polym.Sci. 25, 1427 (1980)
35) S. Reich and E.P. Goldberg, J.Polym.Sci.Physics 21, 869 (1983)
36) R.Tannenbaum, E.P.Goldberg and C.L.Flenniken in "Metal-Containing Polymeric systems" eds. C.Carraher, C.U.Pittman and J.Sheats, Plenum, New York, 1985.
37) H. Hirai, Makromol.Chem.Suppl. 14, 55 (1985)
38) H. Hirai, H.Wakabayashi and M. Komiyama, Bull.Chem.Soc.Jap. 59, 367 (1986)
39) T.W. Smith and D. Wychick, J.Phys.Chem. 84, 1621 (1980)
40) C.H. Griffiths, M.P. O'Horo and T.W. Smith, J.Appl.Phys. 50, 7108 (1979)

RHEOLOGICAL CHARACTERIZATION OF HIGH SOLIDS MAGNETIC DISPERSIONS

Jan W. Gooch
Georgia Tech Research Institute
Georgia Institute of Technology
Atlanta, Georgia 30332

Advancements in magnetic media technology for audio and video tapes have been made possible by investigating the fundamental mechanisms and properties of magnetic dispersions and coatings. In this study, development of magnetic dispersions required novel and quick methods for determining the degree of dispersion during milling in the liquid state since magnetic dispersions degenerate by agglomeration within minutes. Improved formulations were developed by observing compatibility of resins and solvents with acicular magnetic iron oxide powders and results were interpreted in rheological terms for the liquid dispersions since rapid feedback of information was possible. Coatings were characterized by surface gloss, x-ray radiography, optical and electron microscopy together with magnetic measurements. Results show that viscometric trends, degree of dispersion and magnetic properties are interrelated, and that capillary viscometry flow data provide the most consistent method of monitoring the "goodness" of the liquid dispersion before the coating is applied to uniaxially oriented polyester tape.

INTRODUCTION

The magnetic tape industry is continuously searching for better methods to produce magnetic media materials for specific applications including audio, video and computer tapes. The process of manufacturing magnetic tape consists of four major operations:

1. Formulation of the magnetic coating;

2. Dispersing the magnetic pigment in a resin/solvent solution;

3. Coating the liquid dispersion on polyester tape; and

4. Calendaring the coated tape.

It is a long process, even under laboratory pilot plant conditions, to observe the results of modifications to a formulation since steps 1 through 3, at a minimum, must be performed before a coating can be evaluated. The dispersion step is critical and difficult to control since the degree of dispersion in the liquid state has not been satisfactorily determined quantitatively. The rate of particle size reduction is related to the efficiency of dispersing agent and adsorption of resins on the pigment surface. Methods to quickly evaluate the performance of a magnetic dispersion during milling have not been available. Conventional fineness of grind tools are not useful since magnetic pigments form soft agglomerates which deform when stress is applied and give a false reading. These measurements could not be related to magnetic properties. In addition to these ambiguities, the overall dispersion process has not been well-defined,is devoid of a theoretical basis and founded from empirical data.

Each use of a magnetic dispersion requires a different formulation to produce specific physical properties such as hardness, resistance to wear across the recording or playback head and magnetic properties as coercivity and remanence. Coercivity or coercive force[1] is the magnetizing field which reduces the maximum induction to zero in a zero applied field. Remanent induction is that which remains after a saturated magnetic magnetizing field is reduced to zero. Magnetic squareness of a magnetic tape is the ratio of the remanent to maximum magnetic induction. With uniaxial magnetization, i.e. acicular single domain particles, in a tape squareness indicates the degree of alignment of the particles. This alignment can be estimated by comparing the measured value to the theoretical value of 0.5 calculated for randomly oriented particles which ranges from 0.0 to 1.1. Magnetic skewness[2] is a recent improvement over squareness and is about 50% more sensitive and ranges from 1.0 to infinity.

OBJECTIVES

The overall objective of this research was to develop a formulation(s) for a magnetic coating given the application and necessary properties, i.e., design a coating from a theoretical basis or data base without a prohibitive number of man-hours of experimentation.
Specifically, the objective were as follows:

1. Define the mechanism of dispersion with regard to shear stress-shear rate and resolve differences using multiple methods of measurement for high-solid dispersion.

2. Correlate dispersion with coating properties so that measurements can be made in the dispersion stage, but magnetic properties of the coating can be predicted.

3. From the data base generated in 1 and 2, select materials and optimize process parameters for preparing high-solids magnetic coatings.

EXPERIMENTAL APPROACH

Since dispersions exhibit non-Newtonian shear rate trends[3] with shear stress, it was thought that the pseudoplastic and thixotropic curves could generate shear stress-shear rate curves that would be indicative of dispersion quality. By utilizing different methods of viscosity measurement including Haake Rotovisco 2, Brookfield and capillary, differences observed could be useful in interpreting the mechanism of dispersion and give insight into optimizing formulation and dispersion parameters.

The dispersing equipment used for this purpose was the Hochmeyer High Speed Disperser for preparing predispersions followed by milling in the

Chicago Boiler-Red Devil Sand Mill (~40,000 sec.-1) using 3.0 mm silica beads (Potters Mfg. Co.).

The control formulation in Table I was used in the following experiments. Acid-base constants[4] were calculated to select resins and solvents for promoting interaction with pigments, but acid-base values for a range of materials are presently incomplete due a lack of data. Solubility parameters[5] in Table II were evaluated for resin-resin and resin-solvent compatibility/solubility. Wetting agents were screened using solubility parameters and from researcher's[6] experience in this field. After the rheology data base was compiled, a dispersing agent and concentration was selected. Then, a factorial experiment design method[7] was employed for optimizing resin composition in other formulations based on the rate of particle size reduction.

This approach was based primarily on the assumption that "a good dispersion produces a good coating." Although this may not always by true, much work can be eliminated by narrowing the randomness of the tasks. The optimization of coating properties including tensile strength and elastic modulus, etc., is contained in a separate study.

MATERIALS

The materials chosen for demonstrating the results of this research were selected to produce a coating 60-70% by weight of pigment in a thermoplastic matrix requiring no curing, but drying in ten minutes at 25°C.
The materials selected for this purpose were a mixture of "hard and soft" resins to provide film integrity with flexibility. The materials are listed in Table I. Solvents were chosen for a fast initial drying time and slower total film development and the selection was made on the basis of relative vapor pressure;[8] at 25°C, vapor pressure of methyl ethyl ketone = 100 mm Hg and 22.8 mm Hg for toluene.

The magnetic pigment was selected on the basis of past performance in video magnetic media. These materials were used exclusively with the exception of dispersing agents throughout the study.

RESULTS

Coatings were prepared from the formulation in Table I without dispersing agents and at increasing dispersing times; the photo-micrographs of the surfaces are shown in Figure 1. The particles are undispersed agglomerates of the native magnetic particles (1.0 x 0.1 μm). The particle size distribution decreases with dispersing time, and the 360 min. dispersion coating is examined with scanning electron microscopy (samples coated with gold-palladium) in Figure 2. With increasing magnification, the native acicular particles become visible and aligned. The coated tape was drawn through a magnet (1200 gauss) to align the particles in the direction of the tape. Analysis of coatings in Figure 1 by x-ray radiography showed agglomerated particles increasing in density toward the center.

TABLE I.   FORMULATION OF CONTROL MAGNETIC MEDIA DISPERSION.

COMPONENT	WEIGHT PERCENT
A. DeSoto Urethane Resin Azelaic Acid (16.1%) Trimethylol Propane (22.7%) Safflower Fatty Acids (48.1%) Toluene Diisocyanate (13.1%) Mw = 13,017 g/mole Mn = 8,537 g/mole	5.03
B. Vinyl Chloride Terpolymer Union Carbide VAGH Vinyl Chloride (91%) Vinyl Acetate (3%) Vinyl Alcohol (6%) Mw = 39,396 g/mole Mn = 21,338 g/mole	11.73
C. Magnetic Iron Oxide Pferrico 2674, Pfizer Corp. Cobalt surface modified small particle iron oxide	39.89
D. Methyl Ethyl Ketone	19.16
E. Toluene	24.19

TABLE II.   SOLUBILITY PARAMETERS OF MATERIALS.

COMPONENT	SOLUBILITY PARAMETER $(cal / cm^3)^{1/2}$
Methyl Ethyl Ketone	9.3
Toluene	8.9
Cobalt Modified Gamma $-Fe_2O_3$	11.6
VAGH – Vinyl Chloride Terpolymer	7.8 – 9.9
Urethane Resin (DeSoto type)	8.8
Epoxy (EPON 864)	8.5 – 14.7

Gloss and Particle Size

The surface gloss and particle size of the coatings are plotted with dispersion time in Figure 3. Aliquot parts of the dispersion were diluted in urethane resin for particle analysis. The gloss develops logarithmically with dispersion time, and particle size predictably decreases accordingly. The relationship between gloss and dispersing time is described by Equation 1.

$$\% \ Gloss = 66(min) + 25 \ ln(min) \qquad Eq. \ 1$$

Particle size reduction measured with the Leeds & Northrup Microtrac Particle Size Analyzer is related to % gloss as shown in Figure 3, and the rate of size reduction is related to the logarithm of dispersing time. The Gardner Glossgard Glossometer (60x) was utilized for % gloss measurements.

Shear Stress – Shear Rate and Time Relationships

The same dispersions as above were examined with the Haake Rotovisco 2 Viscometer utilizing the MV–I sensor, and the trends are shown in Figures 4 and 5. Viscosity (shear stress/shear rate) decreases with increasing shear stress which indicates pseudoplasticity[9] and decreases with time at constant shear stress which is indicative of thixotropy. More importantly,

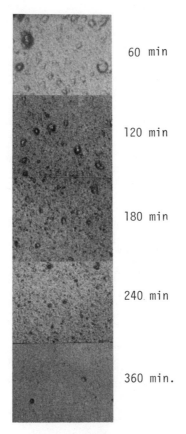

60 min

120 min

180 min

240 min

360 min.

Figure 1. Photomicrographs (x100) of magnetic tape surface at 60, 120, 180, 240, and 360 min of milling,

viscosity increases with dispersion time in each case. However, the time to perform a measurement and clean-up using the Rotovisco 2 is about 30 minutes, which is too long to constantly monitor a dispersion during milling. In any event, the curve would provide a near linear and uneventful curve unlike the curves generated using the Brookfield RVT Viscometer in Figure 6. Shear rate was calculated from viscosity function[10]. These unusual and unexpected curves, generated using the #4 spindle, appear to show the viscosity changes during milling. The curves from 90 minutes to 240 minutes show wide fluctuations due to broad particle size distribution and weak magnetic fields caused by iron oxide agglomerates. At 14.9 sec-1 (10 rpm) the trend is easily observed and finally after 90 minutes the curve stabilizes although realizing that particles continue to decrease in size (see Figure 3). The variation in % nonvolatile is due to a lack of uniformity within the dispersion during milling, but the % nonvolatile becomes more consistent with degree of dispersion.

Capillary Viscometer

Still different trends develop in Figure 7 from the same dispersion using a 0.5 mm diameter capillary viscometer detailed in Figure 8. The initial part of each curve exhibits "peaks and valleys," as shown in Figure 6, but with increasing shear stress as opposed to decreasing shear stress. Also, the curve continues to change with increasing dispersing time, and begins to stabilize only after 180 minutes. The changes are primarily due to the small diameter of the capillary tube, 0.5 mm. The capillary viscometer in Figure 8 was designed using these principles.

Differences in Viscosity Measurement

The capillary viscometer utilizes a different mechanism of detecting viscous flow than that using the rotating-cylinder in the Haake Rotovisco 2, MV–I sensor. The MV–I sensor possesses a "slip" space between cup and moving cylinder of about 2.0 mm. The Brookfield RVT viscometer, #4 spindle,

Figure 2. Scanning electron micrographs of aligned coating at increasing magnifications for a 360 min dispersion.

uses an open reservoir of fluid with at least 1.0 cm slip space. However, the capillary possesses a 0.5mm(500 μm)　for the path of the fluid.

The method[11] of calculating shear stress, s, shear rate, g, and viscosity, μ, for the capillary viscometer are as follows:

$$s = \Delta pr/2L \qquad \text{Eq. 2}$$

$$\gamma = 4Q/\pi r^3 \qquad \text{Eq. 3}$$

$$\eta = \Delta pr^4 \pi/2L4Q \qquad \text{Eq. 4}$$

where   s = shear stress, dynes/cm²

                          p = pressure gradient, dynes/cm²

                          r = capillary radius, cm

                          L = length of capillary, cm

                          γ = shear rate, sec-1

                          Q = flow rate, cm³/sec

                          η = viscosity, dynes . sec/cm²

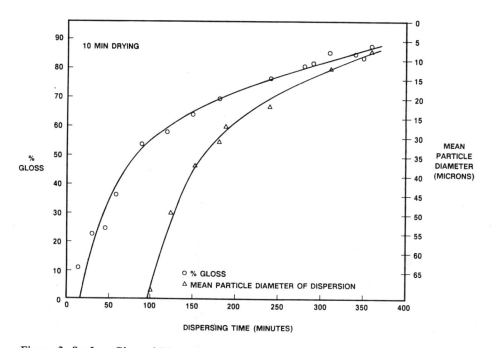

Figure 3. Surface Gloss of Magnetic Coating vs. Dispersing Time of Liquid Dispersion.

Dispersions have been characterized by Rabinowitsch[11] as shown in Equation 5.

$$(-dv/dr)w = [(3 + b)/4] \ (4Q/cr^3) \qquad \text{Eq. 5}$$

where   v = velocity, cm/sec

                       w = "at capillary wall"

                       b = Rabinowitsch correction factor for dispersion
                            dimensionless

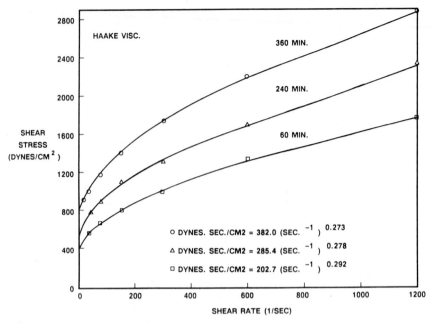

Figure 4. Shear Stress vs. Shear Rate for 60, 240, and 360 Minutes Dispersions.

The correction factor b is determined by plotting log $(\Delta pr)2L$ vs. log $(4Q\pi r^3)$, taking the slope and solving Eq. 5. It is an index of pseudoplasticity. The b correction factor as calculated is 0.6 to 0.8, but the actual measured value is 0.2 to 0.3 for the above magnetic iron oxide pigment dispersion, and therefore is of limited use for magnetic iron oxide dispersion.

Comparing Haake Rotovisco 2 and capillary (0.5 mm x 200.0 mm) shear stress-shear rates curves in Figure 10, it is obvious that the capillary viscometer measures lower viscosity of a milled dispersion. The difference is due to shear measurement only at the capillary wall-liquid interface, but the above capillary was calibrated with standard viscosity silicone oils and shear rate correction factors were inserted in the equation to produce the correct viscosities. However, dispersions are pseudoplastic and therefore shear-stress sensitive which is experienced primarily at the capillary wall compared to the bulk of liquid in the MV-I sensor. The correction factor reduces with decreasing capillary diameter.

Dispersion Aging

A property of magnetic iron oxide dispersions is "fast agglomeration with age" and the effect on capillary flow is shown in Figure 11. Agglomerates cause a decrease in the rate of flow. The formation of agglomerates initiates immediately in the absence of shear stress, and the quality of the dispersion degenerates quickly as described by Equations 6 and 7.

$$1\text{-}30 \text{ min:} \quad cc/min = 4.2 \times 10^{-2} - 1.8 \times 10^{-4} \, min \qquad \text{Eq. 6}$$

$$30\text{--}240 \text{ min}: \quad cc/min = 3.8 \times 10^{-2} - 3.1 \times 10^{-5} \text{ min} \qquad \text{Eq. 7}$$

Therefore, a fast measurement technique (of less than 10 minutes) is imperative due to the rapid formation of agglomerates.

Effect of % Nonvolatile Content in Dispersion

The effect of % nonvolatile in the 210 min. dispersion is to linearly decrease flow rate through the capillary as shown in Figure 12. From 41% to 52% nonvolatile, the flow rate is expressed in Equation 8.

$$cc/sec = 3.68 \times 10^{-1} - 7.07 \times 10^{-3} \qquad \text{Eq. 8}$$

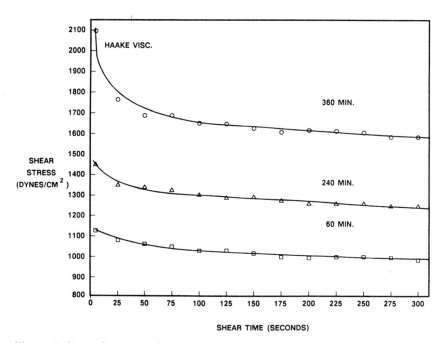

Figure 5. Shear Stress vs. Shear Time for 60, 240, and 360 Minutes Dispersions.

The effect of % nonvolatile is approximately 10 times that of age discussed in Figure 11. Therefore, it is always necessary to correct the flow rate for % nonvolatile to effectively reproduce the flow rate. By using a closed system, this variable would be eliminated since nonvolatile would be constant.

From the rheological findings above, it is apparent that the Haake Rotovisco Viscometer ,using the MV-I sensor, can be used to generate low controllable shear (1198 sec^{-1}) within a dispersion. Since is has been demonstrated in Figure 4 that "viscosity increases with dispersing time," the efficiency of dispersing agents was evaluated by adding dispersing agents to the basic formulation at 0.1 to 3.0% by weight of total formulation weight in Table I. The dispersion was prepared by mixing each agent with the solvent system and pigment followed by mixing with the

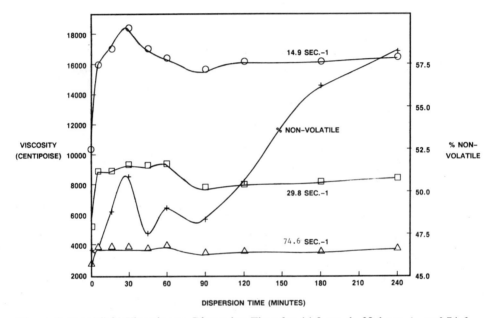

Figure 6. Brookfield Viscosity vs. Dispersion Time for 14.9 sec-1, 29.1 sec-1, and 74.6 sec-1 Shear Rates Demonstrating Nonlinearity in Viscosity During Milling

resins. The predispersion was agitated with an electric shaker in each case. Figure 13 shows the results of each agent at the optimal effective concentration of 0.5%. The initial decreasing slope is due to both drag from large particles and the "yield point" effects, but the consistently increasing slope is due to particle size reduction which is a function of agent penetration and adsorption[12] on the particles which separate with shear.   Disperse Ayd #7 (Daniels Chemical Co.) is clearly the most effective wetting agent in this series. The MV-I sensor was filled with the predispersion and sheared at 512 rpm or 1198 1/sec. This technique was useful for evaluating the compatibility of resins and pigments. Disperse Ayd #7 was added to the formulation in Table I and milling time was reduced producing a similar flow curve.

The general mechanism of dispersion of iron oxide pigments was demonstrated by observing particle size reduction while milling the dispersion, the development of surface gloss of coatings from dispersions, reduction of milling time using dispersing agents, and flow of dispersions through capillary viscometers. The mechanism of dispersion is a stepwise process described as follows

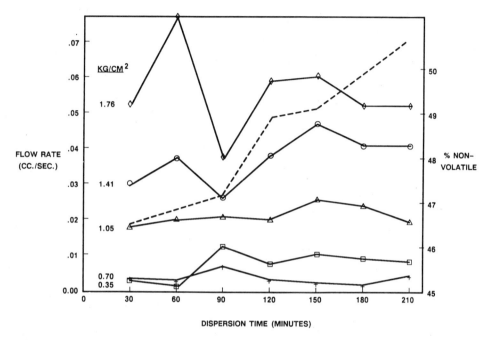

Figure 7. Flow Rate vs. Dispersion Time with % Nonvolatile Content of Dispersion with Capillary Viscometer.

Step 1. Initial adsorption of solvent is more feasible thermodynamically than that of resin due to rate of penetration through particles. Therefore, solvent should be mixed with pigment initially to wet the surfaces while agitating the mixture. Dispersing agents would best be utilized by premixing them in solvent likewise.

Step 2. Solvent and resin solution diffuses through the particles after solvent wetting, but requires agitation to diffuse through the particle agglomerates. Shearing forces are required to disperse the native acicular particles as adsorption of solvent and resin continues. The rate of adsorption is much greater than the rate of particle dispersion; adsorption can only occur after "free surface" is made available. The rate at which this occurs is shown in Figure 3.

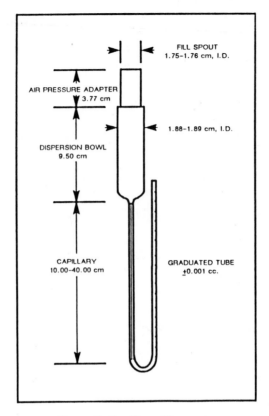

FILL SPOUT
1.75-1.76 cm, I.D.

AIR PRESSURE ADAPTER
3.77 cm

1.88-1.89 cm, I.D.

DISPERSION BOWL
9.50 cm

CAPILLARY
10.00-40.00 cm

GRADUATED TUBE
±0.001 cc.

Figure 8. Capillary Viscometer.

Step 3. The liquid adsorption onto pigment particles is the "rate controlling step" as shown by addition of dispersing agents and development of viscosity. By using these techniques, the milling time was reduced significantly, 150–180 min. compared to 210–360 min.

8

Step 4. The rate of dispersion and the degree of dispersion are directly related to interaction of resin, solvent and dispersing agent with iron oxide, without which an acceptable coating can not be formed.

Step 5. The dispersion mechanism follows a non–linear rheological trend which must be monitored through a pattern of events rather than relying on a single measurement.

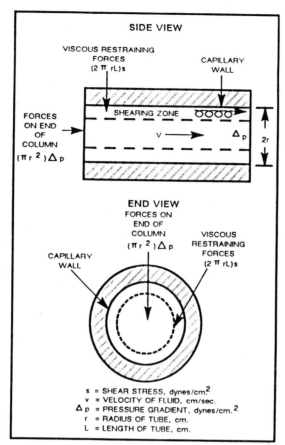

Figure 9. Fluid Through a Capillary

CONCLUSIONS

1. Rheological properties of dispersions were correlated to particle size and distribution. This accomplishment permitted evaluation of magnetic dispersions before preparing the cured coatings on tapes. The most sensitive rheological tool was the capillary viscometer.

2. Optimization of dispersing agents was accomplished with a Rotovisco 2 Viscometer using an MV-I sensor. The same technique was used to evaluate compatibility of resins and solvents.

3. The rate of agglomerate particle size reduction was shown to be directly related to resin and solvent wetting of pigment surfaces. Particle size reduction was more sensitively monitored using a capillary viscometer than rotating cylinder or spindle type viscometers. This was found to be due the sensitivity of the capillary to particle size and distribution.

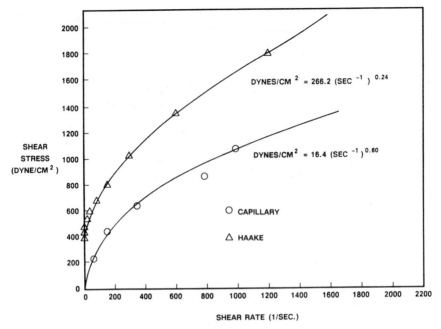

Figure 10. Shear Stress vs. Shear Rate for Capillary and Haake Viscometers for 210 Dispersion.

4. Magnetic skewness is a preferred method of measuring magnetic properties of a cured coating. The percent gloss and magnetic skewness of a cured magnetic coating are plotted in Figure 14 showing a converging set of points which reach an asymptotic zone after 75 percent gloss. However, magnetic skewness continues to increase after percent gloss stabilizes indicating that it is a more sensitive method. Magnetic skewness is approximately 50 percent more sensitive than "squareness" measurements.

5. Rheological properties were correlated to magnetic properties of cured coatings. Using the knowledge gained form the flow of liquid dispersions during the milling operation, the flow rate of dispersions and magnetic skewness of cured coatings were plotted with dispersing time in Figure 15. Following the familiar fluctuations, the experimental curve reaches a constant slope from 180 min. to 210 min. However, by plotting flow rate for a predispersed material corrected for percent nonvolatile, a "control curve" was developed. Correcting the experimental curve for percent nonvolatile and plotting with time produces a curve which converges with the theoretical curve. So, by correcting for percent nonvolatile at 25°C, the degree of dispersion can be determined in the liquid state. The magnetic skewness values of dried coatings from 180 min. to 210 min. dispersions confirm this observation as shown by Figure 15. Using an enclosed pumping viscometer eliminated the fluctuations in capillary flow rate caused by changes in percent nonvolatile content of the dispersions.

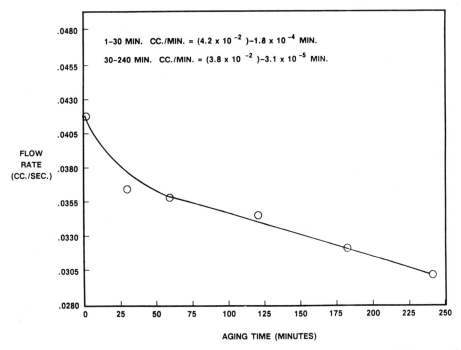

$$1\text{--}30 \text{ MIN. CC./MIN.} = (4.2 \times 10^{-2}) - 1.8 \times 10^{-4} \text{ MIN.}$$

$$30\text{--}240 \text{ MIN. CC./MIN.} = (3.8 \times 10^{-2}) - 3.1 \times 10^{-5} \text{ MIN.}$$

Figure 11. Flow Rate of Dispersion vs. Aging Time Demonstrating Rapid Degeneration by Agglomeration.

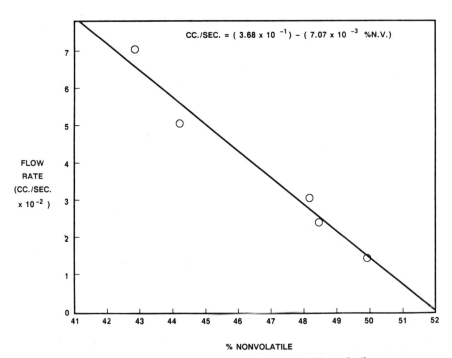

$$\text{CC./SEC.} = (3.68 \times 10^{-1}) - (7.07 \times 10^{-3} \text{ \%N.V.})$$

Figure 12. Flow Rate of Dispersion vs. % Nonvolatile .

287

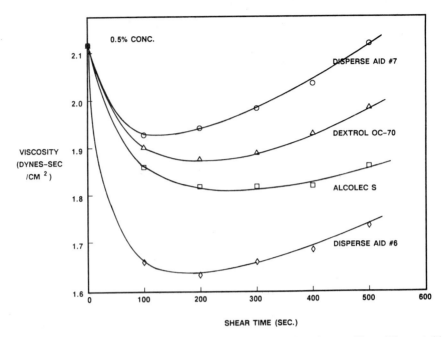

Figure 13. Selection of Dispersing Agents by Observing Viscosity vs. Shear Time at 1198 sec-1.

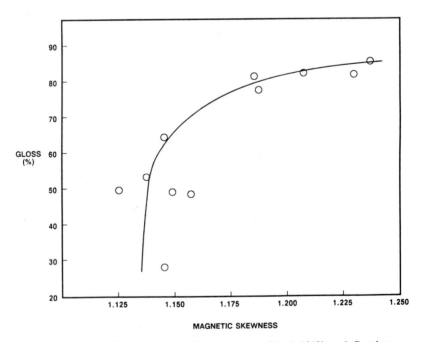

Figure 14. % Gloss vs. Magnetic Skewness of Dried/Aligned Coating.

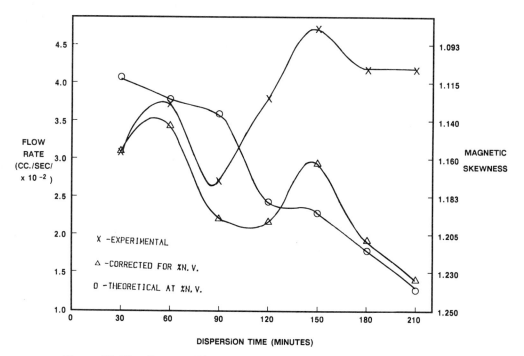

Figure 15. Flow Rate vs. Dispersion Time for Experimental, Corrected for %
Nonvolatile, and Theoretical at % Nonvolatile.

ACKNOWLEDGMENTS

I am grateful to Dr. B. R. Livesay and Mr. J. W. Larson for the magnetic measurements and their innovations.

REFERENCES

1. Magnetic Materials – "A Glossary," Pfizer Corp., Easton, Pennsylvania, 1981.

2. B. R. Livesay, J. W. Gooch, and J. W. Larson, "Studies of Magnetic Particle Dispersion Mechanisms, and the Physical Behavior of Recording Tape Composites," Final Report, Georgia Institute of Technology, Atlanta, Georgia, 1983.

3. T. C. Patton, "Paint Flow and Pigment Dispersion," pp. 10–11, 200. Interscience Publishers, New York, 1964.

4. F.M. Fowkes and M.A. Mostafa, Acid-base interactions in polymer adsorption, I&EC Product R&D, 17, 3 (1978).

5. F. M. Barton, "Handbook of Solubility Parameters," pp. 139–190. CRC Press, Boca Raton, Florida, 1983.

6. L. B. Lueck, Wetting agents and the preparation of magnetic pigment for effective dispersion, Paper presented at the Symposium on Magnetic Media Manufacturing Methods, Honolulu, Hawaii, 1983.

7. D. C. Montgomery, "Design and Analysis of Experiments," pp. 121–159. John Wiley & Sons, New York, 1976.

8. J. H. Perry, editor "Chemical Engineers' Handbook," 7th edition, Section 3, McGraw-Hill Book Company, New York, 1980.

9. G. D. Parfitt, "Dispersion of Powders in Liquids," 3rd edition. pp. 376–377, Applied Science Publishers, Englewood, New Jersey, 1981.

10. P. Mitschka, Simple conversion of Brookfield RVT readings into viscosity functions, Rheol. Acta, 21, 207–209 (1982).

11. R. R. Wazer, "Viscosity and Flow Measurement," pp. 190–191, 207–220, 268. Interscience Publishers, New York 1963.

12. Y. Isobe and K. Okuyama, J. Coatings Technol., 55(698), 23 (1983)

THE ROLE OF POLYMER BINDER IN MAGNETIC RECORDING MEDIA

K. Sumiya, Y. Yamamoto, K. Kaneno and A. Suda

Kyoto Research Laboratory
Hitachi Maxell Ltd.
Oyamazaki, Otokuni, Kyoto, 618 Japan

The mechanical properties of polymer binders for magnetic recording media were evaluated by tensile and fatigue tests. The durability increases with yield stress and strength of repetition fatigue of the coating. These results suggest that the yield point and fatigue properties of polymer binders are very important factors for the improvement of durability. In addition, the results of these tests indicate the relationship between head clogging and the properties of polymer binder.

INTRODUCTION

Extensive developments in the field of magnetic media have taken place[1] and the proper choice of polymer binders offers significant potential for further improvement in magnetic performance. With the recent introduction of high dispersion and high durability binders, it has become clear that particulate media will dominate even in the field of higher density recording media.

To date, many studies[2-8] have focused on the dispersing behavior of polymer binders and/or surfactants in particulate magnetic recording media. These polymer binders are important in the improvement of dispersion and orientation[9] of magnetic particles and the improvement of mechanical properties of magnetic coatings. We have already reported on the dispersion properties in previous papers[10,11] and in this paper, we shall report on the mechanical properties of magnetic coatings. The relationship between the mechanical properties of polymer binders and the durability of magnetic tapes is discussed.

EXPERIMENTAL

Preparation of Free Films

Cellulose nitrate (CN), Polyurethane (PUR), and a crosslinker were mixed as raw materials. The PUR's, which were of the 4,4'-diphenyl methane diisocyanate ester type, varied in the concentration of -NHCOO-. Polyisocyanate, 3TDI/TMP adduct, was used as a crosslinker.

Table I. Composition of Free Films.

		Composition of binder	Concentration of -NHCOO- in PUR (mmol/g)
Free film A	A:	CN/PUR-1/CL	3.3
Free film B	B:	CN/PUR-2/CL	3.1
Free flim C	C:	CN/PUR-3/CL	1.8
Free film D	D:	CN/PUR-4/CL	1.3

CN: Cellulose nitrate   PUR: Polyurethane
CL: Crosslinker

Table I shows the various binders and concentration of the -NHCOO- group. Each PUR and CN was dissolved in the cyclohexanone-toluene (1:1 weight ratio) solvent and mixed with crosslinker. The mixture was coated on a Teflon sheet. The weight ratio of CN/PUR/crosslinker was 2/2/1. The coated film of 20μm thickness, shown as free film in Table I, was then cured at 80°C for 24 hours.

Preparation of Magnetic Tapes

Four different magnetic paints which contained binder A, B, C and D, respectively, were prepared. Each paint was coated on a polymer web, dried, and calendered. The thickness of the magnetic coating was 4 μm. The coated web was slitted to produce magnetic tapes A, B, C and D.

Tensile Test of Free Films

The stress-strain curves of free films were obtained at 25±1°C by using a testing machine manufactured by Simazu Seisakusho.

Fatigue Test of Free Films

The fatigue tests on the free films were carried out at 25±1°C by using the repetition-fatigue test apparatus shown in Figure 1. An important characteristic of the apparatus is that constant stress can be applied until the film ruptures.

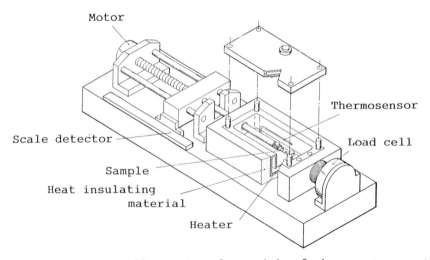

Figure 1. Schematic illustration of repetition-fatigue test apparatus.

Durability Test of Magnetic Coatings

The durability of these magnetic coatings was evaluated at 25°C, 60%RH, by using a durability test apparatus which was especially manufactured for this purpose as shown in Figure 2. The friction force was measured with a semiconductor strain gauge and recorded. The durability of each magnetic coating was evaluated by comparison of the microphotograph of each magnetic coating after the durability test.

Clogging Test of Recording Head

Clogging tests with magnetic tapes A-D were carried out by using a video tape recorder at 25°C and 60%RH. During each test, the tape traversed back and forth over the magnetic head at the speed of 3.33 cm/s. After the 100th traverse, the result was evaluated by output signal and microphotograph of the recording head.

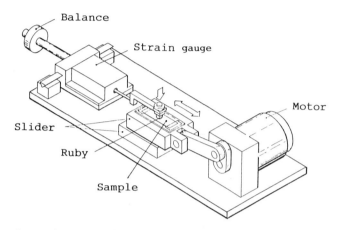

Figure 2. Schematic illustration of durability test apparatus.

RESULTS AND DISCUSSION

Tensile Characteristics of Free Films

The Young's modulus and elongation of free films are important factors for the improvement of the durability of magnetic coatings[12]. Figure 3 shows the stress-strain curves of free films A-D. Film A shows the highest Young's modulus and yield point and Film D has the lowest Young's modulus and yield point. Films B and C have intermediate modulus and show larger elongation than films A and D. The yield point decreases in the order: Film A>B>C>D. This order is related to the concentration of -NHCOO- in the PUR of the films.

Fatigue Characteristics of Free Films

Each fatigue test was carried out under constant stress. As shown in Figure 4, in the case of Film D, it is obvious that there is a relationship between the strain and the number of repetitions when stress is applied from 450kg/cm^2 to 550 kg/cm^2. Film D broke at the 12th repetition under the stress of 550kg/cm^2, and at the 100th repetition under that of 450kg/cm^2, respectively. When the applied stress increased, the film broke at an earlier stage, i.e., the toughness decreased. Figure 5 shows the relationship between the number of repetitions and the applied stress. These results suggest that the toughness is mainly related to the

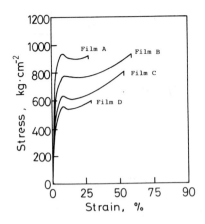

Figure 3. Stress-strain curves of free films.

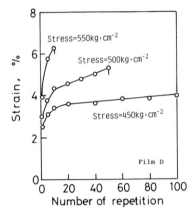

Figure 4. Relation between the strain and the number of repetition (Film D).

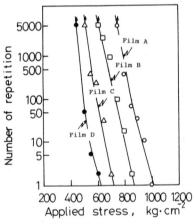

Figure 5. Relation between the number of repetition and the applied stress.

Table II. Durability and coefficient of friction of magnetic coatings.

		COEFFICIENT OF FRICTION	
COATING	DURABILITY	INITIAL	100TH TRAVERSE
Coating A	high	0.19	0.24
Coating B	medium	0.21	0.26
Coating C	medium	0.19	0.25
Coating D	low	0.20	0.27

amount of urethane binding, -NHCOO-, in the PUR of each film.

Durability Characteristics of Magnetic Coatings

The durability of magnetic coatings A-D was evaluated by using a durability test apparatus with a ruby slider. Table II shows the durability and coefficient of friction (COF) of each magnetic coating. Figure 6 shows the photomicrographs of magnetic coatings after the durability test.

The durability of these magnetic coatings decreases in the order: Coating A>B>C>D. Initial COF's of all these coatings were almost the same. After the 100th traverse, however, each COF increased. This result shows the same relationship as the fatigue test of free films. Once again, this order was related to the amount of -NHCOO- in the PUR of these films. The results of all the above tests prove that the amount of -NHCOO- in PUR is a significant factor for the improvement of these magnetic coatings.

Figure 7 shows a typical magnetic head for video recording. Head clogging is a serious problem in magnetic performance. It causes a decrease of output level and/or a loss of output signal. Table III shows the results of tests on magnetic coatings A-D. These results indicate that Coating A is the most suitable for preventing head clogging. Coating A contains the highest concentration of -NHCOO- in PUR. Common PUR's contain a high concentration of -NHCOO-. The -NHCOO- has a particularly high cohesion energy[13]. This cohesion energy affects the mechanical and fatigue properties of free films, and the durability of magnetic coatings. The introduction of -NHCOO- leads to the lowering of the solubility of the PUR in the solvent. Therefore, PUR with -NHCOO- above 3.3mmol/g could not be used as the binder component of magnetic coatings.

CONCLUSIONS

The mechanical properties of polymer binders were evaluated by tensile and fatigue tests. According to the results, durability increases with yield stress and strength of repetition fatigue of the coating. These results suggest that the yield point and fatigue properties of polymer binders are very important factors for the improvement of durability. In addition, the results of these tests indicate the relationship between head clogging and the properties of the polymer binder.

(1) Coating A

(2) Coating B

(3) Coating C

(4) Coating D 100μm

Figure 6. Photomicrographs of magnetic coatings after 100th traverse of ruby slider.

Table III. Head clogging with magnetic coatings.

COATING	HEAD CLOGGING
Coating A	Non-clogging
Coating B	Slight clogging
Coating C	Clogging
Coating D	Clogging

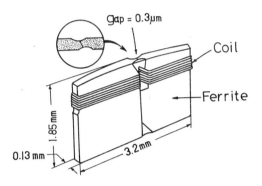

Figure 7.   Schematic illustration of recording head.

ACKNOWLEDGEMENTS

The authors are gratefully indebted to Prof. K. Nakamae for his valuable advice and to Mr. T. Doi for his assistance in performing the durability test.

REFERENCES

1. P. Williams, IEEE Trans. Mag. MAG-24, 1876 (1988).
2. K. Sumiya, N. Hirayama, F. Hayama and T. Matsumoto, IEEE Trans. Mag. MAG-20, 745 (1984).
3. K. Sumiya, T. Taii, K. Nakamae and T. Matsumoto, Kobunshi Ronbunshu (English Edition) 38, 123 (1981).
4. K. Nakamae, K. Sumiya and T. Matsumoto, Prog. Organic Coating, 12, 143 (1984).
5. K. Nakamae, K. Sumiya, T. Taii and T. Matsumoto, J. Polym. Sci. Polym. Symposium, 7, 109 (1984).
6. S. Dasgupta, J. Colloid Interface Sci., 121, 208 (1988).
7. S. Dasgupta, J. Colloid Interface Sci., 124, 22 (1988).
8. A.M. Homola and M.R. Lorenz, IEEE Trans. Mag. MAG-22, /16 (1986).
9. K. Sumiya, N. Nakamae, T. Matsumoto, S. Watatani and F. Hayama, in "Ferrite; Proceedings of the International Conf.", H. Watanabe et al. editors, p. 567, Center for Academic Publications Japan, 1980.

10. K. Sumiya, F. Hayama and T. Matsumoto, J. Adhesion Soc. Japan (in Japanese), <u>17</u>, 155 (1981).
11. K. Sumiya and T. Matsumoto, J. Japan Soc. Colour Material (in Japanese), <u>58</u>, 211 (1985).
12. R.C.F. Schaake, H.A.M. Pigmans, J.A.M. v.d. Heijkant and H.F. Huisman, IEEE Trans. Mag. <u>MAG-23</u>, 109 (1987).
13. C.W. Bunn, J. Polym. Sci., <u>16</u>, 323 (1955).

THE PARTICULATE MEDIA FOR MAGNETIC RECORDING:

CHARACTERIZATION TECHNIQUES FOR PARTICLE

DISPERSION AND ORIENTATION

Myung S. Jhon

Department of Chemical Engineering
Carnegie Mellon University
Pittsburgh, PA 15213-3890

and

Thomas E. Karis

IBM Research Division
Almaden Research Center
650 Harry Road
San Jose, California 95120-6099

Although the particulate magnetic recording industry is mature, there still remains an unsolved problem of dispersion quality, which plagues any coating process involving unstable particle suspensions. We re-examine this issue in light of our recently developed concept for dispersion quality measurement. That is, the particles are oriented in a flow field and, because they are magnetic, we can use a weak magnetic sensing field to detect their orientation. Not only is this a new idea, but also it can be used to derive information about the flocculation, because flocs orient differently in a flow field than the primary particles which are desirable for the coating. Experimental data on the effects of surfactant, dilution, milling, aging, and redispersion are discussed. Qualitative theory describing the relation between the measurement and particle shape is also given to interpret the data. We emphasize that the current method is suitable for process control in particulate media production because it can be placed on-line and measurement can be done in real time.

INTRODUCTION

The most widely used media for information storage are magnetic tapes and disks. Particulate media are manufactured by applying a thin coating of magnetic particle slurry on the substrate [1]. The particles most commonly used for longitudinal recording are various forms of $\gamma - Fe_2O_3$. The length of these rod-shaped particles ranges up to $5\ \mu m$ with diameters equaling 1/6 of the length. They are typically dispersed in resin (polyester, polyurethane blended with phenoxy) and solvent (a mixture of tetrahydrofuran and cyclohexanone) at about 25% by volume of magnetic particles in the final coating [2]. Due to the presence of non-magnetic polymeric binders in the coating, the fundamental limit of the particulate media bit density is, in principle, smaller than that of the thin film media. However, because of other production advantages (for example, reliability and cost) particulate media are presently the dominant media for memory storage.

With the goal of higher density recording, media producers need to use smaller particles (typically metal particles instead of oxide). However, as the particles become smaller, they become more difficult to disperse. Disk shaped particles of Ba-ferrite[3] are a viable alternative to smaller particles through the emerging technology of perpendicular recording (where the particle magnetic moment is oriented normally rather than parallel to the surface of the disk or tape, as is the case with $\gamma - Fe_2O_3$). Despite considerable commercial interest, development of Ba-ferrite media is complicated by dispersion and flocculation. Therefore, the technology of stabilizing the magnetic particle dispersion is an important engineering consideration in particulate media production.

In this paper, we develop a new characterization technique for the concentrated magnetic particle suspension. Even though our focus is on acicular magnetic particles, the technique can be applied to differently shaped magnetic particles by modification of the theoretical interpretation. The most relevant engineering task associated with dispersion technology is to discover a procedure to determine whether the particles were sufficiently separated by the milling or if they were stuck together or agglomerated (due to aging, some further processing condition, or chemical addition). One way of finding out if the particles were sufficiently separated is to evaluate the quality of the final product (using such factors as background noise and missing bits). If the dispersion quality is poor, however, this method results in a low product yield. In the case of large volume production, this is not a cost-effective approach. We will present a new method for measuring the dispersion quality in the early stages of magnetic media production. There already exist several off-line (ideal condition) methods: light scattering, photographic method, and rheology.

Okagawa et. al. [4,5] employed a high-speed photographic technique and then turbidity to study orientation of the particles. Wagner et. al. [6] also developed an apparatus to measure flow orientation of particles by conservative dichroism, while other investigators [7,8] used light scattering to study the particle orientation. However, these optical methods are limited. The particles must be in the range of 0.1 to 100 μm, and the path length must be short enough to allow sufficient light transmission. Therefore, these techniques can only be used to study particle orientation in the limit of infinite dilution.

On the other hand, rheological measurement provides, in principle, an indirect measurement of particle orientation in concentrated suspension. Even though a great deal of published data on zero shear rate viscosity is available (the measured viscosity becomes time dependent and shows hysteresis [9]), there exist virtually no reliable measurements or theoretical interpretation of yield stress, viscosity, storage and loss modulus, etc. Even though the indirect measurement through rheology is as yet undeveloped[10], we believe that rheology will be useful in studying (off-line) dispersion quality.

We introduce an on-line (processing condition) measurement for studying the orientation distribution of particles through magnetic permeability measurement that is novel and complementary to rheological measurement. We believe that the apparatus described in this paper is most suitable for process control because it can be placed on-line and measurement done in real time.

We first derive an approximate theory suitable for specifying the experimental design criteria. The theoretical development is somewhat intuitive and qualitative due to the complexity of the system. The key result is the relationship between the measured quantity (coil inductance) and suspension property (hydrodynamic orientation of the magnetic particles). Later, we present a brief description of our experimental apparatus and a study of the dispersion quality of magnetic inks from the flow orientation data. The effects of surfactant, volume fraction, and milling are considered. Finally, we discuss the suspension hysteresis phenomenon, which is important in the interpretation of our data, and is also vital in correlating the current method with the rheological measurement.

300

THEORETICAL DEVELOPMENT

We illustrate the essence of our idea by studying the single particle dynamics. Consider the motion of a $\gamma - Fe_2O_3$ particle entering a small tube, as depicted in Fig. 1. Far away from the pore entrance, the orientation of the rod-like particle is random due to Brownian motion (it can be characterized by the rotary diffusion coefficient, D_r). However, as the particle enters the pore, the particle tends to line up with its major axis parallel to the flow direction due to the elongational deformation of the fluid element (the deformation is related to the volumetric flow rate, $\dot{Q}$). Therefore, it is expected that the degree of the orientation will depend on the $\dot{Q}$ and D_r . Some details of this analysis are also given in Karis and Jhon[11].

Since the magnetic particle possesses its own magnetic moment along the major axis, the magnetic permeability of the particle suspension depends on the orientation of the particle. This unusual directional magnetic property suggests that the orientation near the pore mouth can be measured by observing the small amplitude, high frequency, inductance of a coil wound around a tube, as shown in Fig. 1.

The inductance of the coil L is related to the coil geometry β and permeability μ by[12]:

$$L = \beta^{-1}\mu,\tag{1}$$

with a coil constant (fixed by the geometry):

$$\beta = 4l/(\pi N^2 d^2),\tag{2}$$

where N is the number of turns, d is the diameter, and l is the length of the coil. In practice, we can measure L from the oscillator frequency f in an L-C circuit by means of the following relationship:

$$f = \frac{1}{2\pi\sqrt{LC}}.\tag{3}$$

The quantity of interest in this paper is the inductance change, δL, (or, analogously, frequency change) due to the magnetic particle, and it is given as[11]:

$$\delta L \equiv L - L_s = \beta^{-1}\mu_m.\tag{4}$$

Here, L_s is the value of L measured with pure suspending fluid, and μ_m is the permeability contribution of the magnetic particles, given by:

$$\mu_m = \mu_0 M_s \int_0^{\pi/2} \frac{dM'}{dH}\Big|_{H\to 0} F(\theta) \sin\theta\, d\theta,\tag{5}$$

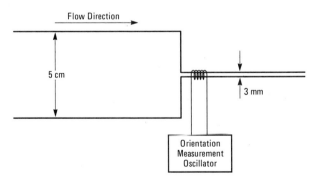

Figure 1. A schematic of the converging flow cell showing the large chamber and the small tube with the coil and oscillator used to measure the particle orientation.

where $M' = M/M_s$, M is the magnetization, M_s is the saturation magnetization, $F(\theta)$ is the orientational distribution of the single rigid rod, and θ is the angle of the particle with respect to the applied field. Equations (4) and (5) imply that δL depends on the magnetic properties of a particle through $dM'/dH\big|_{H\to 0}$ and on the orientational distribution function, F, which depends on D_r and Q.

To calculate μ_m for the given flow geometry, we separately consider the dilute ($\phi < 1/p^2$) and non-dilute ($\phi > 1/p^2$) regions of particle concentration ϕ for the N particle suspension[13].

For the dilute system, one can neglect the interaction between the particles. This implies that the N-particles are statistically independent. Therefore, by using the chain

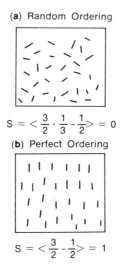

(a) Random Ordering

$$S = \left\langle \frac{3}{2}\cdot\frac{1}{3}-\frac{1}{2}\right\rangle = 0$$

(b) Perfect Ordering

$$S = \left\langle \frac{3}{2}-\frac{1}{2}\right\rangle = 1$$

Figure 2. An illustration of the limiting values of the order parameter, random orientation (a) and perfect alignment (b).

of spheres model[14] to describe the magnetic properties and the Kirkwood-Auer theory [15] to obtain the orientational distribution due to the imposed flow field, we obtain[11]:

$$\delta L/\phi = (1 - S)/\alpha\beta . \qquad (6)$$

in which α is obtained from the magnetic properties of the single-domain particle and $S \equiv \langle (3\cos^2\theta - 1)/2 \rangle$. The angle brackets $\langle\ \rangle$ represent the ensemble average over all orientations of the particles in the system. S is known as an order parameter and it can, in principle, be calculated from the given flow geometry, imposed flow field, and the particle shape. The numerical value of S lies between 0 and 1. The physical meaning of the upper and lower bounding values of S is illustrated in Fig. 2.

In the magnetic recording industry, the objective is to characterize the state of flocculation and agglomeration, or dispersion quality, of the single domain magnetic particle suspension, especially at the non-dilute concentration. However, there is no

302

adequate theory for the non-dilute system, in which the rotary diffusion is hindered by the neighboring particles. The hindered diffusion is due to steric, hydrodynamic, and magnetic interactions that occur between the particles.

To obtain a quick estimate of the dispersion quality, we introduce the concept of *pseudo-particle* explained below. Unlike the non-magnetic particle suspension, the magnetic particles generally form clusters, as shown in Fig. 3. We will call these clusters pseudo-particles. Note that the pseudo-particle possesses a finite lifetime. The shape of the pseudo-particles depends on the flow rate (breaking up of flocs) and on time (by flocculation). Even though the primary particles may be nearly monodisperse, the pseudo-particles are considered polydisperse from our definition. We define the effective diameter and aspect ratio of the pseudo-particles as a_e and p_e.

In the non-dilute region of concentration, the primary particles experience strong magnetic interaction, while pseudo-particles interact weakly with each other. Therefore, one can employ the dilute suspension theory and the pseudo-particle picture to qualitatively describe the non-dilute suspension.

Using the Kirkwood-Auer theory[15] for the dilute suspension of pseudo-particles (with extensional flow) we obtain:

$$S = S(\overline{P}e), \tag{7}$$

and,

$$\overline{P}e = \frac{\gamma \dot{Q}}{d^3 \overline{D}_r}. \tag{8}$$

Here, $\overline{P}e$ is the Peclet number, $\overline{D}_r$ is the rotary diffusion coefficient of the pseudo-particle, and γ is a hydrodynamic coefficient. The approximate functional forms for S in two limiting Peclet number regions are obtained from the dilute suspension theory [11,16] and:

$$S \simeq 0.1 \overline{P}e \quad \text{for} \quad \overline{P}e < 4, \tag{9}$$

and

$$S \simeq 1 - 2/\overline{P}e \quad \text{for} \quad \overline{P}e > 4. \tag{10}$$

Equations (6) through (10) are the key results in designing the experimental apparatus and interpreting the data.

EXPERIMENTAL DESIGN

In the above theoretical development, we employed the extensional flow approximation to simplify the analysis. Nonetheless, most flow fields consist of shear and rotational components[17]. Therefore, considering the practical requirements of laboratory

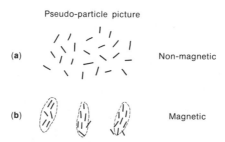

Figure 3. Single primary particles (a) and clusters of primary particles formed into pseudo-particles or flocs (b) due to magnetic interaction forces.

testing, we selected a flow geometry that would have a significant extensional component with little rotational contribution. The flow we selected was the type which occurs when fluid passes from an infinite reservoir into a small orifice. In this case, the fluid velocity is initially very small far from the pore mouth. At or near the entrance, the fluid velocity increases sharply. Especially along the centerline of the convergence, the increase in velocity translates ideally into extensional flow. The sketch of the apparatus is shown in Fig. 1.

In this section we describe the experimental setup, cite relevant data in order to study the variables which affect the dispersion quality, and discuss the potential improvement of the current apparatus.

Experimental Setup

The flow cell shown in Fig. 1 was incorporated into a recirculating flow system holding about 900 ml of suspension. We attached a 3 mm diameter glass tube onto a 5 cm diameter glass chamber. (Glass was used to allow free passage of the magnetic field between the sensing coil and the particles.) The 2 cm long coil was centered 2 cm from the entrance of the small tube, to be as close as possible to the region of extensional flow. The pump speed setting and oscillator frequency measurement used to obtain the $\delta L - \dot{Q}$ plot were performed by an IBM PC with interface cards.

Dry powdered $\gamma - Fe_2O_3$ was combined with ethylene glycol in a paint shaker to obtain a suspension with volume fraction on the order of 0.04. This suspension is hydrodynamically similar to an actual coating formulation used in particulate media for magnetic recording. It was necessary to then pass the suspension through an Eiger mill to further break down agglomerates and disperse the particles. (This was discovered because the suspension density was always below the theoretical density when only the paint shaker was used to prepare the suspensions.) A surfactant (complex organic phosphate ester) was also added to the suspension to alter the floc structure.

Experimental Data

The studies described here include measuring the changes in particle dispersion due to the addition of surfactant, dilution, milling, and aging. One series of tests was carried out to show the ability of our flow cell apparatus to detect differences in the degree of dispersion accompanying various amounts of the surfactant. Interaction of the surfactant with $\gamma - Fe_2O_3$ has also been studied by other methods[18,19]. For each level of surfactant, the suspension was placed in the flow system and recirculated at a high flow rate for 15 minutes in order to obtain a uniform dispersion. Then, the flow rate was decreased from 20.7 to 0.26 ml/sec in 25 decrements with 50 sec at each flow rate, for equilibration, before measuring the oscillator frequency to obtain L. The results of these tests are shown in Fig. 4, where L/ϕ is expressed as a function of $\dot{Q}$ for each of the surfactant levels. The results imply that there is a significant effect of surfactant on the formation of the flocs (pseudo-particles).

Typical data from another test series, in which the volume fraction of the suspension was changed, are shown in Fig. 5, plotted at low and high flow rate for comparison with the theory.

The experimental results in the limit of high $\overline{Pe}$ show that:

$$\frac{\delta L}{\phi} = A + B\left(\frac{1}{\dot{Q}}\right), \tag{11}$$

with $A \neq 0$. Note that the simple theory predicts $A = 0$ [see eqs. (6), (8) and (10)]. This non-zero A can be due to deviation of the flow in the test cell from the ideal extensional flow approximation and from polydispersity of the pseudo-particle.

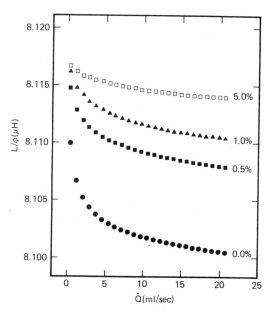

Figure 4. The flow orientation curve for several different levels of surfactant, showing the effect of surfactant on the pseudo-particle shape.

Even though the intercept is non-zero, it is possible to derive an expression for calculating $\overline{D}_r/\gamma$ from the flow orientation data by using the gradient method shown in Karis and Jhon[16]. Because γ and d are constant for a given flow cell, changes in $\overline{D}_r/\gamma$ indicate changes in floc size and shape related to the dispersion quality, independent of any particular model for $\overline{D}_r$. From eqs. (6) through (10), we obtain

$$\alpha\beta = \frac{1}{\sqrt{5}}\left[\left(-\frac{\partial(L/\phi)}{\partial\dot{Q}}\right)_{\dot{Q}\rightarrow 0}\times\left(\frac{\partial(L/\phi)}{\partial(1/\dot{Q})}\right)_{\dot{Q}\rightarrow\infty}\right]^{-1/2} \qquad (12)$$

and

$$\frac{\overline{D}_r}{\gamma} = \frac{1}{2d^3\sqrt{5}}\left(-\frac{\partial(L/\phi)}{\partial\dot{Q}}\right)_{\dot{Q}\rightarrow 0}^{-1/2}\times\left(\frac{\partial(L/\phi)}{\partial(1/\dot{Q})}\right)_{\dot{Q}\rightarrow\infty}^{1/2}. \qquad (13)$$

The gradients are derived from the data by a regression fit to the data in the limit of low (Fig. 5a) and high (Fig. 5b) flow rate.

Another test was performed to study the effect of dilution on $\overline{D}_r/\gamma$. These results are shown in Fig. 6. Here, each symbol indicates a separate dilution experiment. The rotary diffusion coefficient of the pseudo-particles clearly increases as the volume fraction is decreased. Often, we can deduce information about the transition from the dilute to non-dilute region of concentration from these types of data. It does seem as if there is a dilute-concentrated transition ($\phi = 1/p^2$) near $\phi \simeq 0.02$. From this transition we can estimate the aspect ratio of the pseudo-particles p_e by $p_e = (1/\phi)^{1/2} \simeq 7$, which is remarkably close to the primary particle aspect ratio of 6 measured by SEM.

As is well-known in the business of particulate magnetic recording media formulation, an essential part of the process is the milling of the suspension[20]. Even after the suspension has been thoroughly milled, it must be continuously stirred, and sometimes, for an old ink, it is milled again to "redisperse" the particles. We performed several tests to study the effect of milling on the flow orientation curves.

The $\delta L/\phi$ as a function of flow rate $\dot{Q}$ is shown in Fig. 7 for a suspension tested immediately after milling (circles), aged for 16 hours (squares), and the aged suspension

305

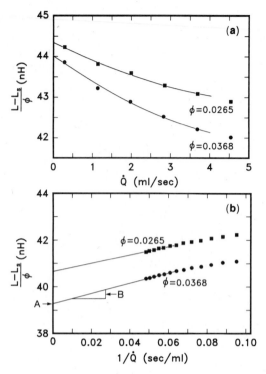

Figure 5. The low (a) and high (b) flow rate orientation curves showing the regression fit used to obtain the slope of the data.

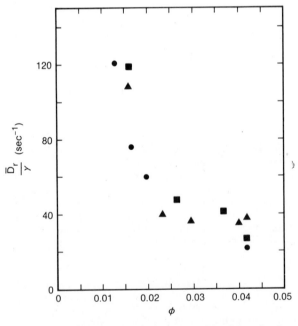

Figure 6. The relative rotary diffusion coefficient as a function of volume fraction. The different symbols indicate separate runs with different batches of $\gamma - Fe_2O_3$.

passed through the mill again (triangles). The effects of aging and redispersion are clearly illustrated. Upon aging, it seems that the pseudo-particle shape changes in such a way that $\overline{D}_r$ decreases, because there is less of a dependence of $\delta L/\phi$ on Q after aging. Redispersion by passing only once through the mill modifies the pseudo-particle shape back towards that before aging.

Another way to look at the dynamics of flocculation and redispersion is to measure $\delta L/\phi$ as a function of time at constant flow rate. This is shown in Fig. 8 for the low flow rate (Fig. 8a) and the high flow rate (Fig. 8b). The aged suspension is initially below the redispersed, but with time of recirculation, the $\delta L/\phi$ of the aged suspension gradually approaches that of the redispersed suspension, while the redispersed suspension changes little. Apparently, even recirculating the suspension in the flow system for a long time has an effect of changing the shape of the flocs.

Finally, we will present an unusual finding called "fluid hysteresis," which is a characteristic of magnetic particle suspension. The unstable suspension generally exhibits a hysteresis curve with respect to shear history in rheological measurements[9]. This is attributed to changes in the state of flocculation of suspended particles. For example, the viscosity of an unsheared suspension is generally a certain high value before shearing. As the shear rate is increased, the viscosity decreases. After some high shear rate, then, if the shear rate is gradually decreased, the viscosity increases along another curve which is below the initial viscosity-shear rate curve. If this is due to changes in the dispersion, and changes in the dispersion are produced by recirculating in our flow system, then we should observe analogous behavior in $\delta L/\phi$ as the flow rate is first increased and then decreased. The results of the test to measure the hysteresis in $\delta L/\phi$ are shown in Fig. 9. Here, the flow rate was gradually increased to $20.7 cm^3/$ sec , maintained at this high value for 30 min. and then gradually decreased, with a previously aged suspension. We do observe the hysteresis in $\delta L/\phi$. The lower curve, during which the flow rate is being increased, is also somewhat unstable, probably due to the breakup of pseudo-particles as the shear forces exceed the floc rupture strength[21]. The upper curve, along which the flow rate is being decreased, is much smoother, although we can see the effect of some flocs reforming as the shear forces decrease sufficiently to allow their growth. This observation suggests that there is a correlation between the current method and rheological measurement. Work along this direction is currently in progress.

Improvement of the Current Apparatus

The entrance flow produced in the current apparatus has certain features in addition to that of extensional flow. After the fluid passes the entrance of the small tube, it becomes an entrance flow which eventually develops into a Poiseuille flow several tube diameters downstream. There is a rotational boundary layer near the wall, starting from the entrance and increasing in thickness through the region of the coil. In addition, the time between the strongest extensional flow at the pore mouth and the center of the coil may allow some relaxation of the particle orientation, since the particles are small enough to undergo Brownian motion.

Both of these deviations from the ideal extensional flow approximation employed to interpret our data can be studied rigorously by numerical simulation, and this work is still in progress. However, some estimates can be made regarding the relative importance of these factors to the operation of our test cell. The highest $\overline{D}_r$ is that for the primary magnetic particle (the $\overline{D}_r$ of the flocculated magnetic particles, or pseudo-particles is lower), for which $\overline{D}_r = 0.147$ sec^{-1}. This gives a time constant for relaxation of the orientation distribution due to Brownian motion of $\tau = 1/6\overline{D}_r = 1.13$ sec . With very slow flow in our apparatus, the worst case for relaxation of the particle orientation distribution, it will take an average of $t = 0.212$ sec for the particle to travel from the entrance to the center of the coil, giving $t/\tau = 0.188$ so that, even in the worst situation, the orientation could only be relaxed by 20% due to Brownian motion. We will see later that very few single particles exist in the non-dilute suspension, so relaxation is not expected to significantly affect the results from the current apparatus.

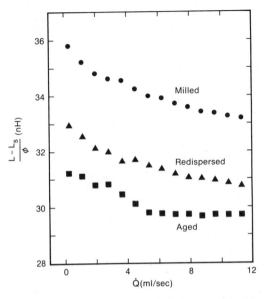

Figure 7. The flow orientation curve for suspension that was freshly milled, aged, and milled again for redispersion.

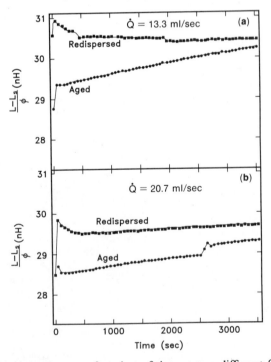

Figure 8. The flow orientation as a function of time at two different flow rates showing the effect of aging, redispersion, and recirculation throughout the flow system.

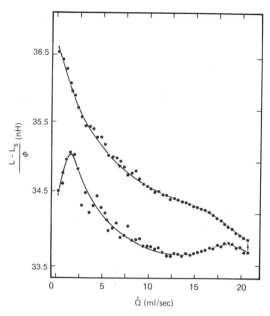

Figure 9. A flow hysteresis curve. The flow rate was first increased and then decreased with aged suspension.

The worst case effect of fluid vorticity, or rotation, on the orientation of the particles as they travel from the entrance to the center of the coil can also be estimated. Consider a fully developed Poiseuille flow, with average vorticity $\overline{\omega} = 8\dot{Q}/\pi d^3$. The average time for a fluid element to travel distance z along the tube is $t = \pi z d^2/4\dot{Q}$. Then, we can estimate the average number of rotations a fluid element will undergo by $\overline{\omega}t = z/\pi d$, or, in the case of our test geometry, $\overline{\omega}t = 3.2$ rotations. Given the approximations that were made in arriving at 3.2, we expect that the vorticity contribution to randomizing the orientation distribution will be small but significant, and less important near the center line, because the actual flow is not fully developed. To minimize the rotational component, we can set the z/d ratio to be small in design of the future apparatus.

DISCUSSION AND SUMMARY

In this paper, we studied characterization techniques for the dispersion quality measurement in the early stages of magnetic slurry production. Dispersion studies are the single most important factor in the quality control of the particulate media. The existing methods are typically off-line measurement. This means we have to take samples from the processing unit to test the dispersion quality. The off-line procedure, therefore, is of limited use in the production line.

Among the off-line methods, the rheological measurement is currently the most promising. Other techniques, such as light scattering, may also be an attractive option; however, it is only possible for the dilute suspension. The technique we described is suitable for the on-line measurement. For example, by passing the stream from the production line through the apparatus we described, one can obtain *in situ* measurement of the dispersion quality. Hence, the current method can be useful in process control.

To interpret the data and to illustrate the design criteria, we presented an approximate theory. For the dilute system, the particles are statistically independent, and one can employ the single particle theory to interpret the data. We have also introduced a *pseudo-particle* concept to qualitatively describe the non-dilute system. Combining the pseudo-particle picture with the dilute suspension theory, we have obtained an approxi-

309

mate, yet useful, relationship between the measured quantity (coil inductance) and the dispersion quality of the non-dilute suspension.

On the basis of the above mentioned ideas, we constructed an apparatus for measuring the orientation through coil inductance with an oscillator and a converging flow cell to simulate an extensional flow. $\gamma - Fe_2O_3$ suspensions were prepared in ethylene glycol to simulate magnetic recording ink formulation flow properties, for a test of the theory with the converging flow apparatus. Tests were performed to study the effects of surfactant, dilution, and milling on the state of flocculation of the suspension. It was found that the gradients of $\delta L/\phi$ with respect to flow rate, in the limit of high and low flow rate, could be used to estimate the relative rotary diffusion coefficient from the data. So far we have studied only the acicular shaped magnetic particles for which $p > 1$ used in longitudinal recording. Currently, we are examining the dynamics of plate-like particles for which $p < 1$ used in perpendicular recording.

ACKNOWLEDGEMENTS

The authors would like to thank the IBM Corporation for permission to publish this study. Thanks are due to Mr. Dave Holmstrom for his technical support. One of the authors (M.S.J.) was supported in part by the IBM General Products Division and Carnegie Mellon University. We are also grateful to Dr. James Lyerla at the IBM Research Division Almaden Research Center and Dr. Irmela Barlow at the IBM General Products Division for making this study possible.

REFERENCES

1. G. Bate, Proc. IEEE, 74 , 1513(1986).
2. H.F. Huisman, J. Coatings Technol., 727 , 49(1985).
3. T. Fujiwara, IEEE Trans. Magn., 21 , 1480(1985).
4. A. Okagawa and S.G. Mason, J. Colloid Interface Sci., 47 , 568(1974).
5. A. Okagawa and S.G. Mason, Can. J. Chem., 55 , 4243(1977).
6. N.V. Wagner, G.G. Fuller, and W.B. Russel, J. Chem. Phys., 89 , 1580(1988).
7. A.J. Salem and G.G. Fuller, J. Colloid Interface Sci., 108 , 149(1985).
8. B. Chu, R. Xu, and A. DiNapoli, J. Colloid Interface Sci., 116 , 182(1987).
9. R.V. Mehta, P. Prabhakaran, and H.I. Patel, J. Magn. Magn. Mater., 39 , 35(1983).
10. M.C. Yang, L.E. Scriven, and C.W. Macosko, J. Rheology, 30 , 1015(1986).
11. T.E. Karis and M.S. Jhon, Proc. Natl. Acad. Sci. USA, 83 , 4973(1986).
12. See standard text on electromagnetism, for example, R. Boylestad and L. Nashelsky, "Electricity, Electronics, and Electromagnetics," Prentice Hall, New Jersey, 1983.
13. M. Doi and S.F. Edwards, J. Chem. Soc. Faraday Trans. II, 74 , 567(1978), ibid., 918.
14. I.S. Jacobs and C.P. Bean, Phys. Rev., 100 , 1060(1955).
15. J.G. Kirkwood and P.L. Auer, J. Chem. Phys., 19 , 281(1951).
16. T.E. Karis and M.S. Jhon, J. Appl. Phys., 64 , 5843(1988).
17. O. Savas, J. Fluid Mech., 152 , 235(1985).
18. F.M. Fowkes, Y.C. Huang, B.A. Shah, M.J. Kulp, and T.B. Lloyd, Colloids Surfaces, 29 , 243(1988).
19. S. Dasgupta, J. Colloid Interface Sci., 124 , 22(1988).
20. P.N. Kuin, IEEE Trans. Mag., 23 , 97(1987).
21. R.C. Sonntag and W.B. Russel, J. Colloid Interface Sci., 115 , 390(1987).

THE EFFECT OF PIGMENT VOLUME CONCENTRATION ON THE MAGNETIC AND MECHANICAL PERFORMANCE OF PARTICULATE DISK COATINGS

H.L. Dickstein and R.P. Giordano

IBM General Products Division
5600 Cottle Road
San Jose, California 95193

W.H. Dickstein

IBM Almaden Research Center
650 Harry Road
San Jose, California 95120-6099

The magnetic and mechanical property changes of a particulate magnetic disk coating were evaluated as a function of the pigment volume concentration (PVC). It was shown that increased PVC improved the magnetic performance of the magnetic coating. The mechanical properties such as impact strength, gloss, adhesion, and scratch hardness also were measured as a function of increasing PVC. It was found that these coating properties all improved up to a critical pigment volume concentration (CPVC) above which they dramatically decreased. The importance of this CPVC is discussed both from the point of view of its determination, as well as its impact as a limiting factor on the ultimate performance of particulate magnetic disks.

INTRODUCTION

Particulate magnetic disks are typically coated with thermosetting organic binders within which magnetic particles have been dispersed[1]. The amount of magnetic particles contained in the organic coating, typically expressed as the pigment volume concentration (PVC), is a critical factor in both the magnetic and mechanical performance of the particulate disk. It has generally been predicted that the signal to noise ratio, a key limiting factor in the magnetic performance of a disk, is directly proportional to the packing density of the magnetic particles in the finished magnetic disk coating[2,3] The PVC is related to the packing density in that the PVC is defined as the total volume of pigment divided by the total volume of solids on the final, cured disk coating, while the packing density is defined as the volume of pigment per total volume of coating on the finished disk. Thus, if air is contained in the final coating, the packing density and the PVC are not identical.

It has long been known that in any filled coating system, there exists a critical pigment volume concentration (CPVC) beyond which any further addition of filler leads to the inclusion of air in the coating[4,5]. At this point, the mechanical properties of the coating generally fall off in a dramatic fashion as the inclusion of air leads to lack of coating integrity.

In this paper, we will describe how the critical pigment volume concentration was determined for an organic thermosetting disk coating. Additionally, we will discuss how the magnetic performance of the disk improved with increasing PVC of the coating. Finally, we will discuss how the CPVC acts as a limiting factor in both the mechanical and magnetic performance of particulate magnetic disks.

EXPERIMENTAL

Materials

The magnetic disk coatings were prepared by ball milling the iron oxide into the organic thermosetting binder as described previously[1]. The degree of grind was monitored and held constant by measuring with a Hegmen gauge[6]. The coatings were then applied to magnetic disks as previously described[1], and further processed to attain similar thicknesses and polished surfaces prior to testing. The coatings of several PVCs also were applied to paint testing panels via a motor driven doctor blade apparatus[6]. The coatings were cured in hot air ovens for two hours at 232°C. Testing on the coated paint panels was carried out without additional surface processing.

Measurements

Magnetic testing results as shown in Table 1 include data on old information, squeeze, and alpha - which is a mathematical combination of the old information and squeeze results. In the old information test the magnetic head is progressively moved off the track center until a bit can not be read; thus high values on this test represent improved magnetic performance. In the squeeze test, adjacent tracks are written and one track is progressively moved closer and closer to the other until the other track can not be read; thus low values from this test are indicative of improved magnetic performance.

The mechanical testing included pencil hardness, direct impact resistance, Arco microknife adhesion, Tabor abrasion wear, and gloss and these were performed as described previously[1,6].

RESULTS AND DISCUSSION

Table I includes the magnetic test data from similarly prepared magnetic disks varying only in the PVC of the as formulated coating. All efforts were made to attain identical degrees of coating dispersion, coating thicknesses, and surface finishes. Additionally, the amount of missing bits on these disks were similar. The data in Table I represent the average of at least three surfaces for each level of PVC. As can be seen, the magnetic performance increases proportionally to the PVC as expressed by improved alpha values with increasing PVC. Old information appears to be the most affected by increases in the PVC.

Table I.   Old Information (IO), Squeeze (SQ), and Alpha
           Values Versus Pigment Volume Concentration (PVC).

| | #ID | | | *MD | | | +OD | | |
PVC	OI	SQ	Alpha	OI	SQ	Alpha	OI	SQ	Alpha
20	119	581	16.9	152	557	21.4	170	565	23.0
30	134	589	18.6	164	561	22.6	179	573	23.8
35	141	582	19.6	167	561	23.0	182	575	24.1

#ID=Inner Diameter   *ML Mid Diameter   +OD=Outer Diameter

Figures 1-4 show the results of the mechanical tests on coatings prepared with six different formulated PVCs. As with the coatings prepared for magnetic testing, the degree of dispersion as measured by a Hegmen gauge was held constant. All coatings were applied to standard, alodined aluminum paint panels using a motor driven doctor blade device with a 2 mil bird applicator. The results shown in the figures represent the mean of at least five tests per PVC.

Clearly, there is a dramatic change in the coating properties beyond a PVC of 33-35. All of the tests show a consistent result indicating that the CPVC is from 33-35. Beyond this value, air is contained in the coating, yielding a coating of decreased gloss and considerably decreased mechanical performance. From a strictly mechanical point of view, the CPVC is a clear limiting factor in magnetic disk performance.

However, at least from a theoretical point of view, the CPVC is also a limiting factor in the ultimate magnetic performance of the disk as reflected in signal to noise and mag-

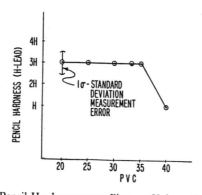

Figure 1. Pencil Hardness versus Pigment Volume Concentration.

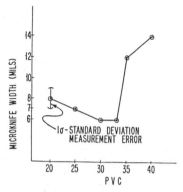

Figure 2. Direct Impact Resistance versus Pigment Volume Concentration.

netic saturation moment performance. As noted in the Introduction, the signal to noise is proportional to the packing density of the magnetic particles in particulate disks[7]. This suggests that increasing the PVC of a magnetic disk coating is always a proper way to improve over-all magnetic performance. The alpha data in Table I support this theory since the magnetic performance goes linearly with the PVC. However, the disks tested for magnetic performance were all formulated from coatings near or below their CPVC, that is where the packing fraction and PVC represent the same value. However, above the CPVC, the PVC and the packing fraction represent different values since included with the binder and magnetic particle is also air above the CPVC. Therefore, above the CPVC the packing fraction of magnetic particles in the coating can not be increased even though one adds more of the magnetic particles into the formulation. A similar argument has been made by Koster and Arnoldussen[7]. They showed that the packing fraction fails to line-arly increase above the mechanically determined CPVC. Similarly, Rasenberg and

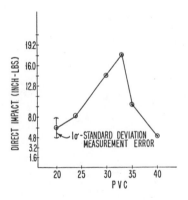

Figure 3. Adhesion Groove Width versus Pigment Volume Concentration.

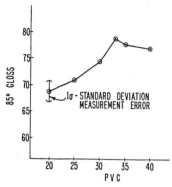

Figure 4. Gloss versus Pigment Volume Concentration.

Huisman have shown that "above the CPVC the magnetic saturation moment (of particulate magnetic tapes) stays constant indicating a constant pigment volume concentration per unit of coating volume"[8]

Therefore, it is clear that the ultimate mechanical and magnetic properties of particulate disk coatings are inherently limited by the CPVC of the given coating. Efforts concentrating towards increasing the CPVC of currently employed magnetic disk coatings via both chemical and mechanical means are a clear avenue towards ultimate properties from particulate magnetic disks.

REFERENCES

1. H.L. Dickstein and W.H. Dickstein, U.S. Patent 4 397 751, 1983 .

2. J.C. Mallinson, in "Magnetic Recording," C.D. Mee, Editor, Ch. 5, McGraw-Hill, New York, 1987.

3. J.C. Mallinson, "The Foundations of Magnetic Recording," Academic Press, San Diego, 1987.

4. W.K. Asbeck and M. VanLoo, Ind. Eng. Chem., $\underline{41}$, 1470, (1949).

5. T.C. Patton, "Paint Flow and Pigment Dispersion," Wiley, New York, 1979.

6. G.G. Sward, "Paint Test Manual," 13th Edition, American Society for Testing and Materials, Philadelphia, Pennsylvania, 1972.

7. K. Koster and T.C. Arnoldussent, in "Magnetic Recording," C.D. Mee, Editor, Ch. 3., McGraw-Hill, New York, 1987.

8. C.J.F.M. Rasenberg and H.F. Huisman, IEEE Transactions on Magnetics. MAG-20 $\underline{5}$, 748, (1984).

VOLATILIZATION MODEL FOR POLYPERFLUOROETHER LUBRICANTS

Steven H. Dillman*, R. Bruce Prime, and Roy B. Hannan

IBM General Products Division
San Jose, CA 95193

The volatilities of two homologous polyperfluoroether lubricants were characterized in a series of isothermal weight loss measurements between 150° and 250°C. A model was derived, based on mass transfer and thermodynamic considerations, which describes the evaporation of polydisperse polymeric liquids with time, temperature, and volume/surface ratio. The model describes the TGA data exceptionally well, and predicts the change in molecular weight distribution of the lubricant with volatility. The model is used to extrapolate the TGA volatility data to operating conditions typical of rigid disk storage devices.

INTRODUCTION

Polymeric liquids are often utilized as disk lubricants in rigid disk storage devices, to minimize friction and wear at the recording head/magnetic disk interface.[1] In some applications, these devices run almost continuously for several years, with internal temperatures approaching 60°C. Lubricant film volume/surface area ratios are on the order of tens of $\mu\ell/m^2$. Under these conditions, potential loss of lubricant to evaporation over long periods of time becomes a significant material consideration. Practicality requires that measurements taken at elevated temperatures be extrapolated to use conditions. In doing so, the change in molecular weight distribution which accompanies, and affects, evaporation must be taken into account.

In this paper the volatilities of two homologous polyperfluoroether lubricants were characterized in a series of isothermal weight loss measurements between 150° and 250°C. A simple time-temperature model was explored but found to be inadequate to describe volatility over wide temperature ranges. A more rigorous model was derived, based on mass transfer and thermodynamic considerations, to extrapolate TGA results (high temperature-high volume/area) to conditions of interest to rigid disk

---

*Present address: Shell Development Co., Westhollow Research Center, P.O. Box 1380, Houston, TX 77251.

$CF_3O\text{-}(CF_2O)_{60.3}\text{-}(CF_2CF_2O)_{35.4}\text{-}(CF_2CF_2CF_2O)_{1.4}\text{-}(CF_2CF_2CF_2CF_2O)_{1.0}\text{-}OCF_3$

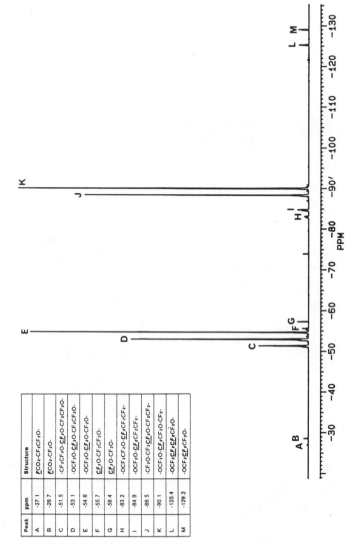

Peak	ppm	Structure
A	-27.1	FCO₂-CF₂CF₂O-
B	-28.7	FCO₂-CF₂O-
C	-51.5	-CF₂CF₂O-CF₂O-CF₂CF₂O-
D	-53.1	-OCF₂O-CF₂O-CF₂CF₂O-
E	-54.8	-OCF₂O-CF₂O-CF₂O-
F	-55.7	CF₂O-CF₂CF₂O-
G	-58.4	CF₂O-CF₂O-
H	-83.2	-OCF₂CF₂O-CF₂CF₂CF₂-
I	-84.9	-OCF₂O-CF₂CF₂CF₂-
J	-88.5	-CF₂O-CF₂CF₂O-CF₂CF₂-
K	-90.1	-OCF₂O-CF₂CF₂O-CF₂-
L	-125.4	-OCF₂CF₂CF₂O-
M	-129.2	-OCF₂CF₂CF₂O-

Figure 1. ^{19}F-NMR spectrum of lubricant A.

files (low temperature-low volume/area). In addition, the model predicts the change in molecular weight distribution of the lubricant with volatilization. This approach to accelerated characterization of the slow evaporation of polydisperse liquids is quite general and can be useful in many applications of polymers. It does not, however, take into account interfacial effects, which come into play with very thin films of liquid.

EXPERIMENTAL

Materials

The polyperfluoroether lubricants studied are random copolymers with the structure shown below. They are homologues, i.e. they differ only in molecular weight distribution. Figure 1 shows the ^{19}F NMR spectrum of lubricant A.   Copolymer ratios and M_n were determined by ^{19}F NMR. Polydispersity was determined by GPC using a viscosimetric detector and Mark-Houwink constants. Viscosities reported are at 1500 sec^{-1}, however these fluids are essentially Newtonian at shear rates up to 10^4 $sec.^{-1}$ The density of both lubricants is 1.9 g/cc. Surface tensions of both liquids were determined to be 19.5 ± 0.5 dynes/cm at 23°C by capillary rise. The refractive indices, n_d(20°C, λ = 632.8 nm), are 1.290 and 1.294 for lubricants A and B, respectively. Comparative molecular weight distributions are shown in Figure 2.

$$(CF_2O)_a - (CF_2CF_2O)_b - (CF_2CF_2CF_2O)_c - (CF_2CF_2CF_2CF_2O)_d$$

	a	b	c	d	M_n	M_w/M_n	$\eta(25°C)$
Lubricant A	60	35	1.4	1.0	8,600	2.0	240 cps
Lubricant B	88	52	2.1	1.5	13,000	1.2	400 cps

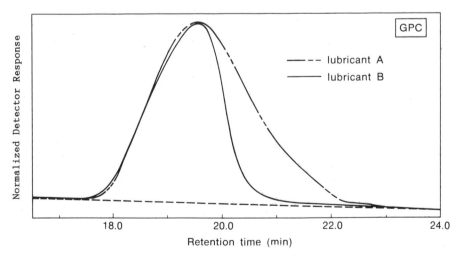

Figure 2. GPC molecular weight distributions. Longer retention times correspond to lower molecular weights. Traces are normalized to 100% at the node.

Methods

Experiments were conducted to investigate the effects of three variables on volatilization: temperature, volume/surface area ratio, and purge rate. Platinum sample pans were used with a diameter of 5.6 mm. Evaporation measurements were performed isothermally, under nitrogen in a Perkin-Elmer TGS-2 thermogravimetric analyzer interfaced to an Omnitherm 35053 Data System. Temperatures of 150, 175, 200, 225, and 250°C were held for periods ranging from 15 to 65 hours. Under these conditions volatility is the only significant mode of weight loss, i.e. the lubricants do not degrade. For these studies sample weights and nitrogen flow rate were held constant at 5.0 ± 0.1 mg. and 100 ± 5 me/min. Because the weight loss of lubricant B was small (<4% after 65 hours at 250°C), all further experiments were performed using lubricant A only. The data in the above table and Figure 2 illustrate the effect of molecular weight on volatility. The effect of volume/area was investigated using samples having initial weights of 2.5, 5, 10, 15 and 20 mg at 250°C for 15 hours. Purge rate was found to have no effect on evaporation at 250°C over the range of 10 to 100 ml/min.; all subsequent data were collected at a purge rate of 100 ± ml/min. GPC measurements of lubricant A were made before and after evaporation at 250±C, to determine the effect of volatilization on molecular weight distribution.

VOLATILIZATION MODEL

The TGA sample is modeled as a liquid vaporizing and diffusing through a cylindrical tube of length z, at the top of which it is carried away by a purge stream (Figure 3). Two phenomena are considered: mass transfer from the lubricant surface to the purge stream (as well as from the liquid bulk to the surface), and the thermodynamics of the phase change at the liquid/vapor interface. There are no significant barriers to heat transfer in the TGA oven, and the evaporation rate of the material is very low, so heat transfer effects are considered negligible.

Mass Transfer

The material is taken as a mixture of components each representing a single molecular weight species. The distance, z, from the interface to

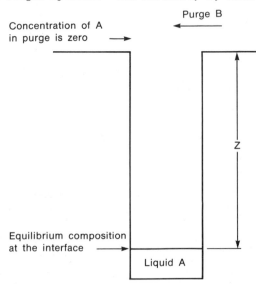

Figure 3. Physical description of the volatilization model.

the purge is constant. As approximations, the vapor and purge are assumed to behave as ideal gases, concentration of the vapor at the top of the tube is assumed to be zero, and the vapor pressure of the lubricant is assumed to be small compared to ambient pressure. Under these conditions the molar flux, N, of component i out of the liquid is given by:[2]

$$N_i = (y_i \times p \times D_i)/(RT \times z) \tag{1}$$

where: y_i = mole fraction of component i at the interface

p = system pressure

D_i = diffusivity of component i in the purge

R = gas constant

T = temperature

z = diffusion distance

The effects of temperature and molecular weight on diffusivity are modeled using the correlation of Fuller[3] which is based on kinetic theory of rigid spheres with an empirically corrected temperature dependence:

$$D_i = C \times T^{1.75} \times (1/M_i + 1/M_p)^{1/2}/[p(V_i^{1/3} + V_p^{1/3})^2] \tag{2}$$

where: M_i = molecular weight of component i

M_p = molecular weight of purge

V_i = collision volume of component i

V_p = collision volume of purge

C = a constant

p = system pressure, as before

Two simplifications may be made to this expression based upon the size of the lubricant molecules. First, since the molecular weight of any component of the lubricant is much greater than that of the nitrogen purge, the expression

$$(1/M_i + 1/M_p) \simeq 1/M_p$$

and may be incorporated into the constant C. Second, since

$$V_i \gg V_p$$

the V_p term may be assumed negligible. Furthermore, the collision volume may be assumed to be proportional to molecular weight for the lubricant. Thus, we obtain an expression of the form:

$$D_i = (C \times T^{7/4})/(p \times M_i^{2/3}) \tag{3}$$

Combining Equations (1) and (3) gives:

$$N_i = (y_i \times C \times T^{3/4})/(z \times M_i^{2/3}) \tag{4}$$

where: M_i = molecular weight of component i

C = a constant

z = diffusion distance, as before

The liquid is assumed to be at uniform composition at all times. This should be a reasonable assumption provided the liquid layer is thin and the rate of evaporation is slow. For 5 mg of lubricant in a TGA pan (ρ = 1.9 g/cc, d = 5.6 mm) the liquid thickness is ~0.1 mm, which should be small enough that concentration gradients may be neglected.

Thermodynamics

Except in the case of very high mass transfer rates, the vapor and liquid phases may be assumed to be at equilibrium at the interface. The agreement between the model and the data, shown below, supports this assumption where typical gas velocities in the TGA are low, on the order of 1 cm/sec. The defining equation for phase equilibrium is:[4]

$$f_i^V = f_i^l \tag{5}$$

where: f_i^V = fugacity of component i in the vapor phase

f_i^l = fugacity of component i in the liquid phase

by definition:

$$f_i^V = \phi_i \times y_i \times p \tag{6}$$

$$f_i^l = \gamma_i \times x_i \times f_i^{ss} \tag{7}$$

where:

$$f_i^{ss} = \phi_i^{sat} \times p_i^{sat} \times \exp[v_i \times (p-p_i^{sat})/RT] \tag{8}$$

ϕ_i = the vapor phase fugacity coefficient

γ_i = the liquid phase activity coefficient

ϕ_i^{sat} = the vapor phase fugacity coefficient of pure i at saturation pressure

p_i^{sat} = saturation pressure of pure i at temperature T

v_i = the average molar volume of pure i over the pressure range p to p_i^{sat}

Both the activity coefficient and the fugacity coefficient are measures of non-ideality, ϕ_i = 1 corresponding to ideal gas, γ_i = 1 corresponding to an ideal solution. Several assumptions must be made at this point. First, the liquid is assumed to be an ideal solution. This should introduce little error since the environment seen by any molecular weight component is nearly identical to what it would see in the pure state due to the similarity between molecules. Second, the liquid densities of all components are assumed equal, which should also introduce very little error since liquid densities for a homologous series are typically quite similar. Third, the saturation pressure is assumed to follow the following relation:

$$\ln(P_i^{sat}) = A + DM_i + (E + FM_i)/T \tag{9}$$

where A, D, E, and F are constants. Equation (9) is a modification of the Clausius-Clapeyron equation which accounts for the effect of molecular weight on vapor pressure. Finally, the ratio ϕ_i^{sat}/ϕ_i is assumed to be constant. While this may not be strictly true, changes in vapor pressure due to molecular weight should be far more significant than changes in fugacity coefficients. Therefore this approximation should introduce only a small amount of error. When Equations (6) - (9) are combined, the following is obtained for the partial pressure of component i:

$$y_i p = C \times x_i \times \exp[E/T + (D + (F + p/(\rho R))/T) \times M_i] \qquad (10)$$

where C is a constant incorporating ϕ_i^{sat}/ϕ_i and A

Finally, Equation (10) may be combined with Equation (4) to give:

$$N_i = C' \times x_i \times M_i^{-2/3} \times \exp[B \times M_i] \qquad (11)$$

where:

$$C' = (C \times T^{3/4})/(z \times \exp[E/T]) \qquad (12)$$

$$B = D + [F + (p/(\rho \times R))]/T \qquad (13)$$

where: x = molar molecular weight distribution function
ρ = liquid density

In terms of mass flux:

$$n_i = C' \times x_i \times M_i^{1/3} \times \exp[B \times M_i] \qquad (14)$$

The total mass flux measured by the TGA may be evaluated as the sum of the mass fluxes for the individual components, or, in the case of a continuous distribution, as the following integral:

$$n = C' \int_0^\infty x'(M) \times M^{1/3} \times \exp[BM] dM \qquad (15)$$

where x'(M) is defined by:

$$\int_{M_a}^{M_b} x'(M) dM = \text{mole fraction of sample with molecular weight M,}$$

such that $M_a < M < M_b$

The volatility model may be used to predict a weight loss curve at any temperature given an initial molecular weight distribution, by determining B and C' from a series of isothermal TGA measurements, as demonstrated in Figure 5, and integrating Equation (15) as a function of time to obtain weight loss.

RESULTS AND DISCUSSION

Figure 4a shows TGA weight loss curves for varying amounts of lubricant A at 250°C. A simple superposition approach was used to extrapolate from TGA volume/area ratios to the much lower values representative of a rigid storage disk. The curves were shifted to a chosen volume/area ratio using Equation (16) which assumes proportionality between the rate of weight loss (expressed in weight percent) and the area/volume ratio.

$$t_2 = t_1 \times w_2/w_1 \times A_1/A_2 \qquad (16)$$

where: t = time to a given percent weight loss
w = initial lubricant weight
A = lubricant area

Figure 4b shows the data shifted to a volume/surface area of 50 μℓ/m², typical for a rigid disk file.

TGA data for lubricant A at five temperatures were fit to the volatility model using a non-linear least squares program to obtain values for B and C' at each temperature. B is plotted versus 1/T in Figure 5a and $\ln(C'/T^{3/4})$ versus 1/T in Figure 5b. These plots follow the linear behavior predicted by the model. Values for B and C' were then taken at each temperature from the least squares regression and the model used to predict the weight loss curves at each temperature. These model predictions are shown in Figure 6 along with the actual TGA data. The change in molecular weight distribution of lubricant A projected by the model to accompany

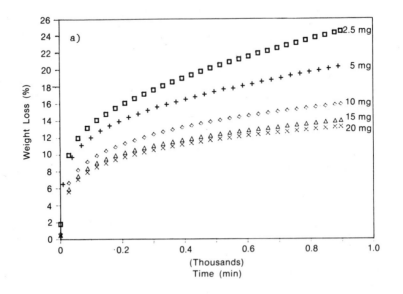

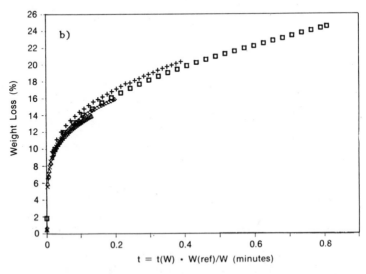

Figure 4. a) Measured TGA weight loss curves for lubricant A at indicated initial sample weights. 15 hours in nitrogen at 250°C.   b) Master weight loss curve predicted from Equation (16) for weight loss at 250°C and a volume/area ratio of 50 μℓ/m². Reference weight = 2.2μg.

evaporation for 15 hours at 250°C is shown in Figure 7, along with the actual GPC measurement. In all cases agreement is excellent.

Projected weight loss curves for lubricant A at 60°C and a volume/surface area of 50 $\mu\ell/m^2$ (conditions of interest to rigid disk files), are shown in Figure 8. Under these same conditions, lubricant B is projected to show negligible weight loss. A simple time-temperature superposition approach[5], which assumes a constant heat of vaporization (molecular weight changes are not accounted for and thus the shape of the

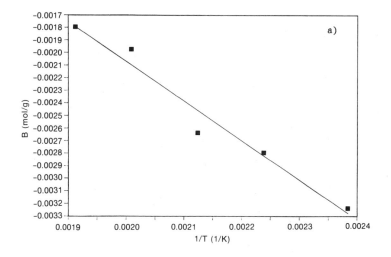

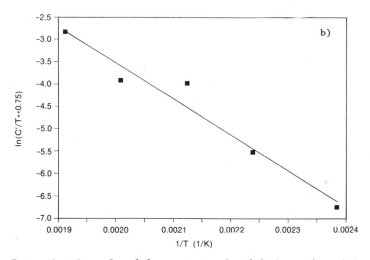

Figure 5. Determination of model constants for lubricant A. a) B vs. 1/T. b) $\ln(C'/T^{3/4})$ vs. 1/T.

volatilization curve is assumed to be independent of temperature), under-
estimates the initial evaporation rate and overestimates the later portion
of the curve, as expected. Such differences can be significant considerations in the selection of an appropriate lubricant. It must be noted that
a number of variables have not been included in this model, most notably
interfacial interaction of the lubricant with the disk surface, which would
retard volatilization, and the case of turbulent air flow (air flow in a
rotating disk file varies from static to turbulent) which would enhance
volatilization. Nonetheless, the model should provide a qualitative description of volatilization in a rigid disk storage device.

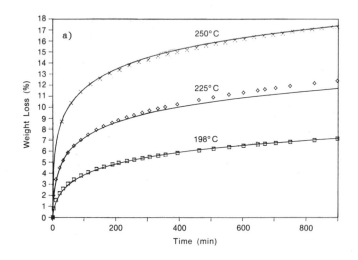

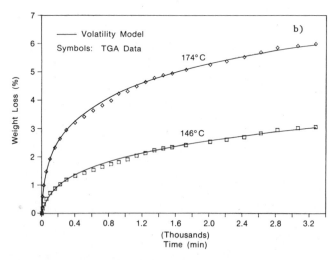

Figure 6. Model predictions and measured weight loss curves for lubricant
A in nitrogen. a) 15 hours at temperature. b) 65 hours at temperature.

SUMMARY

A model was derived, based on mass transfer and thermodynamic considerations, to project volatility of a liquid lubricant from TGA measurements to conditions representative of rigid disk files. The model was shown to describe very well TGA data which varied in volume/area ratio and temperature. The model was able to describe the changes in molecular weight distribution which accompany volatilization, validating the assumptions used in development of the model. Molecular weight distribution of the

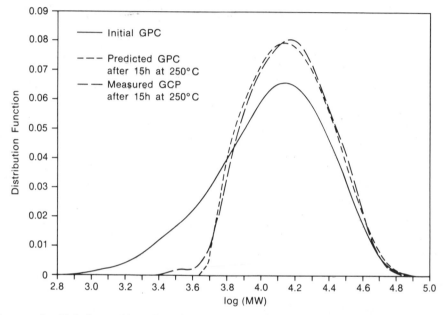

Figure 7. Model predicted and measured molecular weight distribution of lubricant A after 15 hours at 250°C in nitrogen, with initial molecular weight distribution for reference.

lubricant was shown to be a major factor influencing its volatility. The model is believed to give a more accurate projection of evaporation of a polydisperse, high molecular weight liquid than does a simple time-temperature model, which fails to account for the changing shape of the volatilization curve with temperature.

ACKNOWLEDGMENTS

All thermal analysis measurements were very ably conducted by Ms. Connie Moy. The NMR data were kindly provided by Dr. John M. Burns. GPC data were contributed by Ms. Grace K. Cheung and Dr. Manfred J. Cantow. Surface tension data were provided by Dr. Charles J. Mastrangelo, and viscosity data by Dr. Saswati S. Datta.

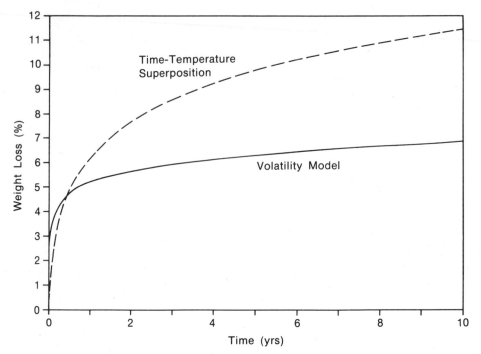

Figure 8. Model predicted and time-temperature superposition predicted weight loss curves for lubricant A at 60°C and 50 μℓ/m² volume/area ratio.

REFERENCES

1. J. M. Harker, D. W. Brede, R. E. Pattison, G. R. Santana, and L. G. Taft, IBM J. Res. Develop. 25(5), 677 (1981).

2. R. B. Bird, W. E. Stewart and E. N. Lightfoot, "Transport Phenomena," Wiley, New York, 1980.

3. R. H. Perry and C. H. Chilton, "Chemical Engineering Handbook," 5th Edition., pp. 3-230 McGraw Hill, New York, 1973.

4. J. M. Smith and H. C. Van Ness, "Introduction to Chemical Engineering Thermodynamics," 3rd Edition, McGraw Hill, New York, 1975.

5. R. B. Prime, Proceedings 14th NATAS Conference, 137 (1985). San Francisco, CA. Available from A-1 Business Service, 219 Park Avenue, Scotch Plains, NJ 07076.

PART V.   PHYSICOCHEMICAL ASPECTS OF MAGNETIC RECORDING

CHARACTERIZATION AND HYDROLYSIS OF MAGNETIC TAPES

T.N. Bowmer, G. Hull and I.M. Plitz

Bellcore Red Bank N.J.

07701-7020

We report the chemical, physical and magnetic properties of a series of virgin tapes and tapes aged in environmental chambers under corrosive and hot/humid conditions. All the tapes were stable under normal operating conditions, but the CrO_2-based tapes deteriorated markedly at high temperature, high humidity or under acidic conditions. Acid-catalysed hydrolysis of the binder and thermally-induced physical disordering of the magnetic particle array were found to cause magnetic property losses of 20-50%. We describe a novel thermogravimetric technique to measure magnetic transitions as well as compositional and thermal stability information.

INTRODUCTION

Magnetic tapes are used to store extensive amounts of information on telephone calls, billing records and other details germane to operating a telecommunications company. Since these records are kept for at least 10 years, long term stability is required under environmental stresses such as indoor pollutants, heat, humidity and other hazards that can threaten the magnetically stored information.

A typical magnetic tape consists of a polymeric base tape coated with an elastomeric binder containing anisotropic particles of magnetic iron oxide (Fe_2O_3) or chromium dioxide (CrO_2). If the magnetic particles have high anisotropies, i.e., needle-like shapes, then a high density of information can be stored on a tape. CrO_2 particles can be prepared with much higher anisotropies than Fe_2O_3 particles and, therefore, are favored for applications that demand high storage densities (e.g., video tapes)[1]. The particles are usually coated on a polymeric substrate with random orientation and then aligned by a magnetic field with their long axis parallel to the tape direction (longitudinal magnetization). Lubricants, dispersants, curative agents or other additives may be present for achieving the desired chemical and mechanical properties[1-6]. Polyethylene terephthalate substrates and polyester-urethane binders are typical of the materials currently used in magnetic tapes. These materials are susceptible to hydrolysis[3,4,7-11] and conditions as mild as 50°C and 50% relative humidity are reported to significantly affect binder materials[3,7].

In this study magnetic tapes are characterized by thermal analysis, solvent extraction, infrared spectroscopy, and magneto-hysteresis. A novel thermogravimetric technique to measure magnetic properties as well as chemical composition and thermal stability is described. We report the chemical, physical and magnetic properties of a series of virgin tapes and tapes aged in environmental chambers under a wide range of corrosive and hot/humid conditions.

EXPERIMENTAL

Tapes from seven different manufacturers were examined with a poly(ethylene terephthalate), PET, film from Martin Corp. used as reference material. Tapes C1-4,C6 and C7 had chromium dioxide as the magnetic pigment while tape F5 had iron oxide. The chemical stability of the tapes was examined by placing one gram of the tape in a flask with 100 ml of aqueous and organic solutions and stirring at room temperature for up to 4 days. The acidic and basic solutions selected cover the pH conditions expected from water damage during a flood or roof leak. The seven organic solvents tested varied in molecular size, polarities and halogen content. Four Freon® cleaners (MS-220, MS-200, MS-180 and MS-165) from Miller-Stephenson Chemical company designed for electronic applications were tested in addition to a typical fluorocarbon (Halon 1211) used in fire extinguishers. After 48 hours in tetrahydrofuran (THF), the soluble fractions were collected and analyzed by Infrared spectroscopy and Size Exclusion Chromatography (SEC).

Infrared (IR) spectra of the tapes (before extraction) were obtained on a Nicolet 10DX Fourier Transform Infrared Spectrometer with a Harrick Attenuated Total Reflectance accessory (Model TMP 220). The soluble fractions from the THF extractions were cast into films for infrared analysis.

A Perkin-Elmer TGS-2/System-4 scanning at $20°C-min^{-1}$ between 50°C and 900°C was used for magneto-thermogravimetric analysis (magneto-TGA). A U-shaped magnet was placed outside the TGA furnace tube producing a magnetic field of 150 Gauss at the sample position. Therefore the measured weight equaled the sample mass (M) plus a magnetic weight (W_m). For reference, a 1 mg sample of iron (Fe) had a W_m of 5.6 mg in this apparatus. For the tapes, W_m was initially 5-10% of the tape mass, but was as high as 60% of the sample mass after cooling from 150-200°C in the magnetic field. In this apparatus a Curie transition was observed as a loss in weight, and a shift in the magnetic particle alignment was observed as an increase or decrease in weight, depending on whether the magnetic particles aligned with the field or against it. The initial magnetic weight was independent of tape orientation in the TGA sample pan. Changes in the magnetic weights were measured during thermal cycling between 50°C and 175°C with the magnetic field both on and off during the cooling step.

Differential Scanning Calorimetry (DSC) profiles were obtained on a Perkin-Elmer DSC-4 system using 6-12 mg tape samples. Two consecutive scans between -60 and 300°C were performed at $10°C\text{-min}^{-1}$ in a nitrogen atmosphere.

The tapes were placed under 1.2 MPa tensile stress along the direction of the tape and their stress/strain behavior measured between -60°C and +220°C with a Perkin-Elmer thermomechanical analysis TMA-2 system. The tapes were stored and measured at $23\pm2°C$ and $50\pm10\%$ relative humidity. The stress relaxation (S_R) at a given temperature (T_1) is defined as

$$S_R = (L_0 - L_1)/L_0$$

where L_0 = initial sample length = 9.0 mm
L_1 = sample length at T_1.

Magneto-hysteresis experiments were performed on a LDJ Electronics Model 9500-VSM Magnetometer. From the hysteresis loop generated by measuring magnetization as a function of applied field, one can determine the coercive force, saturation magnetization and remanent magnetization for a tape (Figure 1). The ideal shape is a square-shaped loop which implies rapid switching and stable information storage.

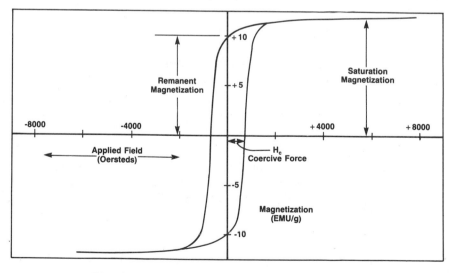

Fig. 1 Typical Magneto-Hysteresis Loop

Aging studies were performed (a) by thermal cycles in the magneto-TGA apparatus, (b) in a flow-through corrosion chamber held at 30°C and 70% relative humidity (RH) with hydrogen sulfide (10ppb), chlorine (10ppb) and nitrous oxide (200ppb) present in the airflow[12], and (c) in temperature/humidity chambers with conditions between 20-100°C and 0-100% RH. The composition, physical and magnetic properties of the tapes before and after aging were examined by the techniques outlined above.

CHARACTERIZATION

Chemical and Physical Properties

All the tapes had a shiny side and a dull side. The infrared reflectance spectra obtained from the dull sides showed the characteristic resonances of poly(ethylene terephthalate), PET. X-ray EDAX analysis of the shiny surface revealed predominantly iron, chromium and oxygen.

The thermal transitions, their quantitative characteristics and compositional information from the DSC experiments are summarized in Table I. DSC profiles obtained for the tapes and for PET showed glass transitions (T_g) at 80-85°C and melting peaks at 248-255°C during the first scan. In the second scan, after cooling from 275°C at 10°C-min^{-1}, the glass transitions remained at 80-85°C while the melting transitions (T_m) decreased by 5-10°C and decreased in magnitude (ΔH_f) by 10-20%. An increase in percent crystallinity will increase ΔH_f, while an increase in size and/or perfection of individual crystallites will increase T_m [13]. The tapes were oriented and stretched during manufacture resulting in stress-induced crystallization[14] that increased ΔH_f and T_m above their equilibrium values. The percentage of PET was determined from the second scan after the 10°C-min^{-1} cooling process had removed previous thermal history. A comparison between the reference PET film and the tapes was then possible[13,15].

Table I. Calorimetry Results.

TAPE	T_g(°C)	T_m(°C) 1st scan	T_m(°C) 2nd scan	ΔH_f (cal/g) 1st	ΔH_f (cal/g) 2nd	Percent PET
C1	82	254	245	12	9.4	78
C2	80	252	245	10.5	10	80
C3	79	255	249	9.8	8	67
C4	83	253	245	9.2	9.8	80
F5	80	255	251	11	10	80
C6	85	253	248	10	8	67
C7	82	254	248	11	9	75
PET	83	255	252	13	12	100

Table II. Solvent Effects on Magnetic Tapes.

SOLVENT	TAPE							PET
	C1	C2	C3	C4	F5	C6	C7	
Water	00	00	00	00	00	00	00	00
Acid(pH 4)	00	00	00	00	00	00	00	00
Base(pH 10)	00	00	00	00	00	00	00	00
Methanol	00	00	00	00	00	00	00	00
Hexane	00	00	00	00	00	00	00	00
Tetrahydrofuran	01	22	22	00	00	00	22	00
Carbon Tetrachloride	01	01	01	01	01	01	01	00
Chloroform	22	22	12	00	00	00	22	00
Acetone	00	00	00	00	00	00	00	00
Toluene	00	00	00	00	00	00	00	00
MS-220	00	00	00	00	00	00	00	00
MS-200	00	00	00	00	00	00	00	00
MS-180	00	00	00	00	00	00	00	00
MS-165	00	00	00	00	00	00	00	00
Halon 1211	01	01	01	01	01	01	01	00

After solvent extraction the tapes were examined for changes in surface reflectivity, shrinkage/swelling, and the adhesion strength between the coating and the PET substrate. The results are shown in Table II where a 0-value means no change, a 1-value means a visible change like a rough and/or dull surface, and a 2-value denotes physical degradation observed in the magnetic coating. The first digit refers to short exposure (5-15 mins) and the second digit refers to long term exposure of greater than one day. Therefore 00 means little or no effect, 01 means no effect initially but deterioration on longer exposure, and 22 means extensive deterioration on either long or short exposures. All these solvents and solutions did not affect the PET substrate.

The standard solvents and cleaners used for electronic equipment present no danger to the magnetic coated tapes as seen from the results for the MS-220, MS-200, MS-180 and MS-165 cleaners. Halon 1211, carbon tetrachloride and acetone produced surface dullness or roughness only after prolonged intense exposure. The magnetic properties of these exposed tapes were found to be the same ($\pm 5\%$) as virgin tapes. Tetrahydrofuran and chloroform were the only two solvents tested that degraded the adhesion enough to cause the coating to peel away from the PET substrate.

Table III summarizes the effect on the tapes of a 48 hour, room temperature extraction in THF. Small amounts (0.2-2.5%) were extracted from all the tapes. The major extractables were low molecular weight binder components with typical M_n's of 25,000-75,000 g/mole. The physical effect of the extraction varied considerably from tape to tape. Tapes C2, C3, and C7 had their magnetic coatings stripped off leaving a bare PET transparent film and metal oxide particles/strips; while tapes C4, F5, C6 and the PET reference film showed no visible changes, and C1 only lost a few isolated flakes of coating. The more the material extracted, as determined by weight loss and SEC peak areas, the greater was the physical damage to the coating.

Infrared analysis showed that the extracts were all poly(ester urethane)s derived from a 4,4′-diphenylmethane diisocyanate and a low molecular weight polyester. The infrared spectra of the extractables were similar for all tapes except for the resonance at 3320 cm^{-1} that is characteristic of >N-H bond stretching. The peak height of the >N-H resonance at 3320 cm^{-1} divided by the peak height of the C-H resonance at 2950 cm^{-1} was defined as the relative concentration of amines in the extracted fraction. A typical polyester-urethane contains hard segments that provide mechanical strength and soft segments that provide elasticity. The urethane (source of amines) is predominately in the hard segments and the polyester is mostly in the soft segments. Coatings that were easily degraded by THF, tended to have higher concentrations of urethane fragments (amines) in their soluble fractions (Table III). Since we do not know the formulations of the uncured binders, it is not possible to determine whether curing conditions or chemical structure differences are responsible for the different solvent resistance of these commercial tapes.

The PET substrate dominated the mechanical behavior of the magnetic tape. Thermomechanical analysis showed flat baselines at low temperatures that are typical of glassy or semi-crystalline polymers below their glass transitions. Under stress, the tapes extended between 20-40°C and reached a maximum strain of 0.002 at 70-80°C. As the tape passed through the PET glass transition (80-90°C), contraction began and was attributed to stress relaxation. Internal stresses are produced in PET during manufacture of the tape and these stresses relax rapidly if the PET is heated above its glass transition[13]. Above 140-150°C, S_R values increased rapidly reaching a maximum of 0.020-0.025 at 220°C for all the tapes. At high temperatures (>220°C), the PET crystallites began to melt, the tapes elongated and fracture occurred at 230-240°C.

Table III. THF Extraction Analysis.

TAPE	% sol by wt.	SEC Area	M_n x 10^{-3}	Comments on magnetic coating	Amine Conc. (relative)
PET	0	0	-	no visible changes	0
C4	0.2-0.5	50	36	no visible changes	0.2
C1	0.2-0.5	150	32	isolated flaking	0.3
C6	1.0	45	48	no visible changes	0.2
F5	1.0	160	54	no visible changes	0.6
C3	1.5	200	24	peeled off in strips	0.5
C7	2.5	220	65	peeled off in strips	0.5
C2	2-3	240	71	stripped as fine particles	0.7

Magnetic Properties

The magneto-hysteresis studies showed that all the tapes had the desired properties for use in standard magnetic tape drives[16]. That is, (a) the hysteresis loops showed ratios of remanent-to-saturation magnetization between 0.82-0.86 (the ideal square loop would have a value of 1.0), (b) the coercive forces ranged from 520 to 690 Oersteds, (c) remanent magnetizations varied from 8.7 emu/g to 12.7 emu/g, and (d) saturation magnetization ranged from 10.9 emu/g to 14.8 emu/g.

Representative magneto-TGA profiles of the magnetic tapes are shown in Figure 2. The initial W_m/M values ranged from 0.068 for C7 to 0.092 for F5. This initial W_m/M is an average over all particles and all tracks of a tape and was independent of the tape's orientation in the TGA sample pan. The magnetic particles in the virgin tapes are, on average, randomly oriented. This was confirmed by the magneto-hysteresis experiments where no initial magnetization was found for zero applied field. Information on individual particles or tracks on the tape is not available from these TGA measurements. When the temperature reached 70-100°C, the magnetic particles moved under the influence of the magnetic field and the W_m values increased. The onset temperature and the peak magnitude (W_{mp}) for this transition is a measure of the mobility of the magnetic particles. W_{mp} is defined as the maximum magnetic weight during this initial TGA scan. If the temperature was held constant at 90-100°C, then this alignment process would continue until the particles were aligned with the magnetic field. The Curie temperature, where all magnetic properties are lost, is 125 ± 5°C for CrO_2 [17,18]. Since the temperature increased at 20°C-min^{-1} during the TGA scan, a competition between particle alignment and the Curie transition was observed. The former increased W_m while the latter reduced W_m. The magnetic properties of the coating decreased above 100°C as the Curie transition dominated and W_m was reduced to zero. No other magnetic transitions were observed for the CrO_2-based tapes. The magneto-TGA profile of the Fe_2O_3-based tape showed no magnetic weight losses until the temperature approached the Curie temperature for iron oxide (585°C)[17,18].

At 380-400°C, the polymer base resin degraded to volatile products (80% acetaldehyde) and a carbon char (13% of initial PET weight)[11,13,19]. Addition of oxygen into the sample volatilized the char when the temperature exceeded 500-550°C. The residue at high temperatures (800-900°C) was the stable oxide of the metallic components of the magnetic coating. Above 350°C chromium dioxide was converted to its stable, green, non-magnetic form of Cr_2O_3[1]. The case for iron oxide is more complicated since magnetic FeO is formed above 650°C in nitrogen. Since the Curie temperature for FeO is 825°C, W_m increased as FeO was formed and decreased as its Curie temperature was approached (Figure 2, curve #5). If heated to high temperatures (800-900°C) in air, the stable form of iron oxide produced was Fe_3O_4.

Table IV. Magneto-TGA Analysis

Tape	W_m/M x 100	Magnetic Transition onset (°C)	Magnetic Transition W_{mp}/W_m	Curie Temp. (°C)	900°C Residue (% of M)	Composition PET %	Composition Coating %
C1	7.5	85	4	122-5	18	76	24
C2	8.0	83	3	120-7	16	77	23
C3	8.6	89	4	122-7	20	70	30
C4	6.7	83	4	122-9	14	80	20
F5	9.2	90	1.5-2	585	18	75	25
C6	7.0	92	4	122-6	18	70	30
C7	6.8	91	4	120-7	17	75	25
PET	0	-	-	-	0	100	0

The residual mass and the mass lost during the poly(ethylene terephthalate) degradation steps determined the compositions shown in Table IV. That is, the tapes consisted of 70-80 wt% PET and 20-30 wt% metal oxide magnetic coating. Assuming a density of 1.4 g-cm^{-3} for the PET and 5.2 g-cm^{-3} for the metal oxide, then the tape consisted of a 30-35μm thick PET film coated with a 3-5μm magnetic layer.

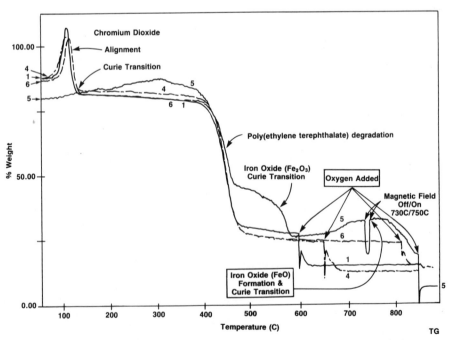

Fig. 2 Representative Magneto-TGA profiles. Numbers adjacent to curves are tape numbers (e.g., profile 4 is from tape C4).

338

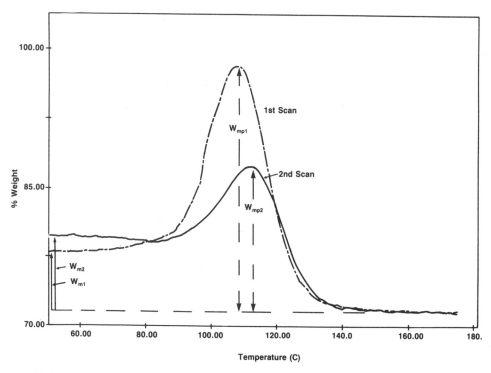

Fig. 3 Representative Magneto-TGA profiles during heating cycles.

AGING STUDIES

A series of cyclic magneto-TGA experiments were performed to examine the thermal durability of the magnetic tapes. A typical experiment is shown in Figure 3 with the results for all the tapes summarized in Table V. The sequence consisted of (i) placing a 1-3 mg (M) sample in the TGA at 50°C with magnetic field in place (W_{m1}), (ii) heating at 20°C-min^{-1} to 175°C (peak maximum = W_{mp1}), (iii) cooling at 20°C-min^{-1} with no magnetic field to 50°C, (iv) reintroducing the magnetic field (W_{m2}), (v) repeating step (ii) measuring (W_{mp2}) and finally (vi) cooling at 20°C-min^{-1} to 50°C with the magnetic field in place that resulted in the maximum magnetic weight observed (W_{m3}).

W_{m1}/M is an average measure of the residual magnetism in the tapes. However, the absolute value of W_{m1} was of limited use since it was highly dependent on the magnetic history of the tape and the geometry of the magnet-TGA furnace-sample apparatus. W_{m2}/M measured the stability of the chemical and magnetic properties after thermal stress. W_{mp1}/M and W_{mp2}/M measured the mobilities of the magnetic particles before and after thermal stress, respectively. The maximum magnetic weight observed (W_{m3}/M) was proportional to the saturation magnetization value found from the magneto-hysteresis experiments.

Table V. Magneto-TGA Thermal Cycle Experiments

TAPE	W_{m1}/M	W_{m2}/M	W_{m3}/M	W_{mp1}/M	W_{mp2}/M
C1	0.075	0.075	0.50	0.28	0.17
C2	0.080	0.070	0.48	0.20	0.18
C3	0.086	0.110	0.59	0.36	0.20
C4	0.070	0.095	0.47	0.25	0.16
F5	0.092	0.095	0.12	0.12	0.12
C6	0.070	0.055	0.44	0.28	0.16
C7	0.068	0.065	0.37	0.19	0.16

Thermal stress does not degrade the average magnetic properties of the particles (Table V). These mild temperatures have not caused the CrO_2 to convert to the thermodynamically more stable form, Cr_2O_3. However, internal stresses in PET substrate relaxed at these temperatures and caused the tape to distort.

Magneto-TGA and magneto-hysteresis experiments were repeated after the tapes were aged in the flow-through corrosion chamber for 17 days which simulated more than ten years in a typical data center office[12]. No changes occurred in physical appearance, overall composition, the Curie transitions, maximum magnetic weights (W_{m3}/M), the thermal degradation onset temperatures or in the magneto-hysteresis coercive forces. There were minor changes (5-10%) in initial magnetic weights and in the magnetic mobilities measured in the magneto-TGA apparatus, as well as 5-10% reductions in the remanent and saturation magnetizations measured in the hysteresis experiments. The capacity of the tapes to store information would not be significantly affected by these small changes.

Aging at 85°C and 85% RH was more severe than the corrosion chamber and resulted in significant changes in all but the Fe_2O_3 tape. These conditions may simulate exceptional occurrences like a nearby fire or a prolonged air-conditioning failure. Although the physical properties and chemical composition of the CrO_2-based tapes were little changed after one month at 85°C and 85%RH, the magnetic properties decreased by 20-50% as shown in Table VI for W_{m3}/M values. Similar results were obtained from the magneto-hysteresis experiments as shown in Table VII. Coercives forces decrease by 5-10%, while remanent and saturation magnetizations decreased by 30-50%. The remanent-to-saturation magnetization ratio remained at 0.8 for all the tapes. The tapes could still store information but with a reduced signal/noise ratio.

It was assumed the magnetic changes in the CrO_2-based tapes arose from either (1) a chemical change in the metal oxide to a non-magnetic state, or (2) a permanent loss in the ordering of the magnetic particles. A color change would be expected if CrO_2 had been converted to a non-magnetic, green chromium-III compound (e.g., Cr_2O_3) or a orange/red chromium-VI compound. A change in

Table VI. W_{m3} Values after Aging in 85°C-85%RH Chamber

TAPE	W_{m3}/M new	aged	percent decrease
C1	0.50	0.25	50
C2	0.48	0.40	20
C3	0.59	0.3	50
C4	0.47	0.27	40
F5	0.12	0.13	0
C6	0.44	0.28	40
C7	0.37	0.19	50

particle shape may also occur with such a change in coordination chemistry. No such large changes in the physical appearance were observed macroscopically or microscopically.

At 85°C and 85%RH, degradation of the adhesive bond between the magnetic particles and the PET substrate could cause disruption of the particle array. Reduced adhesion of the particles would allow relaxation towards a disordered array of particles. High temperatures and humidities can initiate hydrolysis and degradation of the binder[3,7,8,10] and thereby disrupt the ordered array of magnetic particles. On cooling to room temperature, the PET becomes glassy and the particle disorder and any defect sites are frozen into the structure[20]. This would result in decreased magnetic properties and reduced information storage capacity.

Table VII. Magneto-Hysteresis Results
(aged = 1 month in 85°C-85%RH)

TAPE	Coercive Force, H_c (Oersted)		Magnetization Remanent R		Saturation S		Percent change
	new	aged	new	aged	new	aged	(a)
C1	570	520	11.1	5.99	13.6	7.68	-45%
C2	513	487	9.67	6.16	11.2	7.56	-35%
C3	537	515	12.6	7.56	14.8	9.05	-40%
C4	530	500	9.70	5.36	11.5	6.64	-43%
F5	687	681	10.1	9.91	11.9	11.80	- 2%
C6	539	500	11.0	6.23	12.9	7.72	-42%
C7	536	497	8.59	4.74	10.3	5.99	-44%

(a) percent change = average of changes in R and in S.

341

To test the above degradation scheme, aging experiments were performed at a variety of temperatures, humidities and pH values (Table VIII). Higher temperatures, higher humidities and/or more acidic conditions enhanced the deterioration of the magnetic properties and presumably the performance of the tapes (Figure 4). High humidity levels catalysed the degradation, but long-term exposure to temperatures above 60-70°C caused magnetic property losses even in low humidities (<5%RH). Mild acidic conditions (pH 4) lead to 30% losses in magnetic properties, while basic conditions (pH 10) did not affect the tapes noticeably. These results suggest that an acid-catalysed hydrolysis of the binder weakens the adhesive bond between the magnetic particle and the tape, leading to a loss in particle ordering and generation of defect sites. There was no visual, mechanical or calorimetric (DSC) evidence that the PET substrate was affected by these test conditions. Although PET is known to undergo hydrolysis, the conditions used in this study were apparently not harsh enough to cause PET substrate degradation[9].

The CrO_2-based tapes were less stable than the Fe_2O_3-based tape in all of the aging experiments performed. The infrared analysis and the THF extraction results showed the binder used for the iron oxide tape has similar structure and

Table VIII. Aging Results for C2,C3 and F5.

Tape	Conditions			W_{m3}/M after 1 mth	percent loss
	Temp.	%RH	pH		
C2	30	20	-	0.48	0-5
	30	70	-	0.46	0-5
	30	100	4	0.38	20
	30	100	10	0.45	5
	85	<5	-	0.42	15
	85	20	-	0.39	20
	85	85	-	0.40	20
C3	30	20	-	0.64	0-5
	30	70	-	0.63	0-5
	30	100	-	0.62	0-5
	30	100	4	0.44	33
	30	100	7	0.62	0-5
	30	100	10	0.64	0-5
	85	<5	-	0.49	25
	85	20	-	0.38	40
	85	85	-	0.30	55
	100	<5	-	0.35	45
F5	30	<5	4		
	↓	↓	↓	0.13	0-5
	100	100	10		

stability to the binders used in all the chromium dioxide based tapes. If the binders are similar, then either the adhesive bond to iron oxide must be stronger than to chromium dioxide, or the metal oxide catalyses the degradation as was suggested by Bradshaw et.al.[4]. They attributed the enhanced degradation rates to the binder hydrolysis being catalysed by the surface of the chromium dioxide particles.

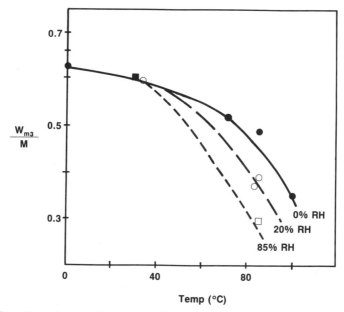

Fig. 4 W_{m3}/M value as function of test temperature after aging tape C3 for one month at 0, 20 and 85% R.H.

SUMMARY

All the tapes examined consisted of a polyethylene terephthalate polymer film with a 3-5 μm magnetic coating of CrO_2 or Fe_2O_3 particles attached to the polymer film with a poly(ester urethane) based on 4,4´ -diphenylmethane diisocyanate and a polyester. We described a novel thermogravimetric technique to measure magnetic transitions as well as compositional and thermal stability information.

Accelerated aging tests predicted that all tapes examined will be stable for greater than 10 years inside a typical data center building. The degradation of CrO_2-based tapes was found to be catalysed by high temperatures, high humidity or acidic conditions. Degradation of the magnetic and physical properties arose from (1) a chemical degradation, i.e., an acid-catalysed hydrolysis of the binder that holds the magnetic particles to the PET substrate, and (2) a thermally-induced physical degradation that disordered the particle array on the tape and created defect sites. A synergy exists between (1) and (2) with high temperatures accelerating the hydrolysis reaction, and degradation of the binder facilitating disruption of the particle array.

ACKNOWLEDGEMENTS

The authors would like to thank R. Schubert, B.T. Reagor, S.M. D'Egidio, C. Fetterman, R.J. Miner and J. Wernick for their assistance and valuable discussions throughout this work.

REFERENCES

1. H. Hibst, *J. Mag. Mag. Mat.*, **74**, 193 (1988)
2. K. Nakamura, K. Monono, H. Kawamura, K. Ota, A. Itoh and C.Hayashi, *IEEE Trans. Mag.*, **MAG-20**, 833 (1984)
3. E.F. Cuddihy, *Org. Coat. Appl. Polym. Sci. Proc.*, **48**, 422 (1983)
4. R.Bradshaw, B. Bhushan, C. Kalthoff and M. Warne ; *IBM J. Res. Dev.*, **30**, 203 (1986)
5. H. Suglhara and Y. Imaoka, *Ind. Eng. Chem. Prod. Res. Dev.*, **23**, 330 (1984)
6. G.Y. Chin and J.H. Wernick in " *Kirk-Othmer Encyclopedia of Chemical Technology*", **14**, 686 (1981)
7. H.F. Huisman, M.T. kroes and R.C.F. Schoake; *Prog. Org. Coat.*, **16**, 177 (1988)
8. D.W. Brown, R.E. Lowry and L.E. Smith; *Macromol.*, **15**, 453 (1982)
9. H. Zimmerman and N. Thac Kim; *Polym. Eng. Sci.*, **20**, 680 (1980)
10. D.W. Brown, R.E. Lowry and L.E. Smith; *Polym. Mat. Sci. Eng.*, **51**, 155 (1984)
11. J.D. Cooney, M. Day and D.M. Wiles; *J. Appl. Polym. Sci.*, **28**, 2887 (1983)
12. R. Schubert in *"ASTM Special Technical Publication on Degradation of Metals in the Atmosphere"*, ASTM Special Publications, **965**, 374 (1987)
13. E.A. Turi, *"Thermal Characterization of Polymeric Materials"*, Academic Press (1981)
14. M.R. Tant and G.L. Wilkes; *J. Appl. Polym. Sci.*, **26**, 2813 (1981)
15. H.W. Starkweather Jr., P. Zoller and G.A. Jones; *J. Polym. Sci., Polym. Phys. Ed*, **21**, 295 (1983)
16. *"Tape and Cartridge Requirements for IBM 3480 Magnetic Tape Drives"* IBM Publication GA32-0048-0 (1984)
17. J.B. Goodenough, *"Magnetism and the Chemical Bond"*, pp 98-110 & 143-150 John Wiley (1963)
18. C. Heck, *"Magnetic Materials and Their Applications"*, pp 22 & 580-90 (Translated by S.S. Hill) Crane, Russak & Co. Publishers (1974)
19. S.L. Madorsky, *"Thermal Analysis of Organic Polymers"*, Wiley Publishers (1964)
20. A.R. Corradi, S.J. Andrews, C.A. Dinitto, D. Bottoni, G. Candolfo, A. Cecchetti, and F. Masoli, *IEEE Trans. Mag.*, **MAG-20**, 760 (1984)

MECHANISM OF CHEMICAL REACTIONS INVOLVED IN MAGNETIC COATINGS

R. S. Tu

Ampex Corporation
Magnetic Tape Division
401 Broadway, Redwood City, CA 94063

The major binders in magnetic coatings are high molecular weight thermoplastic polyurethane and low molecular weight polyisocyanate. This paper reviews the stoichiometry and kinetics of the isocyanate reactions involved in a typical formulation. It was concluded that the dominant reaction is the condensation of the -NCO groups and water in the formulation with the formation of polyurea as an inter-penetrating network in the polyurethane matrix. The degree of crosslinking of the polyurethane molecules was insignificant.

This conclusion was supported by the fact that the removal of the OH groups from the polyurethane, which eliminates the reaction with the polyisocyanate, did not affect the physical properties of the tape. The crosslinking of the cobinders, when used, was not as important as the reaction of the polyisocyanate with water, because of their relatively high molecular weight and low OH functionality.

INTRODUCTION

Magnetic tape consists of a coating of magnetic particles embedded in a matrix of organic polymer on a plastic film. The major binder is polyurethane/polyisocyanate. For certain products, a cobinder, such as modified PVC or phenoxy resin, is used to replace part of the polyurethane. The commonly used polyurethane is Estane[R] 5701 or Morthane[R] CA-280. Both are based on p,p'-diphenylmethane diisocyanate/butanediol-adipate and have a molecular weight of 120,000-150,000 and a Tg of -10°C.

Estane 5701 or CA-280

The commonly used cobinders are Bakelite[R] VAGH[1] and Bakelite[R] PKHH[2]. The former is a hydroxyl modified copolymer of PVC/PVAc with a molecular weight of 23,000 and a Tg of 79°C.

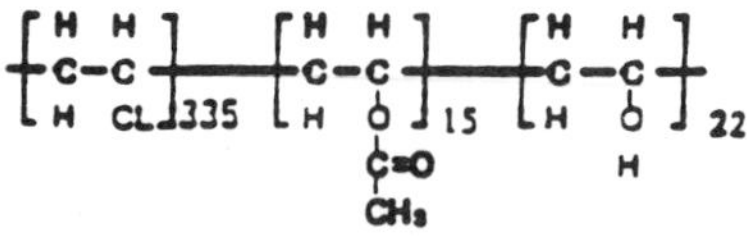

VAGH

The latter is a condensation product of bisphenol A/epichlorohydrin with a molecular weight of 35,000 and a Tg of 98°C.

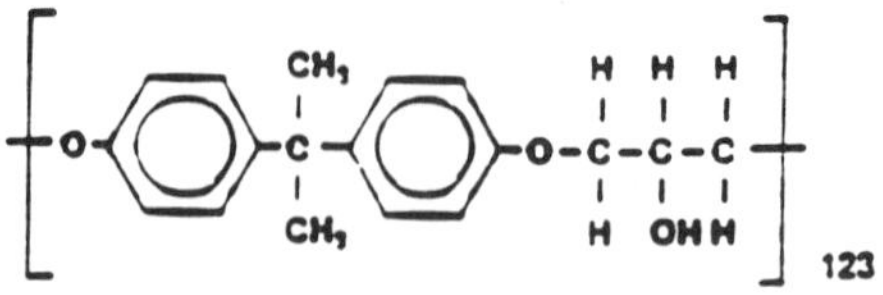

PKHH

The commonly used polyisocyanate is Mondur[(R)] CB-75[3], a 3TDI/TMP adduct.

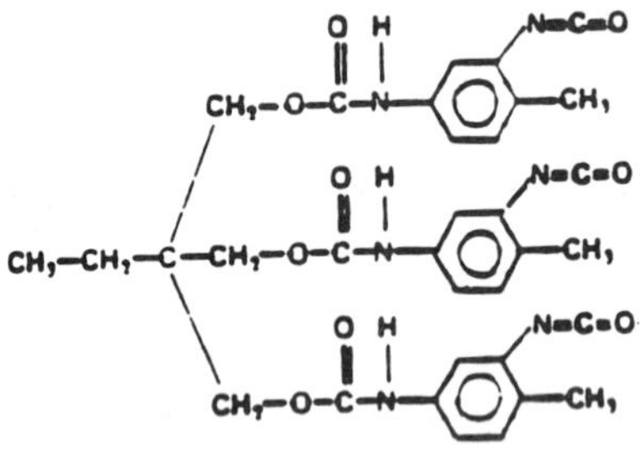

Mondur[(R)] CB-75

As can be seen, the reactions involved in magnetic coatings are actually NCO reactions. Since tape production is carried out practically under ambient conditions, the allophanate and biuret reactions, which are involved with the backbone hydrogens of the polyurethane, are ruled out because these

$$-NCO + RNHCOOR \longrightarrow RNCOOR$$
$$\qquad\qquad\qquad\qquad\quad |$$
$$\qquad\qquad\qquad\qquad CONH-$$

Allophanate

$$-NCO + RNHCONHR' \longrightarrow RNCONHR'$$
$$\qquad\qquad\qquad\qquad\qquad\quad |$$
$$\qquad\qquad\qquad\qquad\qquad CONH-$$

Biuret

reactions require a relatively high temperature ($120-140^\circ$C) and long hours of exposure to crosslink[4]. This is true at least for the uncatalysed systems.

With the exclusion of the allophanate and biuret reactions, the reactivity of the triisocyanate is limited to the hydroxyl groups in the polymers or the formulation. It was the purpose of this paper to identify the reactive groups and to examine the stoichiometry and kinetics of the NCO reactions involved in a typical magnetic formulation.

EXPERIMENTAL AND DISCUSSION

Isocyanate is very sensitive to moisture. Its presence in a formulation presents a competitive reaction with all the OH groups in the polymers. The moisture content of a typical formulation is tabulated in Table I. This data was provided by the suppliers and was reconfirmed by Ampex analysis. For magnetic particles, it was measured by the Mitsubishi moisture analyzer. For solvent, it was measured by the GC method. For others, it was measured by the Karl Fisher method. As can be seen, there were 0.29 parts of water per hundred parts of formulation.

Table I. Typical Experimental Formulation.

Component	Weight %	Moisture Weight %	Moisture in Formulation Weight %
Magnetic Particles	25.67	0.7	0.18
Alumina	1.82	2.0	0.04
Polyurethane	3.65	0.6	0.02
Cobinder	1.22	0.5	0.01
Triisocyanate	1.29	-	-
Dispersant	1.35	1.0	0.01
Lubricant	0.06	0.2	0.00
Solvent	64.94	0.05	0.03
Total	100.00		0.29

In terms of equivalents and in comparison with those of the other reactants, as shown in Table II, there were 10 times more OH groups from moisture than those from VAGH or PKHH,

Table II. Equivalent Weights of Reactants in Formulation.

Reactant	Weight %	Hydroxyl Equivalent Weight	Hydroxyl Equivalents
Moisture	0.29	18	1.6×10^{-2}
Estane 5701 or Morthane CA-280	3.65	67,500	5.4×10^{-5}
Bakelite VAGH	1.22	1,045	1.2×10^{-3}
Bakelite PKHH	1.22	285	4.3×10^{-3}
Mondur CB-75	1.29	219	5.9×10^{-3}

and 1000 times more than those from Estane or Morthane. There was a great excess (10 times more) of OH's from moisture in formulation, not counting those which could be absorbed from the atmosphere, to exhaust the NCO's in the isocyanate.

The kinetics or reactivity of CB-75 in the presence of a typical cobalt doped γ-Fe$_2$O$_3$ and cyclohexanone toward several magnetic tape binder polymers was monitored on a Miran 980 quantitative infrared analyzer using Ampex MTL PCP 251 method and compared with that of water[5]. The oxide has a coercivity of 600 Oe and a specific surface area of 30 m^2/g. Table III shows the order of NCO reactivity of the active hydrogens uncatalyzed at 25°C:

Reactant	$t_{1/2}$, days
Melamine Formaldehyde Resin	0.70
Estane 5701	1.08
PKHH	1.17
VAGH	1.23
Water	1.36

Melamine formaldehyde resin containing primary OH groups was most
reactive. This explains why the pot-life of the formulation containing
it is very short. Estane, also containing primary OH groups, was next
reactive. The rates of reactions of PKHH, VAGH and water containing
secondary OH groups were judged as equal and were least reactive.

The data thus indicated that the dominating reaction in the
magnetic coatings was the reaction of the isocyanates with water. The
reactions take place in both the wet and dry stages. In the wet stage,
there is an excessive amount of water in the formulation. In the dry
stage, the atmospheric moisture cure takes place. The basic reaction of
isocyanate with water is an addition reaction. The intermediate product
decomposes to an amine and carbon dioxide:

$$-NCO + H_2O \longrightarrow [-NHCOOH] \longrightarrow -NH_2 + CO_2$$

The amine formed reacts readily with another molecule of isocyanate
forming a substituted urea:

$$-NH_2 + -NCO \longrightarrow -NHCONH-$$

These reactions represent self crosslinking of the triisocyanate
(CB-75) forming an inter-penetrating polyurea network in the
polyurethane matrix.

The reaction of the isocyanate with VAGH or PKHH was less important
because of the limited numbers of the OH groups and the low mobility of
the molecules. The reaction of the isocyanate with Estane, or Morthane,
was very little for the same reasons and the hydroxyl functionality of
these polyurethanes is even lower than that of the cobinders, although
the OH's in the polyurethanes are primary alcoholic.

To test the reactivity of the polyurethane involved in a magnetic
coating and the effect of the terminal hydroxyl groups in the molecules
on the physical properties of the resulting tape, the performance of
Morthane CA-280 was compared with CA-280EB. The latter is a blocked
polyurethane with the hydroxyl groups replaced by acrylic groups.
Although CA-280EB is radiation curable, tape made for this study was not
irradiated by electron beam. Table IV shows the performance of these
tapes containing CA-280 (CA-280EB)/VAGH/CB-75 at a weight ratio of
59/20/21. Under the same experimental conditions, the performance of
tapes containing the blocked and unblocked polyurethanes respectively
was the same. The removal of the hydroxyl groups in the molecules did
not affect the physical properties of the tape.

Table IV. Performance of Experimental Tapes.

Tape	Unblocked CA-280	Blocked CA-280
Electrical Properties (PAL*), dB		
RF Output	+1.7	+1.6
Video S/N	+1.0	+1.1
Chroma S/N	0	0
Magnetic Properties:		
Coercivity, Oe	768	768
Remanence, Gs	1,227	1,293
Squareness	0.81	0.81
Physical Properties:		
**VPR-1 Still Frame		
Durability, min	60+	60+
**VPR-1 Play Friction, oz		
PF0/PF100/PF200	10/13/14	11/13/14
Layer/Layer Adhesion:		
Unwound, in/Rotate Unwound, in/Del/Blocking	28/36/no/no	36/36/no/no
MEK Solubles, %	4.5	4.5

*Phase Alternating Line
**Video Production Recorder

Therefore, the reactivity of the primary hydroxyl groups in the poly-urethane molecules is insignificant.

In tape testing, the electrical properties were tested by the Phase Alternating Line system (PAL) using 3M's 479 tape as reference. The magnetic properties were tested by the LDJ Electronics BH meter. The still frame durability and play friction were tested on the Video Production Recorder (VPR-1) according to Ampex internal methods. A tape passes the still frame durability test when the reproduced signal of a short recording (7") of the tape maintains at least 75% of the original still frame RF amplitude for a period of 60 min. The layer/layer adhesion was tested according to the Interim Federal Specification W-T-001572. A tape passes the test if the unwound and rotated unwound section of the tape after the oven-heating (130°F. - 85% RH/16 hrs and 130°F./4 hrs) is satisfactory without delamination and blocking. The MEK solubles were determined by soxhlet extraction for a period of 24 hours.

CONCLUSIONS

The reaction rate and stoichiometry of Estane, VAGH, PKHH and water toward Mondur CB-75 in the presence of a typical magnetic pigment were measured.

The dominant reaction involved in a magnetic coating is self-crosslinking of the triisocyanate with water forming an inter-penetrating polyurea network in the polyurethane matrix. The reaction of VAGH or PKHH with the isocyanate was only secondary and the reaction of Estane or Morthane with the isocyanate was insignificant.

REFERENCES

1. Union Carbide, "Chemical and Plastic Physical Properties", 1979-80 Edition (F-44086C, 2/79-40M).
2. Union Carbide, "Bakelite Phenoxy Resins for Solution Coatings (F-41521B, 10/78-6M)".
3. Mobay, "Chemistry for Coatings (3/76)".
4. J. H. Saunders and K. C. Frisch, "Polyurethane Chemistry and Technology", Part I, p. 79, Interscience Publishers, New York, 1965.
5.* J. Curtis, MTD Project Report, "Isocyanate Reaction Chemistry in Magnetic Tape Formulation - IV", May 20, 1981.

* The information could be made available to public upon request.

PLASMA POLYMER FILMS FOR CORROSION PROTECTION OF COBALT-NICKEL 80:20 MAGNETIC THIN FILMS

Hans J. Griesser

Division of Chemicals and Polymers, CSIRO

Private Mail Bag 10

Clayton, Vic. 3168, Australia

Cobalt-Nickel 80:20 alloy thin films suitable for magnetic data storage were protected against atmospheric corrosion by the application of a plasma polymerized protective layer onto the exposed surface of the Co_4Ni film. Coatings were deposited in a custom built reactor onto moving substrate comprising the magnetic thin film on polyester web. Smooth plasma polymer films were obtained from several monomers. Corrosion testing by incubation at 60°C and 90% humidity showed that the plasma polymer overcoat prevented the rapid tarnishing occurring with unprotected Co_4Ni films. Optical microscopy demonstrated absence of localized corrosion spots, indicating that the plasma polymer films were pinhole free. The progress of corrosion was monitored by optical absorption spectroscopy; as corrosion produced an increase in the transmission of light through the semi-opaque films, the extent of the corrosive attack could be quantified by measuring the decrease in optical density of the thin magnetic alloy layer. Protection by plasma polymer overcoats of organosiloxane composition produced a substantial reduction in the rate of corrosion of Co_4Ni films.

INTRODUCTION

Thin continuous alloy films have received much attention for application in high density magnetic data storage [1-4]. The problems associated with thin film media relate to the susceptibility of the alloy coating to mechanical degradation in handling and usage, and corrosion on exposure to severe atmospheric conditions. Cobalt-Nickel 80:20 (Co_4Ni) alloy thin film coatings in particular were observed to corrode rapidly in humid environments and at elevated temperature[5]. We also observed rapid corrosion under the conditions of 60°C and 90% relative humidity (RH), with the shiny metallic coating acquiring a duller, yellowish tarnish within a few weeks. Although unlikely to be encountered in real life usage of magnetic media these conditions nevertheless served to identify, by accelerated testing, concerns with respect to long term stability of experimental media. Surface analysis by X-ray photoelectron spectroscopy showed both segreation of the alloy[6] and production of oxyhydroxides[7] on corrosion in humid environments. In a dry atmosphere, on the other hand, very little corrosion occurred, indicating that the initially formed "dry" oxide allowed only very slow further oxidation. Hence, to prevent catastrophic corrosive degradation it was thought necessary to stabilize the Co_4Ni coating against the rapid damage done by water vapour, whereas the much slower corrosive effects of oxygen and dry air on the magnetic film could be neglected at first.

Co_4Ni coatings may be protected by application of a thin overcoat layer. The overcoat has the functions of alleviating corrosion, providing suitable frictional properties, and mechanical proptection of the easily abraded alloy coating. Ideally an overcoat thickness not exceeding a few tens of nanometres is sought in order to keep the magnetic spacing losses at high recording frequencies minimal across the dielectric protecting layer. Furthermore, the coated thickness and other properties of the overcoat need to be very uniform in order to minimize localized corrosive or mechanical attack through weak spots, which leads to the emergence of "dropouts" in the magnetic signal or abrasion.

Plasma polymerization utilizes organic "monomer" vapour at reduced pressure for vacuum deposition of thin organic-polymeric films[8]. Plasma polymer films are usually smooth, highly conformal to the substrate, extensively crosslinked, dense, and adhere firmly. For application as protective thin film coatings on magnetic media, plasma polymerized films appeared promising not only because their mechanical properties may be superior to those of alternative overcoats, but also because their dense structure may result in low gas permeation[9]. Plasma polymer films have been reported to provide corrosion protection in several applications[10-15]. Among these studies, the closest to ours are those of Harada[11] for protection of plated Co layers and Freitag et al.[13] on protection of Fe-Ni magnetic alloys. Harada obtained sufficient protection only at a plasma polymer thickness excessive for high frequency magnetic data storage.

As the properties of plasma polymer films are strongly dependent on monomer vapour composition, experimental conditions, and the deposition apparatus used, we also investigated the application of plasma polymer films for magnetic media overcoats, using a custom built reactor. This report concentrates on the corrosion protection achieved by the best plasma polymer films, which were deposited from hexamethyldisiloxane monomer gas. The report also presents a simple and inexpensive, stable chamber for corrosion incubation at high temperatures and humidity with avoidance of condensation of water on the samples.

EXPERIMENTAL

Deposition Apparatus

When plasma polymerization is performed in the usual bell jar type chambers with stationary substrates, coating effects on startup and shutdown of plasma are unpredictable and may give rise to interfacial and surface effects and film properties that are not characteristic of those obtainable at equilibrated coating conditions. Moving web coating in a stable, equilibrated plasma is necessary to ensure not only uniformity, but also the commercial viability of production of magnetic media comprising a plasma polymer film. Accordingly, plasma reactors were designed that were capable of experimental semi-continuous plasma coating onto up to 120 metres of thin flexible substrate with a web speed ranging from 0.15 to 6 m/min[16]. A schematic diagram of one reactor model is given in Figure 1. The copper electrodes are 90 mm long and 18 mm wide, and their spacing is 16 mm. The substrate web was transported over the face of one electrode. The reversible web drive improved liberation of entrapped air prior to experimentation, by cycling the web back and forth.

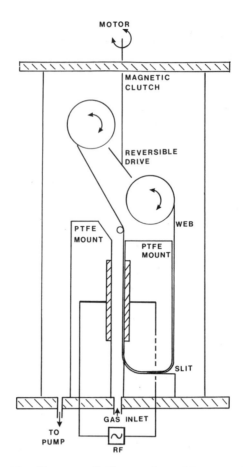

Figure 1. Schematic diagram of the semi-continuous plasma reactor.

RF power was supplied by a custom built oscillator operating at 700
kHz with no matching network. The pressure in the reactor chamber was
monitored by a capacitance manometer. The pumping system included a
throttle valve, a cold trap and a mechanical pump. The "monomer" gas
stream was derived from the liquid, which was held in a round bottomed
glass flask, by using the vapour boiloff at room temperature under
pumping. A manual valve controlled the monomer flow rate, which was
measured prior to deposition experiments by recording the pressure
increase with time in the known volume of the reactor, with the throttle
valve closed. Then the throttle valve was opened partially to stabilize
the pressure at the desired value. Monomer flow rates typically were in
the range 5 to 15 sccm/minute and the pressure 0.3 to 0.9 Torr.

Materials

The monomer liquids were obtained commercially in the purest grade
available and used without further purification, except that prior to

the first deposition, vapour was pumped off the liquid for a few minutes
in the absence of plasma generation, in order to remove more volatile
impurities. Hexamethyldisiloxane (HMDSO) and acrylonitrile (AN) were
used most extensively and their films will be described below. Plasma
polymers from vinyltriethoxysilane and vinyltris-(2-methoxyethoxy)-
silane were similar to those from HMDSO. Methyl methacrylate (MMA) was
investigated for comparison to AN, but offered no advantage.

The substrates consisted of thin film coatings 70 to 120 nm thick[17]
of Co_4Ni alloy deposited by electron beam evaporation onto moving,
12 μm thick polyester web[4]. Rolls of substrates 12.7 mm (1/2 inch) and
8 mm wide were kindly supplied by R.G. Spahn (Eastman Kodak, Rochester,
US). The samples were stored in desiccated bags and XPS indicated that
a surface cleaning step prior to plasma overcoating was not necessary.
For comparison with the plasma polymer protected samples, other Co_4Ni
media were available that were protected by thin proprietary overcoats
not applied by plasma polymerization.

Corrosion Testing and Assessment

The problems involved in producing a stable atmosphere of 60°C and
90% RH relate to the stringent demands placed under these conditions on
stability and the ability to sense and control. The dry bulb
temperature must be very stable as a 1°C change will produce a 4% RH
change, and if the wall or the samples are 2.5°C cooler than the
atmosphere in the chamber, condensation will occur. Furthermore, in
order to monitor the progress of corrosion, we needed to remove and reinsert samples periodically; such transfer to and from room atmosphere
required care to avoid condensation onto the samples. Initial
experiments using a conventional humidity cabinet showed accelerated
localized corrosion in the form of transparent "pinholes", a corrosion
process which was thought to be a result of water condensation on the
Co_4Ni surface. Polarized optical transmission microscopy showed that
the "pinholes" were not empty, but consisted of transparent material,
probably oxidation products from a fast corrosive process occurring
underneath the water droplets. A simple and inexpensive atmosphere
generator was thus developed as shown in Figure 2. The custom built
corrosion chamber provided a continuous flow of the conditioned
atmosphere past the samples. Incoming air (dry, controlled) was split
into two metered streams, one of which passed through a saturator
chamber. To achieve continuous extended corrosion testing, the

saturator was continually topped up by a solenoid controlled, gravity fed line from a distilled water reservoir. The "dry" and "wet" streams were recombined and fed to the sample compartment at a rate of 3 litres/minute. The air condition in the chamber was measured with an EG & G Model 911 humidity analyzer that displayed the temperature and RH. Accurate temperature control was achieved by immersing the entire system in a thermostatted water bath. The standard conditions were 60°C and 90.2% RH. The stability was monitored to be ± 0.1°C and ± 0.8% RH over 2 weeks.

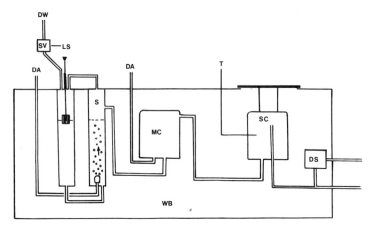

Figure 2.   Schematic diagram of the corrosion incubation chamber. DA: dry air inlet, via metering valve; DW: distilled water inlet; SV: solenoid valve; LS: level sensor; S: saturator chamber; MC: mixing chamber; SC: sample chamber; T: temperature probe; DS: dewpoint sensor; WB: water bath.

Independent control of the "wet" and "dry" streams allowed removal and reinsertion of samples in the course of extended testing with complete avoidance of condensation. Dry air flow replaced the conditioned atmosphere prior to opening the sample compartment. Following reinsertion the compartment was thermally equilibrated with dry air supplied at 60°C before the RH was gradually increased. Typically the time required to reach standard incubation conditions was 30 minutes. Sample strips 50 mm long and 12.7 or 8 mm wide were mounted on cleaned microscope slides by attaching both ends to spacers, 1 mm high, glued to the slide. The spacers enabled access of vapour to the rear (polyester) side of the samples and allowed curling of the samples across their width to develop. Using this equipment and techniques for incubation, optical transmission microscopy confirmed absence of localized, "pinhole" corrosion after 6 weeks of incubation with several

removal and reinsertion cycles over this period. Moreover, optical
reflection microscopy showed that the tarnishing indicative of large
area corrosion from the surface was uniform over the area of the
samples.

The progress of corrosion was periodically assessed by optical
transmission spectroscopy using a Cary 210 UV/VIS spectrometer. Samples
mounted on microscope slides were inserted into a custom built holder
located in the optical beam. This enabled assessment of samples while
they remained mounted on the slides and avoided the problems of
handling. The optical beam measured an area of 16 by 5 mm of the
sample. This area was large enough to minimize the effects of the
occasional localized defects such as small scratches and pinholes from
vacuum coating.

Microscopy

A Zeiss Ultraphot II microscope with an optional Nomarski
attachment was used for optical transmission and reflection
microscopy. Scanning electron microscopy of surfaces was performed
using a Jeol-840 unit. For freeze fractured edge views, samples were
cooled in liquid nitrogen and fractured, a PtPd layer 5 nm thick was
applied by ion beam sputtering and the samples studied in a Jeol 100CX
unit equipped with an ASID-4 scanning attachment. Thin cross-sections
were obtained following embedding of samples in epoxy resin and studied
using a Jeol 100CX transmission electron microscope (TEM).

RESULTS

Corrosion of Co_4Ni Alloy Films

When kept in desiccated bags the unprotected Co_4Ni films retained
their shiny metallic appearance for periods in excess of two years and
XPS showed that only a small extent of additional (to the initial, very
shallow surface oxidation) corrosion resulted[6]. Under humid incubation
(60°C and 90% RH), on the other hand, a yellowish tarnishing and a
generally duller appearance of the films developed within two weeks
typically. The samples also curled to an extent that paralleled the
extent of tarnishing. The tarnishing was uniform over the entire
exposed area except for an enhancement over the outermost area adjacent
to the cut edges, probably due to access of vapour from the side into
the thin film.

In addition, occasional transparent "pinholes" were observed by optical microscopy. The pinholes were often located around coating defects such as incorporated small foreign particles. Their shape and visual appearance often were indicative of silicate dust. Enhancement of corrosion adjacent to defects is consistent with improved access of water vapour along the edge of the particles into the depth of the magnetic film. Several of the particles were dislodged by a fine brush to show the void created and the transparent area around it. Condensation of small drops of water onto Co_4Ni films also gave rise to the development of areas that gradually became more transparent. These pinholes again were observed by microscopy not to be void, but to contain material which was transparent. Films of sub-micrometre thickness of oxides and hydroxides of cobalt and nickel are essentially transparent in tungsten illumination. While it was not possible to analyze these small areas to verify this conjectured composition of the pinholes, it appeared likely that they contained the same oxidized alloy material as that produced by non-localized corrosion of Co_4Ni, identified by XPS as mixed hydroxides and oxides[7]. Nomarski reflection microscopy of nascent pinholes suggested that they developed inwards from the exposed surface.

With avoidance of condensation, corrosive attack progressing from the exposed surface of defect-free areas of films appeared uniform as evidenced by uniform tarnishing. This uniform surface corrosion was of more interest; whereas pinhole corrosion was very infrequent in some coatings and probably can be avoided by clean and defect-reducing coating conditions, the large-area surface corrosion represents the inherent oxidative instability of the Co_4Ni material. We have thus concentrated on studying and attempting to eliminate the large-area corrosion.

The oxide and hydroxide products of large-area corrosion have been discussed elsewhere[7], the present work is concerned with the rate of corrosion. Uncorroded, fresh Co_4Ni thin film coatings were not fully opaque at the thicknesses used; the transmission of visible light through the films was of the order of 1 to 10%, depending on wavelength, thickness and the amount of oxygen incorporated into Co_4Ni during deposition. The pinholes showed the corrosion products to be essentially non-absorbing; accordingly, the progress of corrosion could be monitored directly by the increase in the transmission of light. Absorbance (optical density, OD) was the more suitable parameter as it is directly related to the amount of absorbing material present; the decrease in absorbance Δ OD, was thus a linear measure of the amount of alloy that was converted by corrosive attack over a given time.

As corroded alloy material was contained both in pinholes and in the uniform large-area corrosion, absence of significant pinholes was required to make Δ OD a sensitive and quantitative measure of the large-area corrosion of interest. Absence of pinholes in the area of the samples assessed for OD was ascertained by optical microscopy. Figure 3 shows that the decrease in absorbance due to corrosion was uniform over the visible spectral region. Accordingly, a fixed wavelength of 600 nm, with a spectral bandwidth of 1 nm, was selected for further assessments as at this wavelength the OD values of all samples used were within the range of 0.9 to 2.5, a range which is suitable for good accuracy of OD determinations.

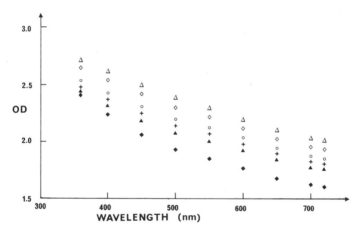

Figure 3. The optical absorbance (OD) of a Co_4Ni thin film as a function of wavelength and time of exposure to the corrosive environment (60°C, 90% RH): Δ, 0 hrs; ◊, 24 hrs; o, 144 hrs; +, 312 hrs; ▲, 456 hrs; ◆, 1008 hrs.

Conversion to an absolute thickness scale required an independent, absolute method of measuring Co_4Ni film thickness; electron microscopy techniques are suitable[17]. From freeze fractured edge views and a TEM cross-section, it was found that the thickness of a film with an OD of 2.5 was 150 (± 15) nm, giving a Co_4Ni film thickness of 60 nm corresponding to an optical density of 1 at a wavelength of 600 nm.

Figure 3 shows a Δ OD (at 600 nm) of 0.43 as a result of 6 weeks exposure to the corrosive environment. The shape of the degree of corrosion with depth is unknown; for simplicity, we define the "thickness of the corroded layer", d_{corr}, as the thickness of a layer with a rectangular step profile containing the amount of fully oxidized oxide/hydroxide necessary to give the same theoretical decrease in absorbance of the Co_4Ni film. It is thus assumed that the shape of the depth profile of the corrosion process is the same for all samples and only the rate of corrosive penetration differs. The above Δ OD of 0.43 corresponds to a value of 25.8 nm for d_{corr}.

Plasma Polymer Overcoating

Among the various monomers tried initially, AN and HMDSO proved most useful. AN showed an outstanding rate of deposition. The thicknesses of the plasma polymer (pp) films and thus the rates of deposition were estimated on-line by visual observation of the brilliant interference colour that the pp film produced on the silvery Co_4Ni coating at lower web speeds. The desired coating thickness values of around 50 nm could then be produced by appropriate speeding up of the web transport. By applying drops of organic liquids of known refractive index it was found that the refractive index of the pp films was $\sim$ 1.5.

For some samples the interference derived thickness estimates were confirmed by electron micrographs. TEM cross-sections suffered from lack of contrast between the epoxy resin and the plasma polymer films, as well as artifacts such as folding in sectioning. Freeze fractured edge views, on the other hand, showed up the pp films well and showed that the majority of the pp films possessed excellent uniformity of thickness, were smooth and had no internal structure; whereas the columnar structure of the Co_4Ni films was clearly observable. The exceptions to smoothness and uniformity were pp-AN films deposited at high deposition rates. Embedded discrete, spherical particles of various sizes and a rough surface with protruding particles indicated a marked sub-structure of many of the films. These spheroidal inclusions were thought to be due to excessive rates of reactions in the gas phase producing oligomeric particles of substantial size before such particles had an opportunity to reach the surface[16]. The extraordinarily high deposition rate of AN films suggests that, in addition to plasma induced reactions, there is a major contribution due to the traditional radical propagation polymerization. Films with such sub-structure were

360

physically inferior in terms of abrasion resistance and coefficient of friction. One magnetic tape sample overcoated with an AN pp film was tested in a video recorder test bed and produced fast clogging of the heads. Although the extremely high deposition rate, of up to 760 nm/min. even in our small reactor, is very attractive for large scale viability, work on pp films from AN was not pursued further for protective overcoating of Co_4Ni media. Methyl methacrylate plasma polymerized somewhat slower, but still quite rapidly; however, as for AN its films were mechanically inferior, due both to a relatively high coefficient of friction and susceptibility to abrasion, and appeared unsuitable for use as protective overcoats. In view of the shortcomings in their mechanical properties, overcoats made from AN and MMA were not subjected to corrosion testing.

Such spheroidal inclusions and other sub-structures were absent in pp films made from the siloxane monomers under all conditions. This observation is different from that of Wertheimer et al.[12] who obtained rough coatings containing speroidal agglomerates when depositing HMDSO pp films onto substrates held at room temperature. Neither did we observe the poor adhesion reported by Schreiber et al.[10] of HMDSO pp films deposited at room temperature; the standard sticky tape test did not lead to removal of our HMDSO pp films. AN films with spheroidal inclusions, on the other hand, failed the sticky tape test in patches, indicating that such sub-structure generally may be unsuitable for good mechanical properties of protective films. We surmise that the equipment of Wertheimer et al. provided a higher plasma reaction rate whereas for HMDSO in our equipment and under our deposition conditions the rates of reactions in the plasma gas phase and the transport to the surface of the growing film were well matched for growth of uniform films at reasonable, but not excessive rates. We only explored a range of relatively low power levels as we wished to retain an organosilicon character of the pp films for tribological reasons. It appears that a fundamental limit to deposition rates for uniform films may exist on account of the onset of excessive gas phase polymerization at higher power producing discrete particles, and that the quality of films from a given monomer depends markedly on the equipment and the process parameters.

Process parameters conducive to deposition of high quality HMDSO pp films were: pressure 0.36 to 0.45 Torr; flow rate 5 to 8 sccm/min; RF power 0.45 to 1.5 W/cm^2 (assuming no transfer losses, although no matching network was available); and web speed 0.2 to 0.6 m/min. This web speed resulted in a residence time of the substrate between 9 and 27 sec in the region between the 90 mm long electrodes, and produced coated thicknesses of 50 to 200 nm. Higher pressure (0.8 Torr) and flow rate (10 to 15 sccm/min) conditions produced somewhat higher deposition rates, but the cross-web uniformity of coated thickness, which was excellent in the former coatings, was inferior under the latter conditions as the outermost ~ 2 mm of the 12.7 mm wide samples showed varying thickness. This is, of course, a consequence of the narrow width of the electrodes.

Analysis by FTIR and XPS of HMDSO plasma polymer films produced under these conditions has been reported elsewhere[18]. In summary, methyl abstraction from the monomer was evident and an organosilicon network resulted, with pendent methyl and trimethylsilyl groups. The chemical composition is thus comparable to that of conventional organosilicon polymers (Silastic, for instance) except that the plasma polymers are three dimensionally, randomly crosslinked. Under the relatively low powers used, the films retained a large degree of organic character. One may thus expect low friction behaviour as in conventional silicones, but higher cohesion and abrasion resistance because of the crosslinked network.

The deposition process of the HMDSO films appeared not to depend significantly on the substrate; with substrates consisting of thin Co_3Cr alloy film on Kapton, and with conventional binder video tape, the same process parameters produced coatings of similar thickness.

The other two, vinyl containing silane monomers gave similar deposition rates with similar process parameters. The vinyl group thus did not give a marked enhancement of deposition rates, an acceleration which might have been expected from a contribution by a free radical propagation mechanism. This observation is in agreement with a recent report[19] disputing the faster plasma deposition of vinyl type monomers suggested earlier[20].

Plasma overcoating with HMDSO produced no noticeable curling of the tape samples additional to the slight curling already present in some of the Co_4Ni coatings from metallization. Whereas plasma polymer films from other monomers were observed under similar conditions in our equipment to give rise to some degree of curling of the composite, HMDSO was exceptionally suitable also in this regard, as production of magnetic tape requires absence of curling stress induced by the overcoat. Our observations parallel those of Yasuda et al.[21] who found several monomers to produce films with curling stress, but tetramethyldisiloxane plasma polymers showed absence of such stress. Hexamethyldisilane films, however, showed stress[14].

Co_4Ni tape samples overcoated with HMDSO plasma polymer films deposited under a variety of conditions were incubated for 6 weeks and the progress of corrosion periodically assessed by transmission spectroscopy at wavelength of 600 nm. Again, the absence of significant

Table I. Decrease in optical absorbance of Co_4Ni thin film coatings on incubation for 6 weeks at 60°C and 90% RH: Samples A and B unprotected; C to J protected by HMDSO pp film; K to M protected by non-plasma overcoat.

Sample	Pressure[a] (Torr)	Power[b] (W/cm^2)	Thickness[c] (nm)	ΔOD	d_{corr} (nm)
A[e]	–	–	–	0.62	37
B[e]	–	–	–	0.43	26
C	0.37	0.83	160	0.13	7.8
D	0.37	0.83	160	0.11	6.6
E	0.38	0.83	80	0.14	8.4
F	0.45	0.83	90	0.14	8.4
G-1[e]	0.39	0.82	50	0.13	7.8
G-2[e]	0.39	0.82	50	0.12	7.2
H	0.40	1.05	105	0.07	4.2
I	0.43	1.44	200	0.14	8.4
J	0.44	1.44	60	0.13	7.8
K	–	–	d	0.22	13.2
L	–	–	d	0.10	6.0
M	–	–	d	0.09	5.4

[a] of plasma gas phase.
[b] consumed by oscillator; not matched into plasma.
[c] of protective layer, from interference colour and web speed.
[d] not known accurately; < 10 nm.
[e] see text.

pinholes in the samples' area assessed by OD measurements was ascertained by optical microscopy, before and after incubation. Table I summarizes results from unprotected Co_4Ni thin film samples, such samples protected by HMDSO plasma overcoats deposited under a variety of conditions, and three samples protected with an overcoat applied by a proprietary, non-plasma technique. Samples A and B were from different Co_4Ni coating runs. Samples G-1 and G-2 were taken from different locations of one extended (30 minutes) plasma coating run onto moving web, and corrosion tested and assessed in independent test runs.

Unprotected Co_4Ni films corroded rapidly, losing about one quarter of the magnetic alloy by oxidation in six weeks. Overcoating provided protection to varying degrees. Visual observations on the tarnishing associated with corrosion paralleled the Δ OD data of Table 1. Unprotected Co_4Ni films had turned from a metallic sheen into a duller yellowish-brown after 6 weeks, and for a few samples, pinkish and bluish hues were also visible, indicating the formation of considerable oxide/hydroxide layers. The poorly protected sample K acquired a

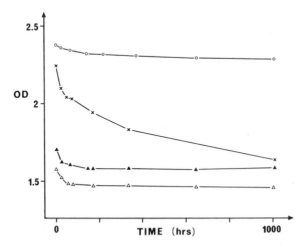

Figure 4. The absorbance (OD) at 600 nm as a function of time of corrosion exposure, of the unprotected Co_4Ni film A (x), samples G-1 and G-2 protected by HMDSO plasma polymer overcoats (▲,△), and sample M protected by a non-plasma overrcoat (o).

brownish-yellow tarnish, again with some dullness. The well protected samples, on the other hand, did not change visibly or acquired only a slightly yellowish tint in the silvery colour. Figure 4 shows the progress of the decrease of OD with time of incubation for selected samples; the behaviour shown is typical. The different initial values of OD are due to different coating thicknesses of the Co_4Ni films.

Optical microscopy after incubation showed that virtually all of the substantial pinholes were associated with coating defects either in the Co_4Ni layer or in the overcoat. The latter defects were not inherent in the overcoat, but consisted of dust particles sitting on the alloy layer, and thus apparently had prevented, in the overcoating process, formation of a sealing plasma overcoat at those spots. Prior web cleaning and clean conditions for the plasma process probably can overcome this in the manufacture of magnetic media. The virtual absence of locally enhanced corrosion in areas devoid of macroscopic imperfections suggested that the plasma polymer layers were uniformly covering, in agreement with the often reported excellent uniformity of thickness under suitable deposition conditions.

DISCUSSION

Assessment of the progress of corrosion by monitoring the absorbance change, Δ OD, relies on the fact that the optical absorbance of the corrosion products is negligible, being orders of magnitude lower than that of the metals. The products are mixed oxides and hydroxides[7]. Hydrated Co^{11} salts are red or pink (for instance, $Co(H_2O)_6^{2+}$ has an absorption maximum at $\sim$ 535 nm), but the absorption strength in the visible spectral region is so low that films of sub-micrometre thickness are essentially transparent. Similar considerations apply for the oxy-hydroxides of Ni. Accordingly, in principle, the oxidation of Co^0 and Ni^0 will lead to a linear decrease in OD, thus making, in the absence of pinholes, Δ OD a quantitative and linear measure of the amount of magnetic material converted by the large-area corrosion process. Reflection optical effects were neglected; the continuing, smooth decreases in OD suggested that they did not exert a substantial influence. On account of the XPS observations of gradual transitions between hydroxide, oxide and alloy rich regions[7] and enrichment of Co in the oxidized region[6] it was not possible to model the optical reflection properties.

The decrease in OD could conceivably be due to the sum of two corroded layers: access of water vapour and oxygen from the "rear", through the polyester base, could also produce a corroded layer growing from the alloy/polyester interface. However, no evidence for substantial corrosion at that location was found on Auger depth profiling through the entire alloy layer[6].

The accuracy of OD determinations was ± 0.003. An independent repeat experiment gave values of Δ OD over 6 weeks of incubation that differed by 0.01 OD units (Table 1). However, some data points in other curves deviated by more from the line of best fit, suggesting that the experimental accuracy in Δ OD is more like 0.03 OD units, corresponding to an accuracy in d_{corr} of 1.8 nm. In addition there is an uncertainty in d_{corr} of 10% due to the electron microscopic thickness calibration, an uncertainty which does, however, not affect comparison of the overcoats for corrosion protection. It is concluded that the incubation and assessment methods provided reliable assessment of the relative depths of corrosion of different Co_4Ni film samples. Furthermore, the ability to remove and reinsert samples enabled following the progress of Δ OD with time, as shown in Figure 4, to elucidate the time law of the corrosion process.

The corrosion process of unprotected Co_4Ni showed a parabolic time dependence as assessed by the decrease in OD with time of corrosion exposure (Figure 5). The decrease in OD of samples protected by

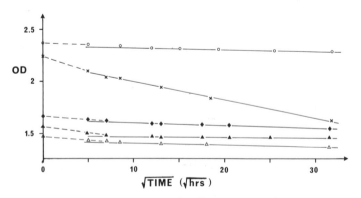

Figure 5. The decrease in absorbance (OD) plotted on a parabolic time scale, for the unprotected Co_4Ni thin film A (**x**), samples D and G-1 protected by HMDSO plasma polymer overcoats (Δ , ▲) and samples L and M protected by non-plasma overcoats (o,♦).

overcoats also followed a parabolic time dependence. Some deviation was
manifested in the initial part (0 to 48 hrs); this could be due to
startup effects and the larger contributions of the startup and shutdown
times for the chamber, relative to the interval between measurements, to
remove the samples for the measurements for the first two data points.
The parabolic relationship enabled extrapolation of d_{corr} for exposure
for one year; for instance for the samples L and M, values of $\sim$ 16 and
18 nm were obtained. The Arrhenius dependence of the corrosion process
of Co_4Ni is not known, but it appears safe to assume that even under
tropical service conditions magnetic media will have a useful service
life of several years before a substantial portion of the magnetic thin
film is corroded.

The data listed in Table I show comparable performances in
corrosion protection by plasma and non-plasma overcoats as assessed by
Δ OD over 6 weeks; this may in some cases be somewhat misleading as the
time dependences within this period varied appreciably. For plasma
polymer overcoated samples there often occurred initially a steeper drop
in OD, but subsequently the decrease in OD with time for some of these
samples became less steep than for other samples and the samples
protected by non-plasma overcoats (Figures 4 and 5). The reason for the
initially steeper drop of OD for some of the plasma overcoated samples
is not known; it may relate to higher susceptibility to the startup
conditions. The slower decrease of OD in the latter part of the
assessment period for the best of the plasma overcoated samples leads us
to expect by extrapolation that in the longer term they may show
performances superior to those of the reference, non-plasma overcoats.
However, this conjecture could not be verified by extended testing due
to practical limitations. It is certainly safe to state that the HMDSO
plasma polymers provided protection at least as good as that of the best
available, at the time, non-plasma coated protective layers. Magnetic
media protected by these HMDSO plasma polymer films can likewise be
predicted to have a useful service life of several years under high
humidity and realistic service temperature conditions.

Finally, possible reasons for the corrosion protection afforded by
plasma polymer overcoats will be discussed. The opinion has been
expressed that unless an extremely low permeability can be obtained,
ultrathin polymer films generally have a limited barrier protection
capability[22]. HMDSO plasma polymers are quite permeable to oxygen as

attested by the proposed use of such films in oxygen enrichment from air[23]. This presented no problems in the present application for protection of Co_4Ni media, because we observed that the alloy did not corrode rapidly in dry air. Water vapour access thus became critical. The permeation of water vapour through HMDSO plasma polymers was found to be low[9]; although those coatings were deposited under plasma parameters quite different to ours, the general tendency probably prevails. Our coatings were deposited at lower power and may, therefore, have retained a larger relative amount of the hydrophobic methyl and trimethylsilyl groups that are pendent on the siloxane backbone. As a result, the water permeability in our plasma polymers is expected to be at least as low. However, even for thin films with very low permeability, water diffusion occurs within a second[13,22]. Furthermore, the permeability might be expected to be related to the crosslink density. XPS data showed that with increasing power input, the extent of methyl abstraction increased, and more Si-O bonds were created to give a more crosslinked network[18]. The data of Table I, however, suggest absence of correlation between plasma power input and protection against corrosion.

Neither did the depth of corrosion appear to relate (inversely) to the thickness of the plasma overcoat; for instance, the pairs D and E, and I and J, were deposited under the same plasma parameters but different web speeds; yet, d_{corr} was the same within experimental error. In other studies, the corrosion protection afforded by plasma overcoats improved with their thickness [11,13,14].

Both these observations suggest that the protective capabilities of the HMDSO plasma overcoats are not primarily a result of the polymers functioning as a water vapour barrier. Winters et al.[22] have stated that in an ultrathin film the contribution of interfacial properties may become as important as the bulk properties of the polymer, and that with a layer of 10 to 50 nm thickness, the major protection must come from the modification of the interface to resist corrosion. In contrast to previous studies on corrosion protection by plasma polymers where the plasma polymer thickness played a role, our HMDSO plasma polymers on Co_4Ni films provide an example where the interfacial properties appear to have assumed a dominant role and surface passivation is the key to the understanding of the corrosion protection. While oxygen in the bulk of the Co_4Ni films is present largely as oxides, XPS indicated that on the very surface hydroxide dominated[7]. Water vapour may thus be expected to show a strong affinity for the alloy film surface and, by hydrogen bonding, to displace a polymer coating that is weakly bonded. On the other hand, hydroxyl bonds react with many silane containing molecules; silylation of hydroxides is a well known derivatization reaction[24]. In the HMDSO plasma, abstraction of methyl groups from the monomer molecules appeared to be the dominant first step[18]. A silyl radical thus generated could, on drifting to the surface, achieve the silylation reaction of surface hydroxide:

$$
\begin{array}{ccc}
& \quad \text{CH}_3 \qquad\qquad & \qquad\qquad \text{CH}_3 \\
& \qquad | \qquad\quad | & \qquad\qquad | \qquad\quad | \\
\text{M} - \text{O} - \text{H} \;+\; & \;\cdot\text{Si} - \text{O} - \text{Si} - \;\;\; & \text{M} - \text{O} - \text{Si} - \text{O} - \text{Si} - \\
& \qquad | \qquad\quad | & \qquad\qquad | \qquad\quad | \\
& \quad \text{CH}_3 \qquad\qquad & \qquad\qquad \text{CH}_3
\end{array}
$$

with the resultant formation of a covalent bond across the interface. The siloxane backbone of the plasma polymer is crosslinked in a random, three dimensional network typical of plasma polymers. Such an arrangement helps to support the interfacial adhesion by providing some rigidity and probably a number of anchorage points not too distant from each other. The presence on the alloy surface of a crosslinked, tight siloxane network with a number of interfacial M-O-Si bonds would make the formation of pockets of water relatively difficult and provide good, water insensitive adhesion.

It thus appears that the corrosion protection of Co_4Ni alloys by HMDSO plasma polymers is related to the theory proposed by earlier workers[25,26] that the dominant property of silicone polymers for use against moisture penetration is their ability to adhere even in the presence of water vapour. Chemical reactions were postulated to deactivate the hydrophilic surface sites, and thus the adsorption of water prevented. The good adhesion of plasma polymers of a silane monomer on aluminium was also ascribed to interfacial Al-O-Si bonds derived from Al hydroxide[27]. As those workers, we only determined the "dry" adhesion, which was excellent for HMDSO plasma polymers on Co_4Ni; whereas for corrosion protection the adhesion under "wet" conditions, which we were not equipped to assess, is important[22]. Nevertheless, it appears reasonable now to postulate a mechanism for the corrosion protection by siloxane plasma polymer films, by invoking that in the initial stages of plasma polymerization of HMDSO, reactive siloxane species drift to the Co_4Ni surface and react with hydroxyl groups under metal-oxygen-silicon bond formation across the interface. Such bonds, coupled with the three dimensionally crosslinked nature of the plasma polymer film, make for a well adhering protective layer. Adhesion is expected to remain strong even under high humidity conditions because such bonds would, unlike hydrogen bonded polymer adhesion, not be broken and displaced by hydrogen bonds to water molecules. Some hydroxyl groups may remain on the surface, but their efficiency in catalyzing oxidative corrosion, by water and oxygen attack, is severely diminished because the formation of substantial water pockets under the dense, well attached plasma coating is made difficult.

CONCLUSIONS

Plasma polymerized films deposited onto moving web, consisting of a Co_4Ni magnetic thin film on polyester, showed, under suitable deposition conditions, smoothness, conformality to the substrate, and uniformity. Under extreme coating conditions, films from acrylonitrile showed a cross-sectional and surface structure with spheroidal inclusions in the film. Films from hexamethyldisiloxane, on the other hand, were of excellent physical quality and adhered well to the substrate. The corrosion of Co_4Ni was reduced greatly by the presence of protective plasma polymer films from the latter monomer. It is believed that interfacial passivation, in this case by the formation of metal-oxygen-silicon bonds from metal hydroxide, is the key factor in the successful protection of the magnetic alloy films by ultrathin protective overcoats of organic-polymeric nature. These bonds also provide strong adhesion of the protective plasma polymer layer.

ACKNOWLEDGEMENTS

This work was done while the author was with the Research Laboratory, Kodak (Australasia) Pty Ltd; generous support and permission to publish are gratefully acknowledged. Thanks are also due to J.W. Chapman and L. Bennett for assistance in corrosion testing, to R.G. Spahn for provision of Co_4Ni-on-polyester coatings, and to T. Poeth for electron microscopy.

REFERENCES

1. H. Sugaya and A. Tomago, in "Proc. Symp. Magnetic Media and Manufacturing Methods", Hawaii, May 1983.

2. T.C. Arnoldussen and E.M. Rossi, Ann. Rev. Mater. Sci. 15, 379 (1985).

3. A. Feuerstein, H. Lammermann, M. Mayr and H. Ranke, Vacuum 35, 277 (1985).

4. J.S. Gau, R.G. Spahn and D. Majumdar, IEEE Trans. Magn. MAG-22, 582 (1986).

5. R.R. Dubin, K.D. Winn, L.P. Davis and R.A. Cutler, J. Appl. Phys. 53, 2579 (1982).

6. N.G. Farr and H.J. Griesser, Mater. Forum, accepted for publication (1988).

7. N.G. Farr, H.J. Griesser, A.E. Hughes and B.A. Sexton, in preparation.

8. M. Shen and A.T. Bell, Editors, "Plasma Polymerization", ACS Symp. Ser., 108, American Chemical Society, Washington (1979); H. Yasuda, "Plasma Polymerization", Academic Press, Orlando (1985).

9. E. Sacher, J.R. Susko, J.E. Klemberg-Sapieha, H.P. Schreiber and M.R. Wertheimer, Amer. Chem. Soc. Coat. Appl. Polym. Sci. Proc. 47, 439 (1982).

10. H.P. Schreiber, M.R. Wertheimer and A.M. Wrobel, Thin Solid Films 72, 487 (1980).

11. K. Harada, J. Appl. Polym. Sci. 26, 3707 (1981).

12. M.R. Wertheimer, J.E. Klemberg-Sapieha and H.P. Schreiber, Thin Solid Films 115, 109 (1984).

13. W.O. Freitag, H. Yasuda and A.K. Sharma, J. Appl. Polym. Sci.: Appl. Polym. Symp. 38, 185 (1984).

14. D.L. Cho and H. Yasuda, Polym. Mater. Sci. Eng. 56, 599 (1987).

15. J.A. van Lier and R.E. Ray, Polym. Mater. Sci. Eng. 56, 603 (1987).

16. H.J. Griesser, Vacuum, in press (1988).

17. H.J. Griesser and L.P. Bennett, Mater. Forum, in press (1988).

18. I.H. Coopes and H.J. Griesser, J. Appl. Polym. Sci., in press (1988).

19. S.P. Swann and S. Srinivasan, Amer. Chem. Soc. Polym. Prepr. 28, 194 (1987).

20. N. Inagaki and H. Yamazaki, J. Appl. Polym. Sci. 29, 1369 (1984).

21. H. Yasuda, T. Hirotsu and H.G. Olf, J. Appl. Polym. Sci. 21, 3179 (1977).

22. H.F. Winters, R.P.H. Chang, C.J. Mogab, J. Evans, J.A. Thornton and H. Yasuda, Mater. Sci. Eng. 70, 53 (1985).

23. M. Yamamoto, J. Sakata and M. Hirai, J. Appl. Polym. Sci. 29, 2981 (1984).

24. D.R. Knapp, "Handbook of Analytical Derivatization Reactions", p. 8, Wiley-Interscience, New York (1979).

25. M.L. White, Proc. IEEE <u>57</u>, 1610 (1969).

26. P.R. Troyk, M.J. Watson and J.J. Poyezdala, in "Polymeric Materials for Corrosion Control", R.A. Dickie and F.L. Floyd, Editors, ACS Symp. Ser. <u>322</u>, 299, American Chemical Society, Washington (1986).

27. A.K. Hays and D.M. Haaland, Amer. Chem. Soc. Org. Coat. Appl. Polym. Sci. Proc. <u>47</u>, 383 (1982).

DYNAMIC MECHANICAL BEHAVIOR OF THERMOPLASTIC POLYURETHANE

IN MAGNETIC COATINGS

H. S. Tseng and E. G. Kolycheck

BFGoodrich Technical Center

Box 122, Avon Lake, OH 44012

Thermoplastic polyurethane has been widely used as the major binder system in the magnetic coating layer along with various magnetic pigments and several other ingredients. For simplicity, the magnetic coating layer can be treated as a composite system with magnetic pigment being well dispersed in the polyurethane matrix. The effect of CrO_2 magnetic pigment and pigment-polyurethane interaction is investigated in this study using the Rheovibron dynamic mechanical analysis technique. Differences in storage modulus and $\tan\delta$ were observed for different pigment/polyurethane ratios for a variety of polyurethane binders. The presence of CrO_2 magnetic pigment resulted in modulus reinforcement. The shift of $\tan\delta_{max}$ toward higher temperature in the pigmented system suggests the possibility of binder-pigment interaction. The addition of dispersant weakens the interface between binder and pigment which then results in lower $\tan\delta_{max}$ temperature and lower storage modulus.

INTRODUCTION

Thermoplastic polyurethanes, especially polyester-polyurethane, have been widely used as binders in magnetic tape. The addition of rigid inorganic magnetic particles to polyurethane results in physical property reinforcement. Since the viscoelasticity of polyurethane is generally an important attribute to the physical aspects of magnetic tape, it is necessary to determine and understand changes in viscoelastic behavior brought about by the presence of magnetic particles.

With regard to the CrO_2-polyurethane system, a thorough study with the emphasis on friction and binder stability was done by Bradshaw et al.[1] In this study, the intent has been to determine the effect of CrO_2 magnetic particles on the dynamic mechanical properties of polyurethane. The influence of dispersant and lubricant in the composite system was also investigated.

A magnetic coating layer which is composed of magnetic particles in a polymeric binder can be treated as a composite system in which the rigid inorganic magnetic particles are uniformly dispersed in the polymeric continuous matrix.

The model systems used to predict the properties of various composite systems were suggested by Charrier[2]. The discrepancy in dynamic properties between predictions and experimentally obtained values is primarily due to the negligence of binder-pigment interface. Hirai and Kline[3], Kardos, et al[4], Lee and Nielsen[5] and Lewis and Nielsen[6] have concluded that the dynamic behavior of a composite system is not only dependent on the nature of polymer and filler but is often strongly influenced by the character of polymer-filler interface. It is, therefore, our interest to investigate the possible polyurethane-pigment interaction in this study.

EXPERIMENTAL

Two ESTANE[R] polyester polyurethanes, ESTANE[R] 5701 and ESTANE[R] 5706, were used in this study. ESTANE 5701 is a linear polyester polyurethane, while ESTANE 5706 consists of sterically hindered species in both soft and hard segments whose structures are illustrated in Figure 1.

DuPont's CrO_2 pigment with aspect ratio of 10:1 and length of 0.3 to 0.5 μm was dispersed into the polyurethane matrix using the Eiger mill mixing technique. Two formulations, shown in Table I, with and without dispersant and lubricant, were used to prepare the dispersions. The dispersions, which contained approximately 80 percent of pigment by weight on dry base, were further diluted down to lower pigment concentrations by adding polyurethane-THF solutions. The diluted dispersions with various pigment levels were then cast on release paper to make thin films of 50-75 μm thickness. Thin strips of length 2-3 cm were cut from the dried cast films for dynamic property measurements. A Rheovibron Model DDV-II-C equipped with a HP minicomputer was used for the dynamic property measurements. The measurements were run at a scanning rate of 1°C/minute. Complex modulus and $\tan\delta$ values determined are related as follows.

$$E^* = E' + iE''$$

$$\tan\delta = \frac{E''}{E'}$$

E' and E'' are the real and imaginary parts of the complex modulus E^*. In practice, the storage modulus E' can be considered as representing the elastic characteristic of polymers; while the loss modulus E'', on the other hand, represents the viscous characteristic of polymers. All the figures were plotted with raw data and no curve fitting was applied through the data points.

Table I. Magnetic Coating Formulas.

	FORMULA A	FORMULA B
Chromium Dioxide	100	100
Polyester-Polyurethane	25	20
Surfactant	0	3
Lubricant	0	2
THF	185	185
Toluene	46	46

Unit: Percentage by weight based on 100 CrO_2.

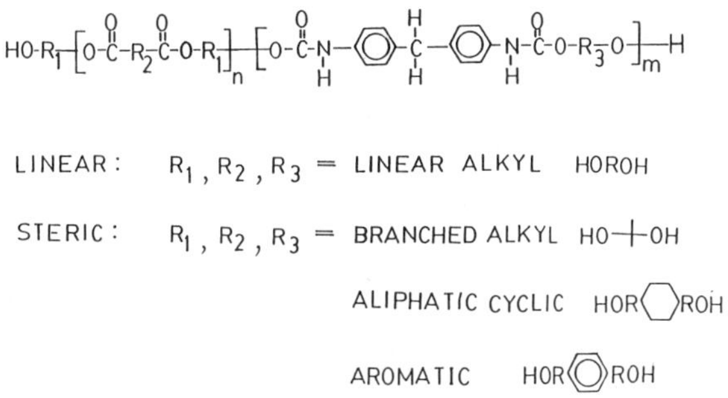

$$HO-R_1\left[O-\overset{O}{\overset{\|}{C}}-R_2-\overset{O}{\overset{\|}{C}}-O-R_1\right]_n\left[O-\overset{O}{\overset{\|}{C}}-\overset{H}{\underset{H}{N}}-\bigcirc-\overset{H}{\underset{H}{C}}-\bigcirc-\overset{H}{\underset{H}{N}}-\overset{O}{\overset{\|}{C}}-O-R_3-O\right]_m H$$

LINEAR: R_1, R_2, R_3 = LINEAR ALKYL   HOROH

STERIC: R_1, R_2, R_3 = BRANCHED ALKYL   HO⊥OH

ALIPHATIC CYCLIC   HOR◯ROH

AROMATIC   HOR◯ROH

Fig. 1.   Structures of thermoplastic polyurethanes.

RESULTS AND DISCUSSION

Effect of Pigment Filling on the Storage Modulus

In general, the pigment presence results in an increase in E' in the glassy region because of the reinforcing effect of a relatively rigid inorganic inclusion. Typical data are shown in Figure 2a and 2b. The reinforcing effect is more prominent in the filled system without the addition of dispersant and lubricant, shown in Figure 2a. With the addition of dispersant and lubricant, the storage modulus decreases to a certain level depending on the amount of these additives and sometimes on the type of polyurethane being used in the system. Table II summarizes the storage modulus of ESTANE 5701/CrO_2 and ESTANE 5706/CrO_2 systems with and without additives.

The storage modulus versus pigment volume fraction ϕ_2 for ESTANE 5701/CrO_2 and ESTANE 5706/CrO_2 systems are shown in Figures 3 and 4, respectively. The log E'- ϕ_2 plot can be approximately expressed by the following equation:

$$\log E' = \log E_1' + A \phi_2 \qquad (1)$$

where E' denotes the storage modulus of a filled system and E_1' the storage modulus of an unfilled system. The dotted line represents the theoretical values for the filler reinforcement only by volume effect. The equation was derived by Guth[7] and can be expressed as follows:

$$E = E_1 (1 + 2.5 \phi_2 + 14.1 \phi_2^2) \qquad (2)$$

where E is the modulus of the composite and E_1 that of the matrix polymer.

Generally, the elastic modulus of a filled system is determined by either the "volume effect" or the "surface effect"[8]. Volume effect means that the increase of modulus is solely from the introduction of a higher modulus rigid particle in the polymer matrix. The surface effect causes an increase by pigment-binder interface interaction. According to this theory, stronger interface between the polymer and the pigment results in higher reinforcement than what is expected by the volume effect. In Figure 3, it is indicated that strong pigment-binder interface exists in ESTANE 5701/CrO_2 filled system. Figure 4 illustrates that the "surface effect" takes place in ESTANE 5706/CrO_2 filled system without the addition of dispersant and lubricant, but the "surface effect" is completely offset by the addition of dispersant and lubricant.

Table II. Storage Modulus of Filled Systems.

P/B	CrO_2 wt.%	ϕ_2	ESTANE 5701/CrO_2 $E' \times 10^{-10}$ (dyne/cm^2)		ESTANE 5706/CrO_2 $E' \times 10^{-10}$ (dyne/cm^2)	
			w/DL	w/o DL	w/DL	w/o DL
Unfilled	0	0	- - -	1.3	- - -	1.6
1/3	25	0.076	1.9	2.2	2.0	2.3
1/2	33	0.11	2.2	2.4	2.4	2.8
1/1	50	0.198	2.6	3.5	2.9	3.4

Symbols and Abbreviations:

P/B:	pigment to binder ratio
ϕ_2:	pigment volume fraction
w/DL:	with dispersant and lubricant
w/o DL:	without dispersant and lubricant

Table III. Tanδ_{max} of Filled Systems at 11 Hz.

P/B	CrO_2 wt.%	ϕ_2	ESTANE 5701/CrO_2 Tanδ_{max} (°C)		ESTANE 5706/CrO_2 Tanδ_{max} (°C)	
			w/DL	w/o DL	w/DL	w/o DL
Unfilled	0	0	--	-17	--	82
1/3	25	0.076	-8	-6	82	85
1/2	33	0.11	-2	0	83	86
1/1	50	0.198	0	2	83	88

Symbols and Abbreviations:

P/B:	pigment to binder ratio
ϕ_2:	pigment volume fraction
w/DL:	with dispersant and lubricant
w/o DL:	without dispersant and lubricant

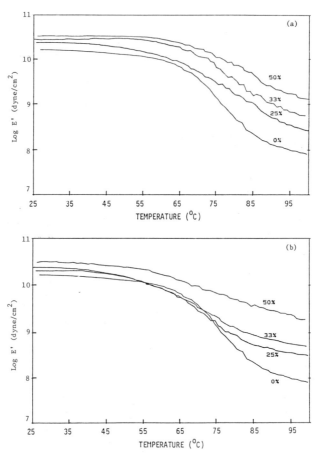

Fig. 2.   Storage modulus E′ vs. temperature of ESTANE 5706/CrO$_2$ filled
system at 11 Hz for various pigment loadings.
(a) Formula A   (b)   Formula B

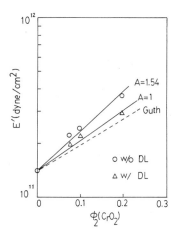

Fig. 3. Storage modulus vs. pigment volume fraction of ESTANE 5701/CrO$_2$
system. The dotted line represents Guth's calculated values.
w/o DL: without dispersant and lubricant
w/DL: with dispersant and lubricant

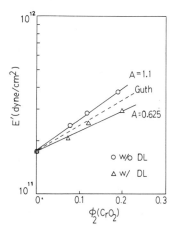

Fig. 4. Storage modulus vs. pigment volume fraction of ESTANE 5706/CrO$_2$
system. The dotted line represents Guth's calculated values.
w/o DL: without dispersant and lubricant
w/DL: with dispersant and lubricant

Effect of Pigment Filler on Tanδ and Activation Energies

The shifting of Tanδ_{max} (maximum in Tanδ peak) has been explained in terms of adsorption phenomena by Seto[9]. The Tanδ_{max} should increase as the interaction between the pigment and polymer increases. Typical data are shown in Figures 5a and 5b. In Figure 5a, without the presence of dispersant and lubricant, tanδ_{max} shifts to higher temperature with higher pigment loading. In Figure 5b, with the presence of dispersant and lubricant, tanδ_{max} is depressed. It is likely that dispersant stays on the pigment surface and loosens the pigment-polymer interface. The lubricant normally disperses in the polymer matrix and results in a lower tanδ_{max}. The amount of tanδ_{max} depends on the levels of dispersant and lubricant.

The summary of Tanδ_{max} at 11 Hz for ESTANE 5701/CrO$_2$ and ESTANE 5706/CrO$_2$ filled systems is presented in Table III. The amount of tanδ_{max} shifting in ESTANE 5701/CrO$_2$ system is higher than that in ESTANE 5706/CrO$_2$ system, indicating that ESTANE 5701 may have a stronger interface interaction with CrO$_2$ pigment.

As shown in Figure 6a and 6b, the dynamic mechanical properties were measured at several frequencies. A plot of frequency vs. the reciprocal of absolute temperature for tanδ_{max} give the apparent activation energy, ΔH^*, for the particular relaxation process.

$$\Delta H^* = -2.303 \ R \ d(\log f)/d(1/T) \tag{3}$$

where R is gas constant, f and T being frequency and absolute temperature of tanδ_{max}, respectively.

The values of ΔH^* for ESTANE 5701/CrO$_2$ and ESTANE 5706/CrO$_2$ systems are given in Table IV. The results of activation energy further suggest that ESTANE 5701 has stronger interface interaction with CrO$_2$ pigment than ESTANE 5706 does.

Table IV. Apparent Activation Energies in Various Systems.

	W/DL* (kcal/mol)	w/o DL (kcal/mol)
ESTANE 5701 (0% CrO$_2$)	----	45.7
ESTANE 5701 (25% CrO$_2$)	56.8	70.6
ESTANE 5701 (33% CrO$_2$)	65.4	73.0
ESTANE 5701 (50% CrO$_2$)	72.6	76.4
ESTANE 5706 (0% CrO$_2$)	----	65.4
ESTANE 5706 (25% CrO$_2$)	59.0	68.0
ESTANE 5706 (33% CrO$_2$)	59.0	69.3
ESTANE 5706 (50% CrO$_2$)	59.0	70.5

*DL: Dispersant and Lubricant

ΔH^* (Apparent Activation Energy) = -2.303 R d(log f)/d(1/T)

CONCLUSIONS

Introducing CrO$_2$ pigment in the polyurethane matrix resulted in modulus reinforcement in the same manner as conventional filler-polymer

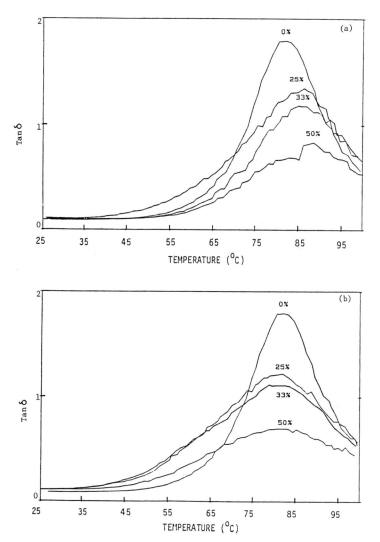

Fig. 5. Tanδ vs. temperature of ESTANE 5706/CrO₂ filled system at 11 Hz for various pigment loadings.
(a) Formula A   (b) Formula B

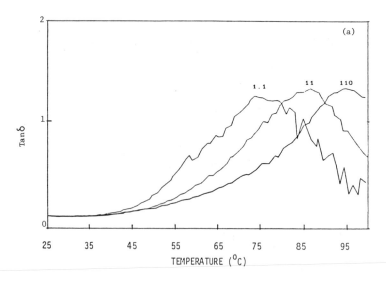

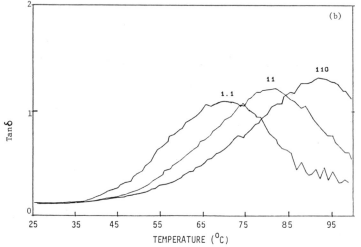

Fig. 6. Tanδ vs. temperature of ESTANE 5706/CrO₂ filled system at
1.1, 11 and 110 Hz at 25 wt.% CrO₂ loading.
(a) Formula A   (b) Formula B

composite systems. Higher pigment loading showed higher storage modulus
in the glassy region.

Without the addition of dispersant and lubricant, both ESTANE
5701/CrO₂ and ESTANE 5706/CrO₂ systems showed pigment-binder interface
interaction. Linear polyester polyurethane (ESTANE 5701) exhibits
stronger interface with CrO₂ pigment than steric polyester polyurethane
(ESTANE 5706) does.

The addition of dispersant and lubricant showed some effect on both
systems. It is believed that dispersant stays mostly on the surface of
the pigment, while the lubricant disperses in the polyurethane matrix as
a plasticizer. Dispersant and lubricant have greater influence on ESTANE
5706/CrO₂ system than on ESTANE 5701/CrO₂ system.

ACKNOWLEDGEMENTS

The authors would like to express their gratitude to the Specialty Polymers & Chemicals Division of BFGoodrich Company for the support of this work and the permission to publish these results.

REFERENCES

1.   R. L. Bradshaw and B. Bhushan, ASLE Trans. 27, 207 (1984).
2.   J. M. Charrier, Polymer Eng. Sci, 15, 731 (1975).
3.   T. Hirai and D. E. Kline, J. Composite Mater., 7, 160 (1973).
4.   J. L. Kardos, W. K. McDonnell, and J. Raisoni, Macromol. Sci., B6, 397 (1972).
5.   B. L. Lee and L. E. Nielsen, J. Polym. Sci., B15, 683 (1977).
6.   T. B. Lewis and L. E. Nielsen, J. Appl. Polym. Sci., B14, 1449 (1970).
7.   E. Guth, J. Appl. Phys., 16, 20 (1945).
8.   K. Sato, Prog. Org. Coat., 4, 271 (1976).
9.   J. Seto, Rubber Chem. Technol., 50, 333 (1977).

POLYESTER-POLYURETHANE INTERACTIONS WITH CHROMIUM DIOXIDE

R. L. Bradshaw and S. J. Falcone

International Business Machines Corporation
General Products Division
Tucson, Arizona 85744

The chemistry of polyester-polyurethanes adsorbed on high surface area, magnetic particles of chromium dioxide is reported. Understanding this chemistry is crucial to the successful use of polyester-polyurethane materials as binders for magnetic recording tape coatings containing chromium dioxide.

The paper discusses the role of chromium dioxide in the hydrolytic degradation of the polyester portions of several polyester-polyurethane materials typically present in magnetic tape coatings. In addition, the oxidative activity of chromium dioxide on the polyurethane segments is presented. While the oxidation was found to occur rapidly at elevated temperatures, no deterioration in the performance of a chromium dioxide tape as a result of oxidation is expected during normal use.

INTRODUCTION

In the course of evaluation and development of polymeric binders suitable for chromium dioxide containing coatings for digital magnetic recording media, a detailed study of the chemistry of the polyester-polyurethane interaction with chromium dioxide (CrO_2) became increasingly important.[1] Previous studies of the degradative reactions that determine the long-term stability of magnetic media clearly established hydrolytic degradation of polyester-polyurethanes as the predominant reaction.[2-5] The active role of chromium dioxide as a catalyst in the hydrolysis reactions leading to degradation of the polyester segments of the polyester-polyurethane binders has only recently been reported.[1,5,6]

A second reaction, the oxidation of the hard-segment (polyurethane) blocks, has only been alluded to in the previous publications.[1,6] Investigations prior to this[2,3] had considered only the oxidative degradation of polyester-polyurethane binders which could arise during aging in air (oxygen) at elevated temperatures. The mechanism of such oxidation was not investigated nor was the involvement of the magnetic particle considered relevant to the oxidation. Oxidative reactions with chromium dioxide have been found to be of critical relevance to the discussions of the resulting impact on the mechanical properties of the coating and on the observed decreasing extractability (or cure) of the coating during thermal aging. The susceptibility of the aromatic portions of the polyurethane hard-segments is well known[7,8], as is the strength of chromium oxides (usually Cr^{+6}) as oxidizing agents for organic materials.[9,10] The potential activity of chromium dioxide as an oxidant for the

385

materials used in typical magnetic-coating formulations made the investigation of such reactions appear a prudent, if not necessary, exercise.

This paper presents some of the experimental evidence that describes the predominant reaction of aromatic polyurethanes with chromium dioxide particles. The discussion concentrates on the chemistry of the hard-segment (polyurethane) portion of the polyester-polyurethane binders as considerable discussion of the role of chromium dioxide on the hydrolytic degradation of the soft-segment (polyester) portion has already been published.[1,6]

EXPERIMENTAL

Commercially available polyester-polyurethanes were used to model conventional formulations used for magnetic-recording-tape coatings. A polyester-polyurethane composed of a butanediol-adipate polyester (pBDA) soft-segment block and a 1,4-butanediol-4,4′-diphenylmethane-diisocyanate (BD-MDI) hard-segment provided an example of a hydrolytically unstable binder for a magnetic-tape formulation containing chromium dioxide. A polyester-polyurethane constructed with a similar BD-MDI hard-segment but with a more hydrolytically stable 1,4-cyclohexanedimethanol-adipate polyester (pCHDMA) soft-segment serves as an example of a more hydrolytically stable binder.

Chromium dioxide magnetic particles manufactured by E. I. DuPont de Nemours & Co. were used to construct model coating formulations and as the reagent for the model-compound reaction studies. The particles possessed a surface area of 25 m^2/g and had been stabilized previously by a proprietary DuPont process. The high surface area ($25 m^2$/grn) of these particles thus provides numerous adsorption sites capable of acting as a topochemical reagent.

Model compounds of the hard-segment portions of a typical polyurethane binder were synthesized from 4,4′-diphenylmethanediisocyanate (MDI) by reaction of recrystallized MDI with methanol to form the bis-methylcarbamate of MDI (MDI-OMe). A model of a potential oxidation product of MDI was synthesized from the bis-methylcarbamate model by refluxing the MDI-OMe model in acetic acid and sodium dichromate to yield the benzophenone analogue. A model of a potential oxidative cleavage product, p-methylcarbamate of benzoic acid, was synthesized from p-aminobenzoic acid with chloromethylformate in pyridine. Precursor materials for other oxidative reactions studied were received from Aldrich and purified as required.

The high-pressure liquid chromatography (HPLC) experiment used a Waters Associates HPLC fitted with a Radial Compression μPorasil column. Samples were passed through the column and eluted with a chloroform and methanol gradient at a l.0 ml/min flow rate. The initial gradient was 2% chloroform, 98% methanol for 18.0 min. After the initial gradient, the eluant solvent for the duration of the run was switched to 100% chloroform. A total run time of 32 minutes was adequate for all the expected reaction products. An ultraviolet (UV) detector was used at 1X sensitivity and a wavelength of 280 nm. Model compounds were used to calibrate the retention times.

The C-13 nuclear magnetic resonance (NMR) spectra were obtained with a Bruker WH-90 FTNMR spectrometer operating at 22.62 MHz. Typically, a total of 16,000 data points were allotted to the free-induction decay (FID) for a sweep of 6024.096 kHz (01 = 8900.000 and 02 = 3400.00). Tetramethylsilane (TMS) was used as the internal reference. Deuterated solvents provided the deuterium lock signal for broad-band proton decoupling. Deuterated dimethyl sulfoxide (D6-DMSO) and deuterated chloroform (CDCl3) were used as solvents. A mixture of DMSO and CDCl3 (60/40) was required for the polymer analyses to obtain useful spectra.

Infrared (IR) spectra were obtained in transmission either as cast films on sodium chloride windows or potassium bromide pellets. A Perkin-Elmer Model 299B grating infrared spectrometer was used to record the spectra of the model compounds as well as extract residues.

Differential scanning calorimetry (DSC) measurements were carried out on the particles with and without adsorbed model compounds as well as before and after aging at 25 and 100°C. A DuPont Instruments Model 910 DSC was used to record the heat flow for 20-25 mg samples as referenced to a pan containing 25 mg of glass beads. Data was collected using a DuPont Model 9900 Thermal Analysis System over the range of -60 to 120°C with a heating rate of 2°C/min under a blanket of dry nitrogen.

Thermogravimetric analysis (TGA) was carried out using a DuPont model 951 Thermogravimetric Analyzer coupled to a model 9900 Thermal Analysis System. Thermal analysis of chromium dioxide with and without adsorbed urethane model compounds was carried out under a dry nitrogen atmosphere using compressed cakes of chromium dioxide powder (typically 20 mg) heated from 25 to 800°C at 20°C/min. The powders were washed with freshly distilled methylene chloride and vacuum-dried at 30°C prior to TGA measurements, with the exception of the initial slurry sample, which was evaporated to dryness without washing. The thermal analyses were routinely run in duplicate to verify the results. Each sample was cooled slowly to room temperature, quenched with liquid nitrogen, and rerun to record the effect of temperature cycling on the coating transitions and pigment-binder wetting interactions.

Reactions of model compounds and polymers were carried out by dispersing with chromium dioxide particles in methylene chloride at 5% by weight of the chromium dioxide using mechanical and ultrasonic dispersion techniques. The particles were then dried either by distillation of the methylene chloride under vacuum using a rotovaporator or by vacuum filtration and air drying in a hood. The particles were characterized by TGA and DSC after drying to establish the initial extent of adsorption and/or reaction. Samples of these particles were then exposed to various temperature and humidity environments to study the effect of these environments on the interaction of the model compounds with the chromium dioxide particles.

RESULTS AND DISCUSSION

The hydrolytic stability of polyester-polyurethanes is significantly reduced by the introduction of the relatively high-volume fractions of chromium dioxide required for high-density, digital tape products.[1] This is demonstrated by the data presented in Figure 1. The curves compare the molecular weights of extractions of polyester-polyurethane materials containing 80% chromium dioxide by weight. An example of a coating containing 80% by weight Fe_2O_3 formulated with the hydrolytically unstable poly(butanedioladipate) (pBDA/BD-MDI) binder is also presented. These results illustrate the sensitivity of polyester-polyurethanes to hydrolytic degradation and the increased rate of molecular weight loss observed for coatings containing CrO_2 as compared to those containing iron oxide magnetic particles.

As was demonstrated in previous publications,[1,6] the extractable binder molecular-weight characterization via gel-permeation chromatography (GPC) is not as straightforward as might be inferred from the data presented here.

The extraction of the polyester-polyurethane is complicated by the competitive adsorption of the binder on the CrO_2 surface. The apparent insolubilization of the binder thus causes fractionation of the polyester-polyurethane during solvent extraction, which gives the appearance of a reduced molecular weight but is not necessarily indicative of molecular-weight reduction due to hydrolytic chain scission.

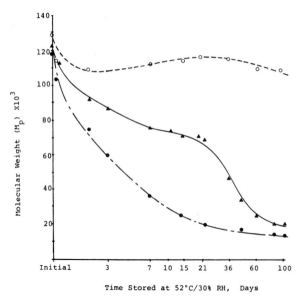

Figure 1. THF Extractable Binder Molecular Weight (GPC) after Hot/Humid Aging.

---o--- γFe_2O_3 + (pBDA/BD-MDI) Binder
·—•—· CrO_2 + (pBDA/BD-MDI) Binder
— ▲ — CrO_2 + (pBDA/BD-MDI) Binder

   The most probable site for adsorption of the polyurethane on the acidic surface of the chromium dioxide particles was anticipated to be through the relatively strong acid-base interaction of the carbamate function (specifically through the carbamate nitrogen). Once adsorbed on the surface of the particle, the benzylic carbon of the MDI was expected to undergo oxidation to the benzophenone analogue, as shown in the reaction scheme I in Figure 2. Further oxidation leading to the possible cleavage of the benzophenone moiety in the presence of humidity was conceivably capable of providing a route to chain scission of the otherwise stable polyurethane hard-segments, as indicated in reaction scheme II in Figure 2. These proposed reactions became the objects of detailed study because the environmental stability of the hard-segment was the remaining issue once a stable polyester had been obtained.

   Model-compound studies soon indicated the unique oxidative properties of CrO_2, as indicated by the reactions and their products summarized in Figure 3. As noted in Figure 3, these reactions were carried out at elevated temperatures (80°C) to effect high yields in a reasonable period of time. The reactions were found to give similar products at room temperature as well but required much longer reaction times (days rather than hours) to produce respectable yields. None the less, it is readily apparent from the scope and yields of these reactions that CrO_2 is a very selective oxidant for benzylic carbons. It is also specific in its preference for producing aldehyde functional oxidation products for primary aliphatic carbon substrates and ketone products from secondary aliphatics. The reactions were highly selective with little or no evidence of further oxidation to the carboxylic acid product found for the series of compounds studied. This observation is quite remarkable given the previous knowledge of chromium oxidation reactions.[9,10]

   We believe the mechanistic explanation involves topochemical oxidation of the reactive benzylic carbon after adsorption of the polyurethane on the acidic surface of CrO_2. The acid-base interaction would be expected to involve the weakly basic carbamate functional groups. Such adsorption would thus disrupt the hydrogen-bonded association of the polyurethane hard-segments, which persists even in polar solvents. Once adsorbed on the

388

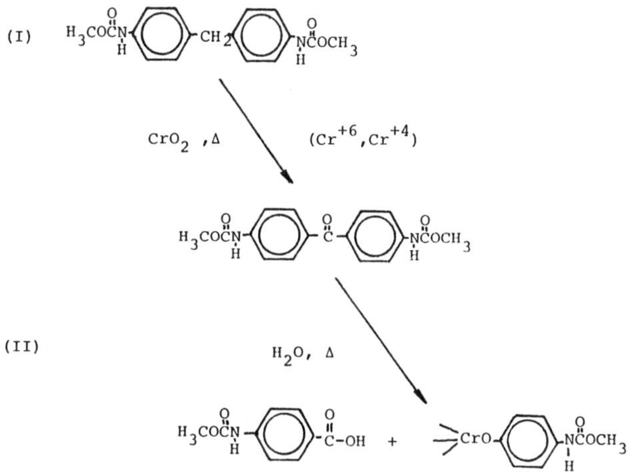

Figure 2. Reaction Scheme for Chromium Dioxide Oxidation of Polyurethane Hard-Segment Benzylic Carbon (I) and Possible Cleavage Reaction (II).

CrO_2 surface, the benzylic carbon participates in the stabilization of the carbamate base through resonance. This places the benzylic carbon in close proximity to CrO_2 such that hydrogen transfer from the benzylic carbon can occur. The reduction of Cr^{+4} to Cr^{+3}(CrO_2 to form Cr_2O_3) involves the concerted reduction of two chromium atoms and the elimination of an oxygen atom. The attack by the oxygen on the benzylic carbon would thus yield a chromate-ester adduct intermediate, typical of chromate oxidations.[10]

The transition state probably resembles the conventional chromate ester adducts,[10] with the notable exception that the reaction occurs on the surface of an insoluble solid surface. A concerted mechanism thus becomes more likely due to the fixation of the reactant species on the CrO_2 crystal surface. Geometric and energetic constraints probably then contribute to inhibition of subsequent addition of water to the intermediate, which would be required to complete cleavage of the chromate-benzylic carbon adduct to produce the corresponding benzoic acid analogue. Proton transfer from the benzylic carbonium ion to the CrO_2/Cr_2O_3 surface appears to be the preferred route leading to aldehydes for primary carbon or to ketones for secondary carbons. Because the reaction produces either the unoxidized starting material or the corresponding aldehyde or ketone, the mechanism proposed seems to be appropriate.

These reactions suggested that while facile oxidation of the benzylic carbon in the polyurethane hard-segment of chromium dioxide tape coatings (reaction scheme I in Figure 2) is probably favored, the cleavage reaction (scheme II) is less likely to occur. Such a prediction is reinforced by the apparent failure of CrO_2 to cleave benzil to two benzoic acid fragments during the oxidation of benzoin (see the bottom of Figure 3).

The study of the relevance of the CrO_2 oxidation reactions to polyester-polyurethane binder degradation in a tape coating was pursued through analysis of tape extracts and comparison to the products obtained for the bis-methylcarbamate (MDI-OMe) model. The possible reactions (shown in Figure 2) were investigated with various analytical and experimental methods. One particularly informative investigation involved quantitative liquid-chromatographic analysis of the products of the reaction of the MDI-OMe model compound with chromium dioxide particles.

389

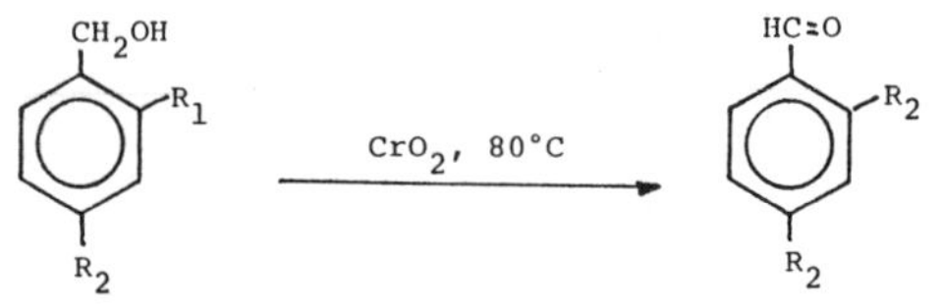

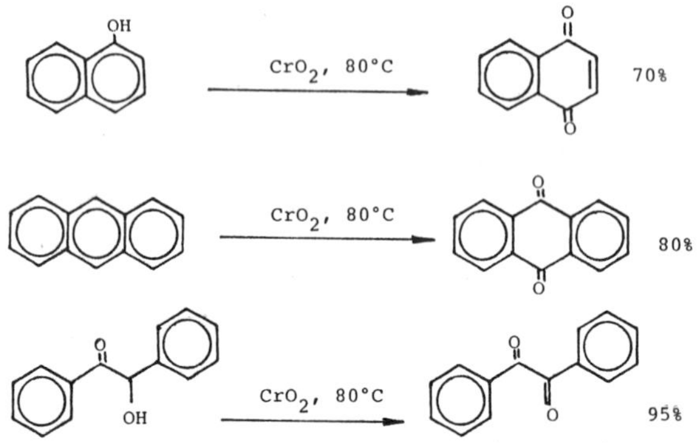

Figure 3. Model Compound Reactions with Chromium Dioxide.

High Pressure Liquid Chromatography Studies

A series of compounds were synthesized to model the expected oxidation products of the bis-methylcarbamate of MDI. These model compounds were used to develop an HPLC method for the separation of the reaction products extractable from CrO_2/MDI-OMe model systems. A representative trace for a mixture of the model compounds at equivalent concentrations (mg/ml) is presented in Figure 4. It is evident that the benzophenone model compound exhibits the greatest UV detector response, while the cleavage product model—p-methylcarbamate of benzoic acid—possesses the weakest UV detector response. The characteristic retention times for these probable reaction products are reproducible to within five to seven hundredths of a minute. The HPLC method thus appeared capable of excellent resolution of the reaction products.

The investigation of polyurethane oxidation by chromium dioxide particles was undertaken using this HPLC method by preparing mixtures of the MDI-OMe model compound with chromium dioxide. A slurry of chromium dioxide with the MDI-OMe model compound in methylene chloride was prepared using mechanical stirring. The slurry was filtered and the filtrate collected. The residue was air dried overnight at 25°C prior to aging or gravimetric analysis.

The methylene chloride was removed from the filtrate by vacuum distillation and the residue taken up into 2 ml chloroform ($CHCl_3$). This sample served as a control for the starting material extractable from the chromium dioxide/model-compound slurry. The HPLC trace is presented in Figure 5. Upon exposure of the dry, chromium dioxide/model-compound powder to humid aging at 35°C, the starting material was noticeably reduced in concentration, as indicated by the LC trace presented in Figure 6. The benzophenone-analogue oxidation product is the dominant species present in the extract. The region in which any cleavage product (the benzoic acid analogue) would have been expected is observed as a weak, broad peak in both the initial extract (Figure 5) and in the aged-sample extract (Figure 6). Little, if any, cleavage is indicated in these results.

390

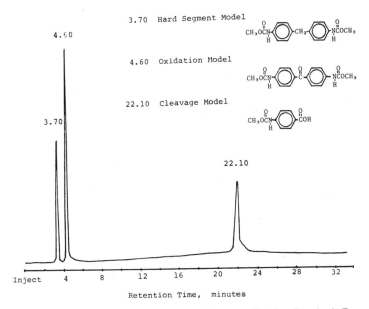

Figure 4. HPLC Analysis of Polyurethane Hard-Segment Models. Standards Traces.

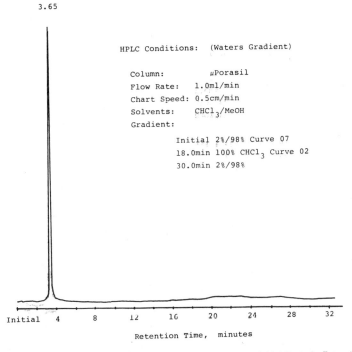

Figure 5. HPLC Analysis of Polyurethane Hard-Segment Model. Initial Sample Extract Prior to Thermal Aging.

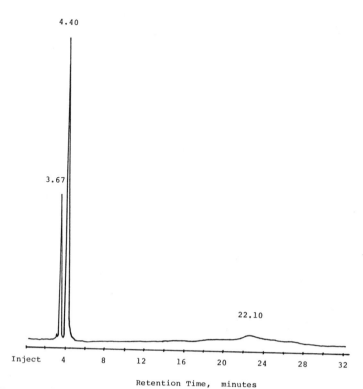

4.40

3.67

22.10

Inject 4   8   12   16   20   24   28   32

Retention Time, minutes

Figure 6. HPLC Analysis of Polyurethane Hard-Segment Model. Extract of Particles Exposed to 35°C/67% RH for 86 Days.

Figure 7 presents the results for the reaction products obtained upon exposure at higher temperature. Only a small amount of the initial model hard-segment is evident in the high-temperature aged sample indicating effective conversion of the benzylic carbons to the corresponding benzophenone. A small amount of the cleavage product from hydrolysis of the benzophenone-chromium dioxide adduct is also possibly present, as indicated by the broad, weak peak observed at 22.10 minutes. This peak was not reproducible, however, even for this high-temperature aged sample.

The HPLC results were correlated with the model-compound retention times and semiquantitated using peak height and area ratios. The results are summarized in Table I below:

Table I. HPLC Analysis of Chromium Dioxide Reaction with Bis-methylcarbamate Model Compound

Compound	Initial	Extract Composition, Area %	
		86 Days 35°C/67% RH	87 Days 52°C/30% RH
Model compound (I)	100.0	13.9	4.7
Model compound (II)	0.0	85.9	95.2
Model compound (III)	0.0	0.2	0.1

It is apparent from the HPLC results that oxidation of the polyurethane hard-segment models by chromium dioxide efficiently converts the benzylic carbon of MDI to a benzophenone. Only a trace of the benzoic acid cleavage product might be present in these extracts.

392

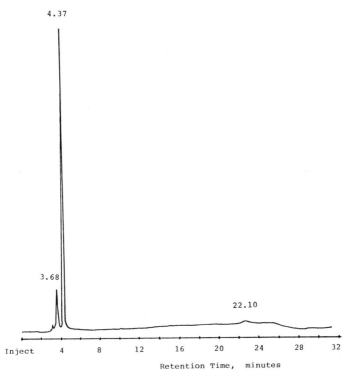

Figure 7. HPLC Analysis of Polyurethane Hard-Segment Model. Extract of Particles Exposed to 52°C/30% RH for 87 Days.

Note that the HPLC studies were carried out on the extracts of the chromium dioxide-model compound slurried residue. It is likely that additional material remained bound to the chromium dioxide particle surface. Any material not present in the extract would not be detected. The HPLC results would thus incorrectly describe the distribution of the actual reaction products. Since the amount of material relative to the weight of the chromium dioxide reagent was very small, microscale thermalgravimetric analysis of the extracted residues was carried out to obtain a mass balance of reaction products.

Thermal Analysis Studies

The TGA trace for chromium dioxide particles slurried in methylene chloride without the model compound present is indicated by the solid line trace in Figure 8. The slurry was dried by vacuum filtration, followed by additional vacuum drying at 30°C to achieve a constant weight prior to TGA analysis. Little weight loss other than adsorbed volatiles is observed prior to -450°C, at which the CrO_2 conversion to Cr_2O_3 with loss of oxygen is observed. The region of interest from 25 to 400°C is presented on an expanded scale in Figure 9. The CrO_2 particles thus appear to retain roughly 1% volatiles even after drying, with little else noted in the TGA traces upon further heating. The fact that the volatile material boils off below 100°C suggests that methylene chloride rather than water is present.

Slurries of CrO_2 with the MDI-OMe model, on the other hand, indicate considerable additional material (~4%) present on the particle surface prior to exhaustive extraction with methylene chloride. The TGA results for the methylene chloride extracted particles, presented in Figure 10, illustrate the tenacity with which the oxidation products are held to the CrO_2. This retained organic layer on the surface of the CrO_2 would be expected to render the particle surface less polar, which might be the reason for the reduced weight loss for the low temperature ($<$ 100°C) region of the TGA traces (Figure 10) after reaction, as compared to the solvent washed particles alone.

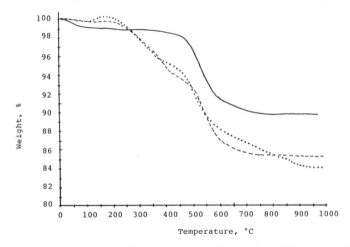

Figure 8. Thermogravimetric Analysis (TGA) of Chromium Dioxide. (——) CH₂Cl₂ slurried, dried; (------) MDI-OMe/CH₂Cl₂ slurried, dried, aged 20 days at 25°C; (······) MDI-OMe/CH₂Cl₂ slurried, dried, aged 20 days at 100°C.

In the region of the TGA trace between 100 and 200°C, little weight loss is observed; between 200 and 325°C, a significant weight-loss region is observed for the CrO_2/MDI-OMe model-compound slurries. The amount of material lost in this region (Figure 10) increases with increased oxidation. This behavior probably indicates an increase in the amount of the bound organic residue on the CrO_2 particle surface. This unextractable, tightly bound organic material does, however, indicate the need for correction of the HPLC product ratios, because for the extensively oxidized sample obtained after 20 days at 100°C, approximately 2.5% by weight of the sample is surface-bound organics.

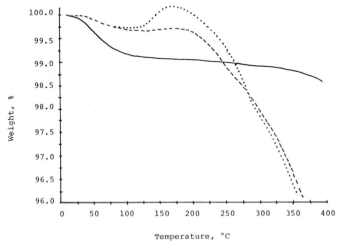

Figure 9. Thermogravimetric Analysis (TGA) of Chromium Dioxide. (——) CH₂Cl₂ slurried, dried; (------) MDI-OMe/CH₂Cl₂ slurried, dried, aged 20 days at 25°C; (·······) MDI-OMe/CH₂Cl₂ slurried, dried, aged 20 days at 100°C.

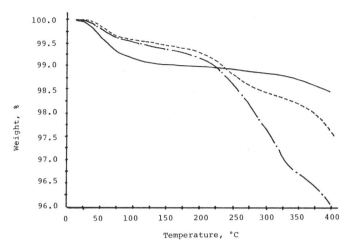

Figure 10. Thermogravimetric Analysis (TGA) of Chromium Dioxide. (———) CH₂Cl₂ slurried, dried; (------) MDI-OMe/CH₂Cl₂ slurried, dried, aged 20 days at 25°C and extracted with CH₂Cl₂ (4X); (_ ._) MDI-OMe/CH₂Cl₂ slurried, dried, aged 20 days at 100°C and extracted with CH₂Cl₂ (4X).

These results indicate that as much as 40% of the material initially adsorbed onto the particle surface is not recovered during the solvent-extraction process used to obtain the HPLC samples. The identity of the unextracted organic is important to any attempt to correct the product mass balance. Additional analyses carried out with the TGA for the various slurried chromium dioxide samples provide considerable insight as to the probable composition of the unextracted organic material.

The TGA traces presented in Figures 9 and 10 are consistent with the expected behavior of such topochemical oxidations as conceived by the proposed mechanism. These tightly bound organic residues are, therefore, probably the benzophenone oxidation products rather than the starting material or cleavage product residues. All the traces reported represent the TGA results obtained for the unextracted slurry after drying or hot and humid aging, as indicated. Note that prior to hot and humid aging (that is, oxidation of the MDI-OMe benzylic carbon to the benzophenone), the amount of the adsorbed organics is relatively small as detected by TGA (Figure 10). All of the adsorbed unaged organics are boiled-off below 300°C. After oxidative aging, the material requires temperatures up to 350°C to complete the volatilization of this material. In addition, considerably more material is present on the surface of these high-temperature aged particles after methylene chloride extraction than on the corresponding ambient-aged particles.

Furthermore, slurries of the model of the cleavage product, the p-methylcarbamate of benzoic acid, with CrO₂, did not leave detectable amounts of material on the methylene chloride extracted particles. Based on these observations and the HPLC results described previously, it seems reasonable to ascribe the behavior of these unextracted organic residues to the oxidized product rather than to retention of the starting material or the cleavage product.

The HPLC experiments explored the extractable products while TGA experiments permitted quantitation of the amounts of extracted and nonextracted material. Differential scanning calorimetry (DSC) experiments were also carried out to observe the expected adsorption and reaction with chromium dioxide for the dried slurries. The data presented in Figure 11 compares the traces obtained for the initial heating of three samples of chromium dioxide slurries. Figure 12 compares the results obtained for the same samples after slowly cooling and rescanning the sample after cooling. This approach was found to provide additional details of the interaction of the adsorbed species on the chromium dioxide particles.

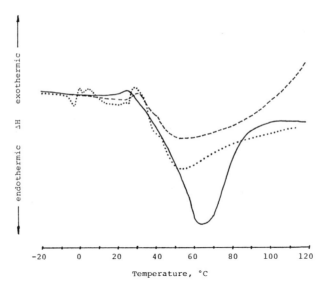

Figure 11. Differential Scanning Calorimetry (DSC) Traces of Chromium Dioxide. (———) slurried with CH₂Cl₂, dried, aged 20 days at 25°C; (------) slurried with MDI-OMe in CH₂Cl₂ dried, aged 20 days at 25°C; (······) slurried with MDI-OMe in CH₂Cl₂, dried, aged 20 days at 100°C; Initial Heating.

The DSC traces recorded for the CrO₂ particles after slurry mixing with methylene chloride were expected to be relatively clean. The results presented in Figure 11, on the other hand, show the presence of an adsorbed material that releases the heat of adsorption as a weak exotherm around 26°C. A broad endothermic region is observed from 30 to about 90°C, which is typical of loss of a volatile material. Because adsorbed water would appear well above this temperature, it is more likely retained methylene chloride (bp ~40°C).

The persistence of solvent is surprising since the particles were subjected to aggressive vacuum drying. Methylene chloride, furthermore, was not expected to display such a strong affinity for the CrO₂ surface. This affinity is further demonstrated by the results obtained for the same sample upon reheating after cooling, as shown in the analogous trace presented in Figure 12. Despite heating to 120°C, the endothermic peak is still visible, although weaker. As the DSC experiments were carried out under a nitrogen atmosphere using closed (but not hermetic) aluminum pans, the reappearance of the endotherm is probably the result of recondensation of methylene chloride on the particles during cooling.

When the CrO₂ particles were slurried with the polyurethane hard-segment model, the endothermic peak observed from 35 to 80°C is still evident, but noticeably shallower and rising in an exothermic direction above 80°C. The trace obtained for the model compound slurried over CrO₂ aged at ambient is also noticeably shifted to higher temperature. This presumably results from the competition afforded by the MDI-OMe model compound for the particle surface. A strong exothermic region beginning just above 54°C is evident in the initial heating trace for the MDI-OMe model system. This is believed to be associated with the oxidation of the adsorbed model compound to the benzophenone analogue during the heating of the sample. This explanation is consistent with the observation (Figure 12) that upon cooling and reheating, the solvent adsorption is little affected while the exotherm is markedly weaker and shifted to above 100°C. This is presumed to be evidence of the reaction of the benzylic carbons with the CrO₂ surface at temperatures near 100°C.

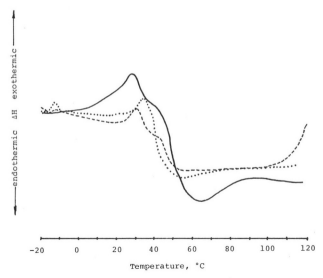

Figure 12. Differential Scanning Calorimetry (DSC) Traces of Chromium Dioxide. (——) slurried with CH₂Cl₂, dried, aged 20 days at 25°C; (------) slurried with MDI-OMe in CH₂Cl₂ dried, aged 20 days at 25°C; (······) slurried with MDI-OMe in CH₂Cl₂, dried, aged 20 day s at 100°C; Cooled from 120°C, quenched to -60°C and reheated.

When a sample of the MDI-OMe model slurry on CrO_2 was aged at 100°C, the expected oxidation was almost complete by HPLC analysis. The DSC data presented as the dotted traces in Figures 11 and 12 are consistent with this expectation. The sharp exothermic region near 100°C observed for the ambient-aged sample is absent in the DSC traces obtained for the 100°C aged sample. The adsorbed solvent peak (35°C) persists even for this sample. Apparently, the sample-preparation methodology is reproducibly ineffective in removing the methylene chloride, or another feature of the CrO_2 not related to the solvent, persists through these heating cycles.

By plotting the results we obtained for these model-compound slurries on the same scales in Figures 11 and 12, the changes imparted by exposure to aging are more evident. As shown in the plots of the initial heating curves presented in Figure 11, the solvent endothermic peak from 30 to 90°C is dominant only in the absence of the model compound, which reduces the amount of solvent retained. In addition, the slurries with the polyurethane model require temperatures near 100°C to effectively complete the oxidation. This is indicated by comparing the two traces in Figure 11 for the model polyurethane adsorbed on CrO_2 stored at ambient and at 100°C for the same time interval. Only a weak exothermic peak is observed in the region above 80°C for the sample exposed to high temperature, while a significant exotherm is observed for the ambient-aged sample. These results, not unexpectedly, suggest that the benzylic oxidation is incomplete after 20 days at 25°C, while exposure to 100°C increases the extent of the reaction. Subsequent heating, therefore, yields a significantly lower exotherm than that observed for the ambient-aged sample. This explanation is further supported by the DSC results obtained on the second heating of these samples, as shown in Figure 12.

The DSC plots for the second heating of the samples shown in Figure 12 permit a view of the samples after removal of a significant portion of the adsorbed solvent and a short period of high-temperature heating. The two traces obtained for the particles slurried with the polyurethane model compound are very similar, with the exception of the exothermic region present in the ambient-aged sample above 80°C. The sample aged for 20 days at 100°C does not exhibit such an exothermic behavior. Conversion of the benzylic carbon to the ketone was essentially complete for this sample. The TGA for this 100°C aged sample showed an appreciable amount of retained material (2.5%, Figure 10) even after solvent extraction and the DSC indicates the material to be an oxidized product, the binding to the CrO_2 surface might be indicative of the stability of the CrO_2 complex with the oxidized benzylic carbon of the polyurethane hard-segment. The precise identity of the material, however, could not be unequivocally established with these analyses.

Spectroscopic Studies

Only the HPLC results presented earlier support the assignment of the oxidation product as the benzophenone analogue rather than the cleavage product. Infrared (IR) analysis of the extracted residues was used in an effort to identify the materials further. Resolution of the characteristic absorption bands in the IR spectra for models of the oxidation products provided a reference set for the interpretation of spectra obtained for the CrO_2 slurry extract residues. These spectra were obtained as KBr pellets, while most of the extract residue spectra were obtained as thin films cast from the methylene chloride, acetone, or chloroform extraction solvents on sodium chloride (NaCl) disks.

The carbamate carbonyl band (amide I) at 1700-1710 cm⁻¹ was expected to dominate the carbonyl region of the spectra for all of the expected oxidation products.[11,12] The ketone carbonyl absorption band was expected to occur in the same region of the spectra as the amide I.[11,12] The benzoic acid analogue of the potential cleavage product was also expected to exhibit carbonyl absorption bands in the same region. Two bands were expected for the acid as a solid cast on NaCl. Monomeric acid was expected to exhibit a weak band near 1740 cm⁻¹, while the associated dimer was expected to exhibit a strong absorption near 1700 cm⁻¹ This dimer absorption would thus not be readily distinguished from the ketone or carbamate carbonyl absorptions for mixtures of the products.

The amide I band for the unoxidized model was measured at 1700 cm⁻¹. The ketone oxidation product model exhibits both the amide I band at 1710 cm⁻¹ and the ketone band at 1690 cm⁻¹. The model of the cleavage product, p-methyl carbamate of benzoic acid, was found to contain bands in the same region. The band at 1700 cm⁻¹ is probably the amide I band, while the absorption at 1750 cm⁻¹ is associated with the benzoic acid carbonyl stretching vibration. It is thus apparent that mixtures of these possible reaction products would contain broad bands in this region of the IR spectrum making conclusive identification difficult.

As shown in Figures 13 and 14, the extract residues obtained for the extracts of the MDI-OMe model-compound slurry on CrO_2 were broad and poorly resolved in this region. The extract residues thus appear to be mixtures. Comparison of the relative peak intensities for the absorption bands in the region from 1720 to 1500 cm⁻¹, however, do suggest that the ambient-aged extract (Figure 13) contains considerably more of the unoxidized material than the extract obtained for the slurry aged at 100°C shown in Figure 14. Note also that less material was extracted (that is, weaker IR intensity for an equivalent aliquot of extract) from the 100°C aged sample compared to the ambient-aged sample. This is consistent with the results of the HPLC analysis discussed earlier.

The IR analysis results obtained for the extracts of the model-compound slurries with CrO_2 thus proved to be inadequate to identify the mixtures obtained for the extracts. Also note that methylene chloride, although suitable for the preparation of the CrO_2 slurries, is not a particularly good solvent for the model compounds. Acetone is a better solvent for the oxidation products. Tetrahydrofuran (THF) or dimethylformamide (DMF) are efficient extraction solvents but not particularly useful for carrying out the reactions on CrO_2. When sequential extraction was carried on the model-compound slurries, very small amounts of material were recovered in acetone, THF, or DMF extracts. The IR of these extracts more closely resembled the benzophenone oxidation product than the corresponding benzoic acid model, as shown in Figure 15.

398

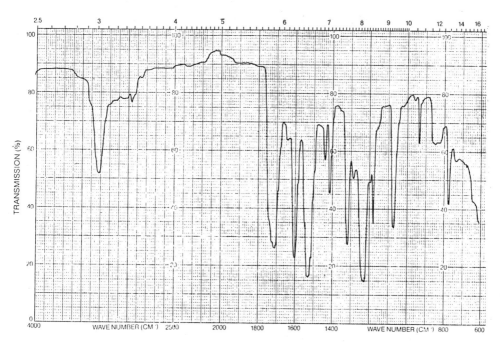

Figure 13. Infrared Spectrum (Grating), Thin Film Cast on NaCl from Methylene Chloride. Methylene chloride extract of CrO_2 slurry with MDI-OMe model after 20 days aging at 25°C.

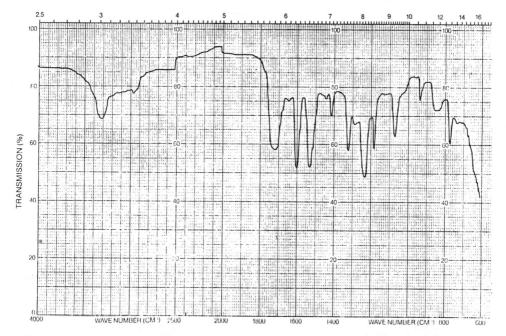

Figure 14. Infrared Spectrum (Grating), Thin Film Cast on NaCl from Methylene Chloride. Methylene chloride extract of CrO_2 slurry with MDI-OMe model after 20 days aging at 100°C.

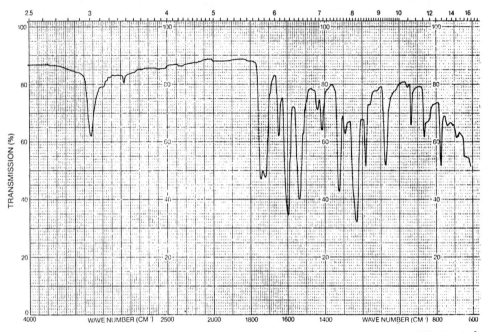

Figure 15. Infrared Spectrum (Grating), Thin Film Cast on NaCl from Tetrahydrofuran.. THF extract of CrO₂ slurried with MDI-OMe model in CH₂Cl₂ after aging 20 days at 100°C. Sequential extraction with THF after CH₂Cl₂ extraction.

The material retained on the CrO₂ after methylene chloride extraction thus appears to be the oxidation product or products. The unequivocal identification of this material, however, was not resolvable through IR spectral analysis of the extracts alone. HPLC analysis of the chloroform and THF extracts, on the other hand, resolved the possible reaction products, as was demonstrated earlier. HPLC analysis of the THF extract of the CrO₂ slurry with the MDI-OMe model aged 20 days at 100°C did not detect peaks associated with either the unoxidized MDI-OMe model or the benzoic acid analogue. Only a peak corresponding to the benzophenone oxidation product was observed. It seems relatively certain then that the material held tenaciously on the CrO₂ particle surface is the benzophenone oxidation product.

To further characterize the model systems and facilitate the characterization of the polyester-polyurethane coatings containing chromium dioxide, C-13 NMR studies of the model compounds were also undertaken. The general C-13 literature gave useful guides to the assignment of the model compound C-13 resonances.[13,14] A particularly relevant paper by Delides et al.[15] gave a great deal of guidance in the assignments of the C-13 NMR resonances for polyurethanes. That paper provided much of the corroboration in the otherwise tentative assignments of the carbon resonances for the CrO₂/polyurethane model systems. However, the paper was in error in its assignments for bis-methylcarbamate of MDI when it assigned one of the downfield resonances of the four aromatic carbons to the nitrogen substituted carbon. We undertook an additional effort to ensure that the assignments of this critical model compound were correct.

The C-13 NMR spectrum of the methyl carbamate model is indicated in Figure 16. The benzylic carbon, à=41.9 ppm, (marked by (g) in the structure in Figure 16 is an important resonance which is absent in the benzophenone model C-13 resonances shown in Figure 17. The resonance of the methyl group of the carbamate (52.2 ppm) is present in both spectra. It is designated a (f) in Figure 16 and by a (g) in Figure 17. The conspicuous absence of a resonance near 42ppm for the benzylic carbon, the shift in the aromatic ring carbon resonances, and appearance of a new resonance line (b) at 144 ppm (C=O) in the spectrum of the benzophenone model suggested the potential utility of C-13 NMR analyses for the oxidation products extractable from the CrO₂ tape extracts. Comparison of these simple model compound C-13 NMR spectra with more complex, polymeric models of a typical polyester-polyurethane was therefore undertaken.

In the spectrum presented in Figure 18, a low molecular weight polyurethane (MW~20,000 by GPC in THF) derived from 1,4-butanediol and 4,4′-diphenylmethane diisocyanate is used to model the hard-segment portions of the polyester-polyurethane binders used in conventional magnetic-tape-coating formulations. The assignments are indicated on the accompanying structure and agree with the assignments made for the model in Figure 16 and reported in the literature.[15] These models and the structural

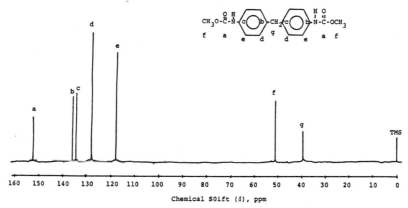

Figure 16. Broad Band Decoupled, C-13 NMR Spectrum of Bis-methylcarbamate of MDI in D₆-DMSO/CDCl₃.

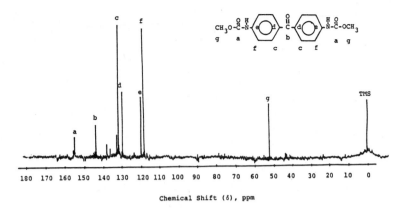

Figure 17. Broad Band Decoupled, C-13 NMR Spectrum of Bis-methylcarbamate benzophenone in D$_6$-DMSO/CDCl$_3$.

assignments present in the literature provided the elements required to characterize the extracted material obtained for a typical magnetic-tape polyester-polyurethane binder.

The next step was to use C-13 NMR techniques to study the extracts from a polyester-polyurethane coating containing the typically high concentrations of CrO$_2$ (80% by weight) normally employed in 80% CrO$_2$ by weight. The presence of particles of CrO$_2$ or even chemically bound chromium species in the binder extracts, however, presented some experimental difficulties due to the interferences introduced by the magnetic properties of these chromium species. The use of a cation exchange resin was found to minimize the interference, but some shift of the resonances for the degraded coating extracts may persist.

In the NMR trace shown in Figure 19, the extract obtained by methyl-ethyl-ketone (MEK) extraction of a hot and humid-aged polyester-polyurethane coating containing 80% by weight chromium dioxide is presented. The sample was not sufficiently concentrated to produce the desired spectrum quality even after a large number of scans, but does illustrate the information obtained through C-13 NMR analyses of degraded tape extracts. The appearance of a strong resonance line at 144 ppm in this extract supports the presence of oxidation of at least some of the benzylic carbons of the polyurethane hard-segments. Because the resonance at 42 ppm is associated with the benzylic carbon of unoxidized polyurethane segments, it appears that a significant portion of the extracted material has not been oxidized. Even the relatively severe aging environments used for these tapes samples appear to be ineffective in oxidizing all of the available benzylic carbons.

Because the model-compound studies had indicated that temperatures well above 80°C were required to achieve efficient benzylic carbon oxidation, it is not surprising that extracts of tapes aged at temperatures well below 80°C exhibit incomplete oxidation. Recall,

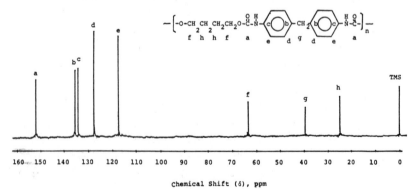

Figure 18. Broad Band Decoupled, C-13 NMR Spectrum of Poly(1,4-butanediol-4,4'-diphenylmethanediisocyanate) in D$_6$-DMSO/CDCl$_3$.

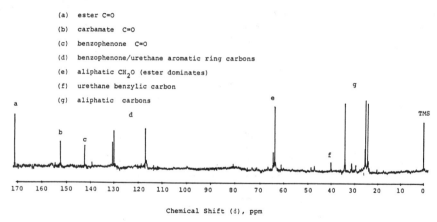

(a) ester C=O
(b) carbamate C=O
(c) benzophenone C=O
(d) benzophenone/urethane aromatic ring carbons
(e) aliphatic CH₂O (ester dominates)
(f) urethane benzylic carbon
(g) aliphatic carbons

Chemical Shift (δ), ppm

Figure 19. Broad Band Decoupled, C-13 NMR Spectrum of Methylethyl Ketone Extract of CrO₂-Polyester-Polyurethane Coating. Aged 25 days at 60°C/30% RH.

furthermore, that the oxidized model compounds displayed a tenacious adsorption and reluctance to be completely extracted from the CrO₂ particles. It seems likely, therefore, that additional oxidized polyurethane might have been formed during the hot and humid aging in a polyester-polyurethane coating containing chromium dioxide. This oxidized material was probably not completely extracted by the ketone solvent used for these tape extracts.

Additional sharp resonances in the aliphatic carbon region of the C-13 NMR spectrum (20-70 ppm, Figure 19) indicate that a considerable amount of polyester soft-segment is present in the extract as well. Although not detailed here, appropriate resonances for the end-groups associated with hydrolytic degradation of the polyester were indicated in the extracts, based on published assignments.[13-15] The IR spectrum of the extract, presented in Figure 20, also indicates hydrolytic degradation of the polyester segments by the dramatic increase in the bands associated with alcohol and carboxylic acid functional groups, which are produced during hydrolysis of a polyester.

When a polyester-polyurethane coating containing CrO₂, typical of magnetic-coating formulations, is aged severely at high temperature and humidity, the molecular weight is reduced through hydrolytic degradation of the polyester.[1-4] The potential for oxidative reactions of the polyurethane structural units to participate in degradation of a magnetic-tape coating has been investigated as well. The evidence obtained through this

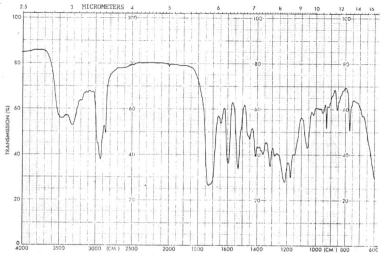

Figure 20. Infrared Spectrum (Grating), Thin Film Cast on NaCl from Acetone. Acetone extract of CrO₂ tape coating (pBDA-pMDI-BD) aged 25 days at 60°C/30% RH.

study indicates that oxidative reactions do occur, preferentially at the benzylic carbons of the polyurethane hard-segments. Furthermore, the oxidation appears to require temperatures sufficiently in excess of those to which magnetic tapes should be exposed, even under extreme use or storage conditions. The participation of the oxidative reaction in chain scission of the polyester-polyurethane binder does not appear to be operable to any significant extent.

CONCLUSIONS

The oxidation of MDI containing polyurethane hard-segment blocks by chromium dioxide involves oxidation of the benzylic carbon of the MDI moiety after adsorption of the hard-segment onto the chromium dioxide particle surface. The adsorption and subsequent oxidation produce a tenacious interaction of the hard-segment with the chromium dioxide surface. The oxidative reaction does not give appreciable chain scission, but does effectively hold the polyurethane portions of the polymer on the particle surface resulting in significantly decreased extractability of the oxidized polyurethane.

Extracts of aged polyester-polyurethane tape coatings possess evidence of hard-segment oxidation and considerable polyester hydrolysis. Cleavage of the hard-segment portions is not, however, a significant contributor to molecular-weight reduction of tape coatings containing chromium dioxide. Prevention or inhibition of molecular-weight reduction in a polyester-polyurethane binder for these tape coatings is a critical parameter governing the long-term utility of high-density, high-performance, digital, magnetic-recording tapes.

ACKNOWLEDGMENTS

The authors wish to recognize the cooperation and assistance of R. D. Stacy, E. A. Bartkus, T. R. Martin, R. M. Phelan, D. R. Smith, and J. R. Koch, of the General Products Division, International Business Machines Corporation, Tucson, Arizona. The assistance of Dr. J. Witchell, formerly of Arizona State University, Tempe, Arizona, in performing the C-13 NMR experiments is especially appreciated. The support provided by the the International Business Machines Corporation and its community of engineers and scientists provided the means and the motivation for this and ongoing activities critical to the future of magnetic storage technology.

REFERENCES

1. R. L. Bradshaw, B. Bhushan, C. H. Kalthoff, and M. D. Warne, Chemical and mechanical performance of flexible magnetic tape containing chromium dioxide, IBM J. Res. Develop., 30(2), pp.203-216 (1986).

2. E. F. Cuddihy, Aging of magnetic recording tape, IEEE Trans. on Magnetics, MAG-16,4, pp.558-568 (1980).

3. E. F. Cuddihy, A Chemical aging mechanism of magnetic recording tape, Org. Coat. Appl. Polym. Sci., 43, pp.422-425 (1983).

4. D. W. Brown, R. E. Lowry, and L. E. Smith,"Prediction of the Long Term Stability of Polyester-Based Recording Media,"NBSIR 84-2988, December 1984.

5. B. Bhushan, R. L. Bradshaw, and B. S. Sharma, Friction in magnetic tapes II: Role of physical properties, ASLE Trans., 27(2), pp.89-100 (1984).

6. R. L. Bradshaw and B. Bhushan, Friction in magnetic tapes III: Role of chemical properties, ASLE Trans., 27(3), pp.207-219 (1984).

7. A. Potter, H. G. Schmelzer, and R. D. Baker, High-solids coatings based on polyurethane chemistry, Prog. Org. Coat., 12, pp.321-338 (1984).

8. N. S. Allen and J. F. McKellar, Photochemical reactions in an MDI-based elastomeric polyurethane, J. Appl. Polym. Sci., 20, pp.1441-1447 (1976).

9. L. Fieser and M. Fieser, "Reagents for Organic Synthesis," Vol.1, John Wiley & Sons, New York, pp.142-153 (1968).

10. H. O. House, Oxidations with chromium and manganese compounds, "Modern Synthetic Reactions," 2nd Edn., W. J. Benjamin, Inc., Menlo Park, California, pp.257-291 (1972).

11. L. J. Bellamy, "Infra-red Spectra of Complex Molecules," Vol.2, John Wiley & Sons, New York, pp.149-201 (1975).

12. M. Avram and G. Mateescu, "Infrared Spectroscopy, Applications in Organic Chemistry," John Wiley & Sons, New York, pp.341-462 (1972).

13. W. W. Simons, Ed., "Sadtler Guide to Carbon-13 NMR Spectra," Sadtler Labs, Philadelphia, Pennsylvania, pp.63-90, 603-607 (1983).

14. R. M. Silverstein, G. C. Bassler, and T. C. Morrill, "Spectrometric Identification of Organic Compounds," 4th Edn., John Wiley & Sons, New York, pp.263-266 (1981).

15. C. Delides, R. A. Pethrick, A. O. Cunliffe, and P. G. Klein, Characterization of polyurethane elastomers by C-13 NMR spectroscopy, Polymer, 22, pp.1205-1210 (1981).

ROLE OF ACTIVE FUNCTIONAL GROUPS AND CONFORMATION OF ADSORBED POLYMERS

IN THE DISPERSIBILITY OF MAGNETIC PARTICLES

Katsuhiko Nakamae, Satoshi Tanigawa, Kenji Sumiya
and Tsunetaka Matsumoto

Department of Industrial Chemistry, Faculty of Engineering
Kobe University, Rokkodai, Nada, Kobe 657 Japan

The Dispersibility of inorganic particles in coatings depends
on the conformation of polymer adsorbed onto the particles. The
effect of active functional groups in polymers on their adsorption
behavior on γ -Fe$_2$O$_3$ and performance of the magnetic tape was
investigated. The adsorption behavior of polymers having the active
functional groups is Langmuir type. A small amount of the
functional groups in polymers allowed to improve the dispersibility
of γ -Fe$_2$O$_3$ in the magnetic coatings. The effect of the
functional groups on the dispersibility follows the order.

$$-PO_3H_2 > -SO_3H > -COOH > -OH > -N< > -C-C = -CN$$

This order corresponds with the interaction force between the
functional groups and the γ -Fe$_2$O$_3$. The dispersibility was
greatly dependent on conformation of adsorbed polymers on
γ -Fe$_2$O$_3$. The conformation of adsorbed polymers was investigated
by various interfacial methods.

INTRODUCTION

The need for high density recording has grown along with the
development of recording methods[1].The high magnetization retentive
Co-epitaxial γ -Fe$_2$O$_3$ powder or α -Fe powder have been developed as
magnetic particle materials in the place of conventional γ -Fe$_2$O$_3$[2, 3].
The properties of the coating materials for the magnetic recording depend
not only on the properties of the inorganic powder itself, but also on
the interfacial properties between the magnetic powder and the binder
polymers. For high density recording, good dispersion, high packing and
high orientation of acicular magnetic particles in a magnetic tape are
important factors. However, only few fundamental studies have been
published on polymeric binders to provide an optimum dispersion of the
magnetic powder[4].
In previous papers[5-10], we have applied a combination of
physicochemical methods to establish an approach to evaluate quantitatively
the interaction forces between a polymer and an inorganic powder. We

found that the surfaces of most inorganic powders were covered with water molecules adsorbed both chemically and physically. These water molecules were not able to be desorbed by the usual drying method until 500 C. If the surface of inorganic powder covered with chemically and physically adsorved water can be substituted by water phase, a distinct relationship is observed to exist between the absorption of a polymer on a solid surface and the interfacial tension ($\gamma_{w/o}$) at the water-polymer solution interface[4,11]. Figure 1 shows the schematic model of our concept for the adsorption of a polymer.

It is possible to evaluate the interaction force between a polymer and an inorganic powder by means of the measurement of interfacial tension. A polymer having a strong interaction with an inorganic powder than with another polymer gives a lower interfacial tension $\gamma_{w/o}$. This relationship is very useful for quantitative evaluation of the interaction forces between a polymer and an inorganic powder. With this method, it is possible to explain systematically the relationship between the chemical structure of polymers and the interaction force[12-14].

In this paper, the role of active functional groups and the conformation of adsorbed polymer molecules on γ -Fe$_2$O$_3$ was investigated, as a function of the nature and content of the active functional groups, by the surface pressure(π)-area(A) curve, the dispersibility and orientation of magnetic particles.

EXPERIMENTAL

Adsorption Isotherms of Polymers on γ -Fe$_2$O$_3$ Particles Surfaces

Glass tubes containing 20 cm^3 of the polymer solution and 2.0 g of γ - Fe$_2$O$_3$ were subjected to 29 kHz ultrasonic waves for 30 minutes. Afterwards the samples were shaken for 24 hours and allowed to stand for another 24 hours. Adsorption was then determined by measuring the change in the concentration in the supernatant solutions. All measurement were carried out at 30 $\pm$ 0.5 °C.

Interfacial Tension at Water-Polymer Solution Interface

The interfacial tension $\gamma_{w/o}$ was measured by the du Nouy ring method. The adsorption of polymers at an interface between water and a polymer solution was achieved by placing 50 cm^3 of water into each of several du Nouy dishes and carefully adding a similar volume of the polymer solution of the desired concentration. All measurements were carried out at 25 °C. Since the interfacial tension at the water-polymer solution interface is slightly time dependent, $\gamma_{w/o}$ was measured after 3 hours of standing .

Surface Pressure(π)-Area(A) Curves

A surface balance of the auto-recording Langmuir-Adam type was used. The monolayer was prepared by spreading 1.0 mg cm^{-3} of a polymer solution in benzene on the water surface. Further detail are described in previous publications[15,16].

Preparation of Magnetic Coatings and Their Magnetic Properties

The magnetic coating material was prepared by using a ball milling method. The vessel contained 20 g binder polymer, 46.7 g of γ -Fe$_2$O$_3$, 125 g of chosen solvent and stainless steel balls. The dispersing process was carried out for over 20 hours. The magnetic coating was cast on a poly(ethylene terephthalate) film. The magnetic properties were measured by Toei Kogyo Vibrating sample Magnetrometer(VSM).

Polymers

Poly(vinyl butyral) having a hydroxyl group content of 24 mol% and a molecular weigh(Mw) of 43,000 g/mole was obtained from Sekisui Chemical Co.Ltd. Partially sulfonated polystyrene was prepared by sulfonating of polystyrene with sulfonic chloride, and polystyrene used had a molecular weight(Mw) of 220,000 g/mole. Partially sulfonated polyester had a molecular weight (Mw) of 16,000 g/mole and sulfonic content of 4.8×10^{19} number/g-polymer. PMMA and its derivatives were obtained by copolymerizing MMA monomer with monomers having functional groups.

RESULTS AND DISCUSSION

Adsorption of Polymers having Various Kinds of Functional Groups on γ -Fe$_2$O$_3$ Particles Surfaces

Particle dispersibility in a polymer solution is generally affected by the adsorption behavior of binders polymers on particle surface. The adsorption behavior of polymers having various kinds of functional groups on γ -Fe$_2$O$_3$ will be considered.

Figure 2 shows the adsorption isotherms of polymers having various kinds of functional groups on γ -Fe$_2$O$_3$. The adsorption of polymers increases with increase in the interaction of the functional groups in polymers to γ -Fe$_2$O$_3$.

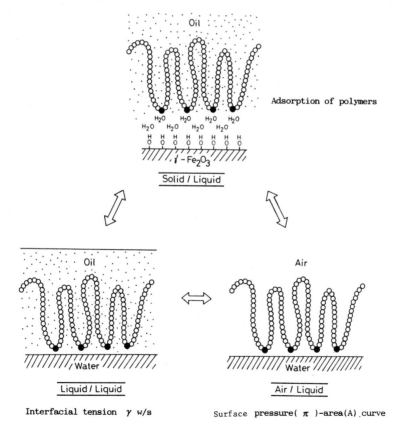

Figure 1.   Schematic representation of the adsorbed polymer model.

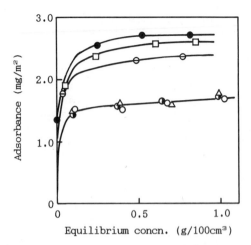

Figure 2.   Adsorption isotherms of various polymers on γ -Fe$_2$O$_3$ from benzene solution at 30 °C. Polymer: ◯ , PMMA; △ , P(MMA-AN) [-CN]; ◐ , P(MMA-GMA)[-C≟C]; ⊖ , P(MMA-DM)[-N<]; ☐ , P(MMA-HEMA)[-OH]; ● , P(MMA-MMA)[-COOH]: Functional group, 5mol%.

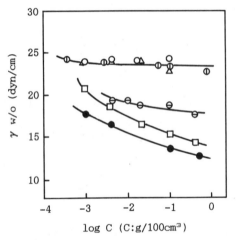

Figure 3.   Interfacial tensions at water/benzene solution of various polymers. Polymer: ◯ , PMMA; △ , P(MMA-AN)[-CN]; ◐ , P(MMA-GMA)[-C≟C]; ⊖ , P(MMA-DM)[-N<]; ☐ , P(MMA-HEMA) [-OH]; ● , P(MMA-MMA)[-COOH]: Functional group , 5mol%.

Figure 3 shows the effect of the functional groups of polymers on the interfacial tension at water-benzene. The polymer having functional groups with a strong interaction to γ -Fe$_2$O$_3$, i.e. high adsorbance of polymer on γ -Fe$_2$O$_3$ gives a lower interfacial tension. The polymers having functional groups (GMA, AN) could lower interfacial tension by only a few dynes and the interfacial tension reached almost constant, the same as PMMA.

Figure 4 shows the interfacial tension at water/toluene solution of polymers having the other functional groups. The interfacial tension for PES-S was lowered with increase in the PES-S concentration. PES-S is

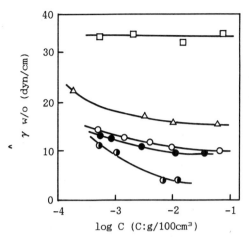

Figure 4. Interfacial tensions at water/toluene solution of polymers.
◑ , PES-S(16000); ● , PVB(M_w:77000); ○ , PVB(43000);
△ , PVA_c(42000); □ , PS(310000).

strongly adsorbed and fixed at the interface because of the strong
interaction between sulfonic groups and water. The interfacial tension
depends on the chemical structure of polymers and the nature of active
functional groups introduced in the polymer chains
      From these results, the order of interaction for the active
functional groups was obtained as follows.

$$-SO_3H > -COOH > -OH > -N< > -\underset{\underset{O}{\diagdown\diagup}}{C{-}C} \doteqdot -CN$$

      Figure 5 shows the adsorption isotherms of polymers having a sulfonic
acid or a phosphoric acid group on γ -Fe_2O_3 particles and Langmuir
plots (CA^{-1} vs. C), where A is the adsorbance of polymer (mg m^{-2}) and C
is the equilibrium concentration of polymers (g/100cm³). The adsorption
isotherms have the characteristic plateau and the CA^{-1} vs. C plots for
each polymer are straight lines. Therefore, these isotherms are of the
Langmuir type adsorption, which meanes monolayer type adsorption.
      Figure 6 shows the effect of the functional groups in polymers on the
saturated adsorbance on γ -Fe_2O_3. The adsorbance increases with the
increase in the content of the functional groups in polymers. The
phosphoric acid group shows stronger effect on adsorption than the
sulfonic acid group.
      Figure 7 shows the relationship between the interfacial tension at
the water/toluene solution interface(γ $_{w/t}$) and the functional groups
contents in polymers. The γ $_{w/t}$ for these two kinds of polymers
decreases abruptly with the increase in the functional group content. The
effect of the phosphoric acid groups on the γ $_{w/t}$ is larger than that
of sulfonic acid groups. This means that the phosphoric acid groups more
effectively take part in the adsorption of the polymer at the
water/toluene interface, similar to that at the γ -Fe_2O_3 surface.

Adsorption of PES-S Monomolecular Film at the Air/Water Interface

      The adsorption morphology of partially sulfonated polyester (PES-S)
on γ -Fe_2O_3 is considered based on the surface pressure-area curves of
PES-S monomolecular film. The reasons for such attempt are as follows.
1) Since the γ -Fe_2O_3 surface has -OH groups and adsorbed water, the

411

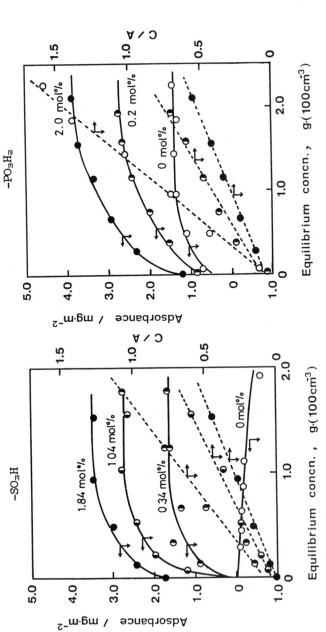

Figure 5. Adsorption isotherms and Langmuir plots of PS-S and P(MMA-PO₃H₂) adsorbed on γ-Fe₂O₃ from benzene at 30 °C.

412

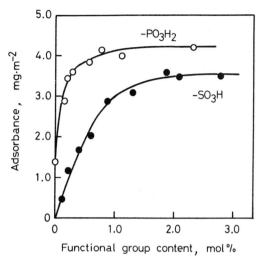

Figure 6. Relationship between the saturated adsorbance of polymers on γ -Fe$_2$O$_3$ and the functional group content of polymer.

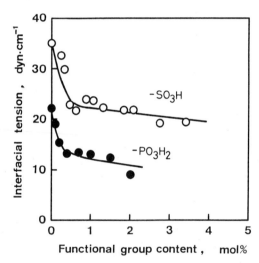

Figure 7 Relationship between the interfacial tensions at water/toluene solution of polymers and the functional group content of polymer.

events at the γ -Fe$_2$O$_3$ surface would seem to the same as at the water surface, from standpoint of polymer adsorption. 2) The adsorption of PES-S on γ -Fe$_2$O$_3$ is monomolecular layer adsorption as shown by the adsorption isotherms.

Figure 8 shows the adsorption and desorption isotherms of polyester (PES) and partially sulfonated polyester (PES-S) and the π -A curve for PES-S. These two polyesters also show Langmuir type adsorption. PES-S was not desorbed from the γ -Fe$_2$O$_3$ surface because of strong ionic interaction of the sulfonic groups with the γ -Fe$_2$O$_3$ surface. We proposed that the main factor for good dispersion and stabilization of particles should depend not only on the adsorbance of polymer but also on the conformation of polymers at the solid-liquid (liquid -liquid) interface. In order to clarify the conformation of PES-S at the γ -Fe$_2$O$_3$ surface, a π -A curve was measured. A monomolecular film which covers the water surface completely is formed in region I. Subsequently,

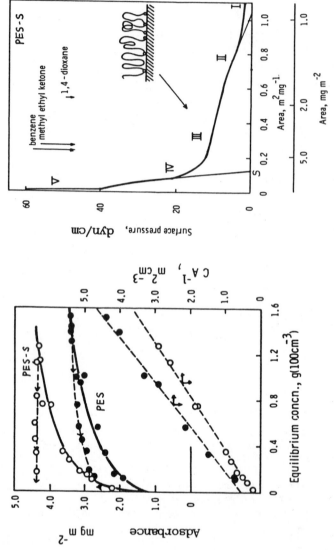

Figure 8 Adsorption isotherms and Langmuir plots of partially sulfonated polyester, and surface pressure(π)-area(A) curve.

the hydrophobic segments of polymer which have weak interaction with water surface desorb from the water surface, as shown in region II. The hydrophobic segment in polymer forms loop chain toward the air from water surface in region III. The monomolecular film is broken by further compression and an aggregated film is formed, as in region IV.

The arrows in figure indicate the points which correspond to the saturated adsorbance of the polymers on γ -Fe$_2$O$_3$ from various solvents. If PES-S adsorbs on γ -Fe$_2$O$_3$ similar to that on water surface, PES-S is assumed to have an adsorbed morphology as in region III in all these solvent systems. The loops with a higher segment density are formed in solvent systems with lower solvent power (dioxane < methyl ethyl ketone < benzene). PES-S is reoriented to provide for the sulfonic groups attachment on γ -Fe$_2$O$_3$ surface. The segment density of the hydrophobic region changes in different solvent systems.

Packing and Particle Orientation in the Magnetic Coatings

Table 1 shows the magnetic properties of the coatings prepared from different combinations solvents and polymers. PES, which has no functional group, gives a lower squareness and a lower packing density, despite a large saturated adsorption, where the squareness is defined as the ratio of residual magnetization (Mr) over the saturated magnetization (Ms). However, PES-S, which has a small amount of the sulfonic group, give a higher squareness and a higher packing density, despite having almost the same saturated adsorbance as that of PES. This is the conformation effect of polymer molecules adsorbed on the γ -Fe$_2$O$_3$.

From these results, it is reasonable to conclude that the polymer-solvent system having the most compact train loop type adsorption gives the highest dispersion of γ -Fe$_2$O$_3$ in magnetic paint and, consequently, gives the highest packing and orientation of γ -Fe$_2$O$_3$ in magnetic film.

Figure 9 shows the effect of the functional groups content in polymers on the squareness. It is clear that each functional group acts in a different way, and the optimum functional group content is different depending on the interactive force.

Table 1. Physical properties of magnetic films.

Polymer	Solvent	Ms(G) [a]	Rv(%) [b]	Mr/Ms [c]	Adsorbance(mg/m^2)
PES	benzene	1240	26.7	0.613	3.5
PES-S	benzene	1470	12.9	0.832	4.4
PES-S	MEK [d]	1450	14.1	0.845	3.9
PES-S	1,4dioxane	1380	18.3	0.832	2.1

a) saturated magnetization, b) Volume fraction of voids,

c) Squareness, d) methyl ethyl ketone.

Polymer : Polyester or partially sulfonated polyester

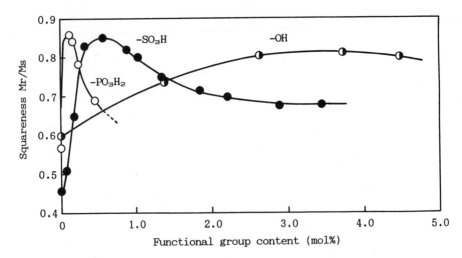

Figure 9. Relationship between the squareness of magnetic coating and
the functional group content of polymer.

The Determination of the Dispersibility of γ-Fe$_2$O$_3$ Particles in Magnetic Coatings

An attempt was made to evaluate the quality of dispersion of the
particles in the magnetic coatings material under a low magnetic field.
The rotation of particles in the coating material can be traced by
measuring the initial magnetization (M_i) under a magnetic field of 50 Oe,
where most of the moments of particles do not switch.

Figure 10 shows the relationship between the ratio of the initial
magnetization (Mi) to the saturated magnetization (Ms) and the holding time
for the applied field of +50 Oe. The Mi/Ms of coating material A increased
with time, but that of the coating material B remained constant and was
equivalent to that of cobalt-containing γ-Fe$_2$O$_3$ powder. This means that the
particles in coating material B were not well dispersed. In the case of A the
Mi is given by

$$M_i = M_{im} + M_{io}$$

where M_{im} is the magnetization associated with removing the magnetic
moment from the easy axis (=long axis of particles), and is the quickly
responding component. M_{io} is the magnetization associated with the
orientation of particles, and is the slowly responding component. Because
the particle should orient in the viscous magnetic coating material.

Figure 11 shows the vibration curve of M_i/M_s for paint 1 and 2 over a
1 minute interval. The curve of paint 2 did not involve the component M_{io}.
The curve of paint 1 shows a large M_{io} component and a slower decay
vibration (dotted line). The degree of decay is related to the stability
of the coating material.

Figure 12 shows the amplitude and the decay curve for partially
sulfonated polystyrene(PS-S). The γ-Fe$_2$O$_3$ in the paint for polystyrene
(PS) is not well dispersed. The amplitude of PS-S (0.61 mol%) is very
high and the decay curve has a steep slope. However, with the increase in
the functional group content, the amplitude decreases as shown in figure
12. Figure 13 shows the effect of the functional group content on the
dispersibility in the magnetic paint. The optimum amount of functional
groups in a polymer chain depends on the nature of functional group.

Figure 14 shows a schematic representation of the model of adsorbed
polymer having functional groups at the interface. It is reasonable to

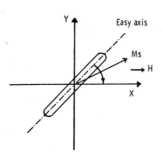

$$Mi = Mim + Mio$$

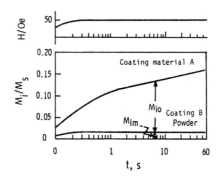

Figure 10. Relationship between the M_i/M_s and time of applied field of +50 Oe.

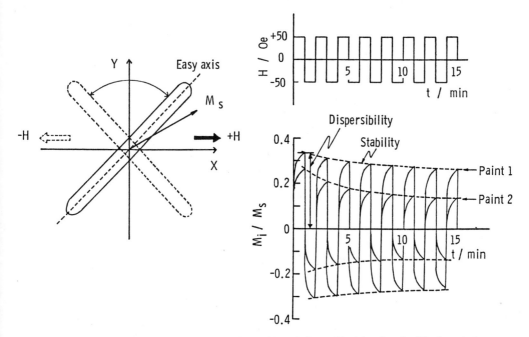

Figure 11. Vibration curve of M_i/M_s with a field of ± 50 Oe at 1 minute interval.

Partially-sulfonated polystyrene

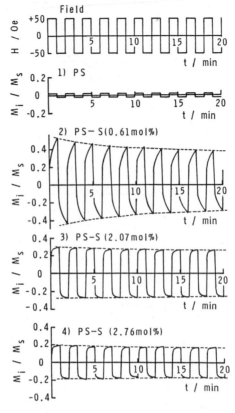

Figure 12. The M_i/M_s curves for various coatings.

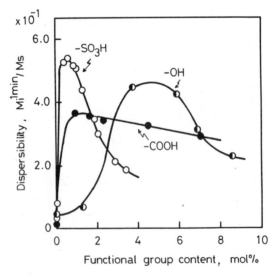

Figure 13. Effect of the functional group content of polymers on the dispersibility of coatings.

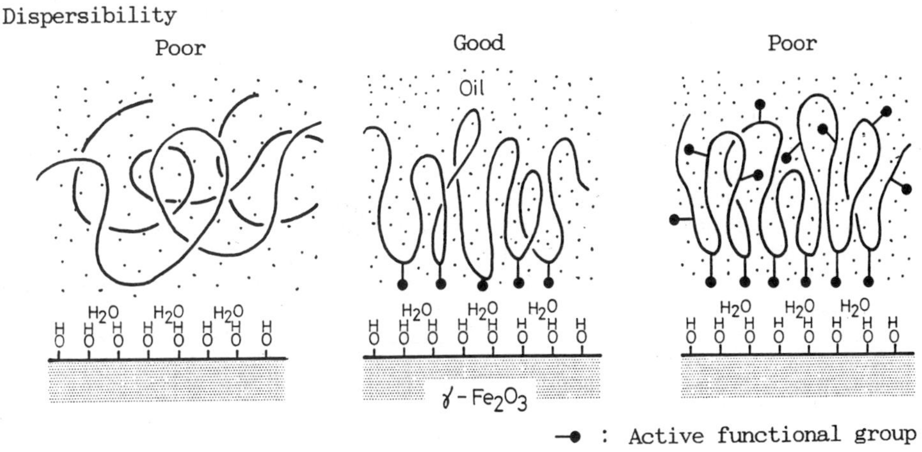

Dispersibility

Figure 14.   Schematic representation of the model of adsorbed polymer chain at the interface.

conclude that the one of the ideal binder polymer molecules should be the polymer having proper amount of functional groups and an ability to adsorb in a "loop-point" type conformation.

CONCLUSION

The effect of active functional groups in polymers on their adsorption behavior on γ-Fe$_2$O$_3$ and the performance of the magnetic tape were investigated. A small amount of active functional groups in polymers improved the dispersibility of γ-Fe$_2$O$_3$ in the magnetic paints and also the magnetic properties of magnetic tapes. The polymer molecules having proper amount of functional groups adsorb in a "loop-train" or "loop-point" mode conformation on γ-Fe$_2$O$_3$ surface because the active functional groups act as anchor to the surface. The order of interaction force of the functional groups to γ-Fe$_2$O$_3$ is as follows;

$$-PO_3H_2 \; > \; -SO_3H_2 \; > \; -COOH \; > \; -OH \; > \; -N< \; > \; -\underset{\backslash O \diagup}{C-C} \; \doteqdot \; -CN$$

REFERENCES

1. A.L.Oppegard, F.J.Darnell and H.C.Miller, J.Appl.Phys., 32, 184s (1961).
2. K.Sumiya, T.Matsumoto, S.Watatani and F.Hayama, J.Phys.Chem.Solids, 40, 1097 (1979).
3. T.Uehori, A.Hosaka, Y.Tokuoka, T.Izumi and Y.Imaoka, IEEE Trans.Magn., Mag-14, 852 (1978).
4. K.Nakamae, K.Sumiya and T. Matsumoto, Prog.Organic Coat., 12, 143 (1984).
5. T.Matsumoto, K.Otsuki and I.Sakai, J.Adhesion.Soc.Japan, 2, 391 (1966).
6. T.Matsumoto, T.Doi and K.Otsuki, ibid, 3, 65 (1967).
7. T.Matsumoto, T.Doi, K.Otsuki and T.Nakamura, ibid, 4, 1 (1968).
8. T.Matsumoto, K.Nakamae and K.Nonaka, Kobunshi Ronbunshu, 31, 7 (1974).

9. T.Matsumoto, K.Nakamae, K.Nonaka, S.Miyoshi and I.Sakai, ibid, <u>31</u>, 522 (1974).

10. I.Sakai, K.Nakamae, T.Okawa, H.Inubushi and T.Matsumoto, ibid, <u>34</u>, 661 (1977).

11. K.Nakamae, K.Sumiya, T.Taii and T.Matsumoto, J.Polym.Sci.Polym. Symp., <u>71</u>, 109 (1984).

12. T.Matsumoto and K.Nakamae, in "Adhesion and Adsorption of Polymers", L.H.Lee(Ed), p.707, Plenum Press, New York (1980).

13. K.Sumiya, K.Nakamae, T.Matsumoto, S.Watatani and F.Hayama, Ferrites, Proceeding of the 3rd international conference, 656 (1980).

14. S.Watatani, K.Sumiya, F.Hayama, H.Naono and T.Matsumoto, Kobunshi Ronbunshu, <u>35</u>, 565 (1978).

15. K.Sumiya, K.Nakamae, S.Watatani, F.Hayama, H.Naono and T.Matsumoto, ibid, <u>37</u>, 49 (1980).

16. Y.Fujimura, I.Sakai, K.Nakamae, M.Hatada, T.Okada and T.Matsumoto, ibid, <u>37</u>, 29 (1980).

RAPID DETERMINATION OF COATING PERCENT PIGMENT

FROM MAGNETIC RECORDING INK DENSITY AND

VOLUME FRACTION

Thomas E. Karis

IBM Research Division
Almaden Research Center
650 Harry Road
San Jose, California 95120-6099

A magnetic recording ink formulation consists of mixed solvents, a polymeric binder, and magnetic pigment. This is coated onto a disk or tape substrate and cured. In the process, the solvent evaporates, and the binders polymerize, embedding the pigment to form the magnetic film. For quality control of the process, it is important to maintain the composition of the film. Conventionally, this is done by an off-line method of Thermo Gravimetric Analysis (TGA), in which the solvent is first evaporated and then the binder is pyrolized, leaving behind the magnetic pigment. The TGA method is time-consuming, and often too slow for process control. The advantage of the new method described here is that it can be used on-line, as well as for rapid measurements, therefore it can be applied to process control. The ink volume fraction and density are related to the percent pigment by a mass and volume balance, and the method is illustrated with data from a typical recording disk ink.

INTRODUCTION

In the conventional quality control of magnetic recording ink, the weight percent pigment to be expected in the cured film is routinely monitored by thermogravimetric analysis (TGA). In TGA, the solvent is evaporated, leaving behind a film consisting of binder and pigment. The film is then weighed, followed by heating that pyrolizes the binder. The weight ratio of the pigment residue to the composite film before pyrolysis is the mass fraction of pigment in the cured film. However, TGA is time consuming and often too slow for adequate process control.

This paper describes a new, alternative, method for predicting the percent pigment in the cured film from the measured volume fraction of magnetic pigment particles in the ink and the ink density. The relation between the amount of pigment, binder, and solvent in the ink and the ratio of pigment to binder that will result in the cured coating is derived from a mass and volume balance. The result is an equation for the pigment (oxide) concentration in terms of the component densities, the volume fraction of oxide in the ink, and the bulk density of the ink.

An electromagnetic method for rapidly measuring the magnetic oxide volume fraction in the ink, based on the principle that the inductance of a coil is linearly proportional to the amount of magnetic material in its core, is described[1]. The inductance is easily measured by constructing an oscillator with the coil as the inductor. The oscillator frequency can be used to measure the volume fraction of oxide in a tube placed inside the coil. Oscillator stability and calibration with reference standards are addressed. This method provides a way to rapidly and accurately measure the volume fraction for use in estimating the final oxide concentration in the cured film.

Experiments done with an actual disk coating formulation are presented. A magnetic disk ink was prepared and then diluted with binders or solvent, changing both the volume fraction and density of the ink. The percent pigment in the coating is unaffected by dilution with solvent and is decreased by dilution with binders. Comparison with TGA evaluation of the inks shows that the derived relation can be used to estimate the percent pigment from the ink density and volume fraction.

THEORY

This section presents the derivation of the percent pigment equation. The derivation relies on the principles of conservation of volume and mass going from the ink to the cured binder. If we know what the volume fraction of the pigment is in the ink, and the ratio of solvent to binder, then it follows that we know the ratio of pigment to binder in the ink. Since this ratio will be the same once the volatile solvent is driven off during the curing process, we then know the ratio of pigment to binder expected in the cured film.

Derivation of the Percent Pigment Equations

The following derivation presents a means to predict the percent pigment from the volume fraction of magnetic particles in the ink ϕ and the ink density ρ. Here the ink refers to the mixture of magnetic particles, binders, and a variable amount of solvent (due to solvent evaporation). The percent pigment is defined as $100 \times PMC$, where PMC is referred to as the pigment mass concentration, or weight fraction of pigment in the cured film. In general, the PMC is the ratio of the mass of pigment M_p to the total mass of the cured film:

$$PMC = \frac{M_p}{M_p + M_b}.$$ (1)

Here M_b is the mass of binder fluid.

Several operations are now done to more conveniently express the PMC as a function of the pigment volume fraction ϕ. Equation (1) is inverted to get:

$$\frac{1}{PMC} = 1 + \frac{M_b}{M_p}.$$ (2)

From eq. (2), it is apparent that the PMC is a function of the ratio of binder to pigment in the coated film M_b/M_p. The key to the following steps in the derivation is to express the binder to pigment ratio in terms of ρ and ϕ. This is done by considering a volume and mass balance for the ink. A volume of ink V is made up of some volume of pigment V_p, binders V_b, and volatile solvent V_s according to:

$$V = V_p + V_b + V_s.$$ (3)

Since volume is mass divided by density, eq. (3) can be expressed in terms of the masses of ink, pigment, binders, and solvent in the ink M, M_p, M_b, and M_s, and the densities of the ink, pigment, binders, and solvent ρ, ρ_p, ρ_b, and ρ_s, respectively:

$$\frac{M}{\rho} = \frac{M_p}{\rho_p} + \frac{M_b}{\rho_b} + \frac{M_s}{\rho_s}.$$ (4)

(It will be shown later that the only density value which needs to be measured is the ink density ρ.)

The mass balance gives:

$$M = M_p + M_b + M_s . \tag{5}$$

Further, the mass fraction of pigment in the ink can be expressed in terms of the volume fraction of pigment in the ink and the ink density. Both quantities are easily measured. That is:

$$\frac{M}{M_p} = \frac{\rho V}{\rho_p V_p} = \frac{\rho}{\rho_p} \frac{1}{\phi} . \tag{6}$$

Here $\phi = V_p/V$ is the volume fraction of magnetic particles.

From eqs. (4) through (6), the PMC can be written as:

$$\frac{1}{PMC} = a_0 + a_1 \left(\frac{1}{\phi} \right) + a_2 \left(\frac{\rho}{\phi} \right) . \tag{7}$$

Here a_0, a_1, and a_2 are expressed in terms of component densities only:

$$a_0 = \frac{\rho_s}{\rho_p} \frac{(\rho_p - \rho_b)}{(\rho_s - \rho_b)} , \tag{8}$$

$$a_1 = \frac{\rho_b \rho_s}{\rho_p (\rho_s - \rho_b)} , \tag{9}$$

and

$$a_2 = \frac{\rho_b}{\rho_p (\rho_b - \rho_s)} . \tag{10}$$

Inverting eq. (7) gives the final expression for the PMC as a function of three coefficients for the binder system, the ink density ρ, and the volume fraction of pigment in the ink ϕ:

$$PMC = \frac{1}{a_0 + a_1 \left(\dfrac{1}{\phi} \right) + a_2 \left(\dfrac{\rho}{\phi} \right)} . \tag{11}$$

It is also of interest to consider the alternative definition, the pigment volume concentration PVC, in the coating[2]. That is:

$$PVC = \frac{V_p}{V_p + V_b} = \frac{1}{1 + \dfrac{V_b}{V_p}} = \frac{1}{1 + \dfrac{\rho_p}{\rho_b} \left(\dfrac{M_b}{M_p} \right)} . \tag{12}$$

From eqs. (1) and (12), the PVC is related to the PMC by:

$$PVC = \frac{1}{1 + \dfrac{\rho_p}{\rho_b} \left(\dfrac{1}{PMC} - 1 \right)} . \tag{13}$$

Conversely:

$$PMC = \frac{1}{1 + \dfrac{\rho_b}{\rho_p} \left(\dfrac{1}{PVC} - 1 \right)} . \tag{14}$$

423

Test data verifying eq. (11), and the procedure for calculating coefficients a_0, a_1, and a_2 (given in eqs. (8) through (10)) from measured PMC, ϕ, and ρ for a particular binder system are given later.

Now that the relation between the experimentally accessible values of density and volume fraction has been established, we consider the method for practical measurement. The density can be measured with an on-line density meter. For off-line testing, it is desirable to measure the density at the time of the test in case some volatile solvent has evaporated in the process of sampling from the line. The volume fraction measurement is achieved by a rapid electromagnetic technique described below.

Derivation of the Volume Fraction Equation

The inductance is defined as $L = 1/C(2\pi f)^2$ where f is the frequency of an oscillator attached to the test coil, and C is the capacitance [3]. For practical purposes, the inductance is related to the volume fraction of magnetic pigment in the ink by a straight line, as shown schematically in Fig. 1, in which L is the inductance with volume fraction ϕ of pigment in the coil. Without any pigment, that is, pure solvent in the coil, $L = L_0$ and $\phi = 0$. For the volume fraction of interest in magnetic recording, L is linearly dependent on ϕ, that is:

$$L = L_0 + B\phi . \tag{15}$$

The slope B can be easily obtained from a known volume fraction $\phi = \phi_1$ in the coil giving $L = L_1$ and:

$$B = \frac{L_1 - L_0}{\phi_1} . \tag{16}$$

Equations (15) and (16) are inverted to provide a measurement of the unknown volume fraction ϕ with:

$$\phi = \left(\frac{L - L_0}{L_1 - L_0} \right) \phi_1 , \tag{17}$$

or

$$\phi = -\left(\frac{L_0}{B} \right) + \left(\frac{1}{B} \right) L . \tag{18}$$

Equation (18) is the working equation used to obtain the ϕ value from the measured L. The coefficients must be determined from a calibration with pure solvent and a sample with known volume fraction ϕ_1, as described below.

Calibration of the Volume Fraction Measurement

Calibration is necessary for each device to measure the coefficients given in eq. (18). These coefficients are determined by the coil geometry, temperature, circuit components, and the interaction of the magnetic field with surrounding materials. The calibration can be separated in two parts: (i) an initial measurement of the slope coefficient B, and then routinely specified (probably weekly) calibration to verify the current value, and (ii) a reference calibration done at the beginning of each measurement with pure solvent.

In the initial calibration of a new measurement cell, as well as in the routinely scheduled maintenance calibration, the slope coefficient B is assumed unknown and is measured using two reference points as illustrated in Fig. 1. That is, the solvent value of $L = L_0$ when $\phi = 0$, and the value with a sample of known volume fraction ϕ_1, because $L = L_1$ when $\phi = \phi_1$. (Oxide can be embedded in casting resin to make up a solid volume fraction reference standard that is free of evaporation.) The slope coefficient is then calculated from these calibration data and eq. (16).

For the calibration at the beginning of every measurement, the operator inserts a sample of pure solvent in the cell to obtain the current value of L_0. The pure solvent in

424

the coil gives a reading $L = L'_0 \neq L_0$. This means that new coefficients are required. The adjustment to the coefficients is represented by a lumped parameter K, such that the new working equation becomes [modified version of eq. (15)]:

$$L = K[L_0 + B\phi],\qquad(19)$$

or,

$$L = L'_0 + B'\phi.\qquad(20)$$

Notice that the slope must be adjusted. Since the previous value of $L = L_0$ with pure solvent is known, as well as the current value $L = L'_0$, it is possible to calculate $K = L'_0/L_0$. Therefore the new slope becomes $B' = BL'_0/L_0$, and the currently calibrated equation for the volume fraction is:

$$\phi = -\frac{L_0}{B} + \left(\frac{1}{B'}\right)L.\qquad(21)$$

This type of correction, rather than just a level shift in L_0, seems to provide less sensitivity to drift in the oscillator frequency. The above sequence of calibrations is iteratively programmed and made transparent to the operator.

EXPERIMENTAL

A magnetic recording ink system can be thought of as being made up of three components (pigment, binder, and volatile solvent), even though the binders are a combination of resins and the solvent may be a mixed solvent system. Data were obtained to verify this concept, which is fundamental to eqs. (8) through (11).

The test consisted of diluting magnetic disk ink with either binders or solvent. Diluting the ink with solvent should not change the PVC, while diluting with binders will decrease the PVC. Also, since the density of the binders is higher than the density of the solvent, we expect more of a decrease in ink density for dilution with solvent than with binders.

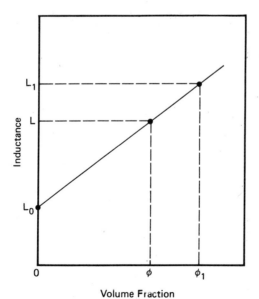

Volume Fraction

Figure 1. An illustration of the relation between the calibration points. With the pure solvent $L = L_0$ when $\phi = 0$, with the reference standard $L = L_1$ when $\phi = \phi_1$, and the inductance L corresponding to arbitrary volume fraction ϕ. The slope is B.

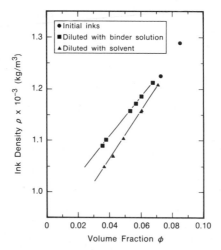

Figure 2. The ink density as a function of volume fraction for disk inks diluted with
binders or solvent, illustrating the density difference between binders and solvent.

This test was carried out twice, each time starting with an independently formulated
ink and diluting with either binders or solvent. As shown in Fig. 2, dilution with the solvent lowered the ink density more than dilution with binders.

In the next part of the test, the *PVC* of each ink sample was measured by TGA. This
data for each of the initial inks is shown in Fig. 3. The PVC was unaffected by dilution
with solvent and decreased by dilution with binders in both cases.

To test the ability of the percent pigment equation to fit these data, coefficients
a_0, a_1, and a_2 were calculated by a regression fit to the *PVC* from TGA, the volume fraction of pigment in the ink ϕ, and the ratio of the ink density to the volume fraction ρ/ϕ.
The correlation of the *PVC* calculated from eqs. (11) and (13) with the measured *PVC* is
shown in Fig. 4.

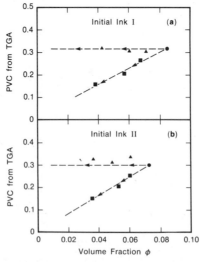

Figure 3. The PVC as a function of the volume fraction of oxide for two disk inks diluted
either with binder or solvent, showing how the PVC is unchanged by dilution
with solvent and is decreased by dilution with binders. The symbols correspond to those in Fig. 2.

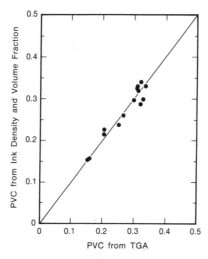

Figure 4. A correlation between the PVC measured by TGA and the PVC predicted from the percent pigment equation with the coefficients calculated from a regression fit to the measured PVC, ϕ, and ρ/ϕ for the disk ink.

Overall, this provides a good prediction of the *PVC* regardless of whether the ink is diluted with binders or solvent. The coefficients can then be used together with ϕ and ρ to predict the final *PVC* or *PMC* from measurements on the ink.

DISCUSSION

The coefficients a_0, a_1 and a_2 in the percent pigment equation [eq. (11)] can, in principle, be calculated from the component densities if they are known. Table I lists typical density values for a disk and a tape ink, and the corresponding coefficients.

Notice that the numerical value of the coefficients is quite sensitive to any changes in the density. To eliminate the effect of uncertainty in the component density, it is preferable to calculate the coefficients of the percent pigment equation by a regression fit to the data (*PMC*, ϕ, and ρ/ϕ). These data measured for the disk ink are listed in Table II.

Table I. Ink component densities and the coefficients of the percent pigment equations [eqs. (8) through (11)] calculated from the densities for a disk and a tape ink.

Quantity	Disk Ink	Tape Ink
ρ_s (kg/m^3)	870.6	870.92
ρ_b (kg/m^3)	1182	1250
ρ_p (kg/m^3)	4650	4649.2
a_0	-2.0851	-1.6797
a_1	-0.71066	-0.6177
a_2 $(m^3/kg) \times 10^4$	8.1629	7.09254

Table II. Measured *PMC* in the cured film, pigment volume fraction ϕ, and density ρ for recording disk inks. The last three columns show the quantities calculated for input to a linear regression fit to determine the coefficients in the percent pigment equation.

PMC	ϕ	$\rho \times 10^{-3}$ (kg/m^3)	$1/PMC$	$1/\phi$	$\rho/\phi \times 10^{-3}$ (kg/m^3)
0.627	0.07127	1.226	1.594	14.03	17.2
0.649	0.0851	1.289	1.54	11.75	15.15
0.638	0.07127	1.21	1.568	14.03	16.98
0.667	0.06102	1.16	1.499	16.39	19.01
0.639	0.06056	1.158	1.566	16.51	19.12
0.642	0.04904	1.105	1.556	20.39	22.53
0.651	0.04255	1.071	1.537	23.5	25.17
0.658	0.03688	1.05	1.519	27.12	28.47
0.591	0.06773	1.213	1.693	14.77	17.91
0.571	0.06056	1.186	1.751	16.51	19.58
0.507	0.05699	1.172	1.973	17.55	20.57
0.506	0.05341	1.158	1.978	18.72	21.68
0.428	0.0382	1.102	2.339	26.18	28.85
0.416	0.03596	1.09	2.404	27.81	30.31

A linear regression fit with two independent variables is carried out to calculate the values of the coefficients (Fig. 4). Here $1/PMC$ is the dependent variable, and $1/\phi$ and ρ/ϕ are the independent variables. The results are listed in the last three columns of Table II. The best fitting coefficients for the disk ink are:
$$a_0 = -1.5535,$$
$$a_1 = -0.5576,$$
and
$$a_2 = 6.418 \times 10^{-4} \ (m^3/kg).$$
The resulting equation for the *PMC* is therefore:

$$PMC = \frac{1}{-1.5535 - 0.5576\left(\dfrac{1}{\phi}\right) + 6.418 \times 10^{-4}\left(\dfrac{\rho}{\phi}\right)}. \tag{22}$$

In eq. (22), ρ is in kg/m^3.

The coefficients from the regression differ substantially from those calculated with the disk ink component densities shown in Table I, which may be attributed to the difficulty in measuring the solvent and binder density, and, possibly, binder shrinkage on curing. The significance of this to calculating the *PMC* is illustrated in Table III. The fourth column in Table III is calculated from the coefficients in Table I, and the fifth column is obtained from eq. (22). The *PMC* difference using the densities is up to 20%, while using the regression fit the difference is less than 6%. Therefore, in application of the percent pigment equation to a particular binder system, the best coefficients should be determined from a regression fit to the data.

SUMMARY AND CONCLUSIONS

A magnetic recording ink system can be thought of as being made up of three components: pigment, binder, and volatile solvent. Thus, a material balance was used to obtain an equation for the percent pigment in the coating as a function of the ink density and volume fraction, referred to as the percent pigment equation [eq. (11)].

Table III. Measured PMC, ϕ, and ρ in the first three columns, the percent PMC difference between the PMC calculated from the approximate component densities in Table I, fourth column, and the PMC calculated from the coefficients determined by a regression fit to the data for a disk ink, fifth column.

PMC	ϕ	$\rho \times 10^{-3}$ (kg/m^3)	Percent PMC Difference from Densities	from Regression
0.627	0.07127	1.226	-19.7	-4.1
0.649	0.0851	1.289	-20.1	-4.7
0.638	0.07127	1.21	-13	3.3
0.667	0.06102	1.16	-16.1	0.6
0.639	0.06056	1.158	-12.5	3.6
0.642	0.04904	1.105	-14.3	1.3
0.651	0.04255	1.071	-12.6	2.8
0.658	0.03688	1.05	-19.5	-5.0
0.591	0.06773	1.213	-17.1	-0.9
0.571	0.06056	1.186	-19.2	-3.2
0.507	0.05699	1.172	-11.6	6.0
0.506	0.05341	1.158	-14.3	2.9
0.428	0.0382	1.102	-18.2	-1.1
0.416	0.03596	1.09	-17.0	0.4

A test was done with magnetic disk coating formulation and expressed in terms of the pigment volume concentration PVC. However, the form of eq. (11) is the same regardless of whether one is considering the PVC or PMC as the process variable. The test was done twice, each time starting with an independently formulated ink and adding either binders or solvent. The solvent lowered the ink density more than the binders. The measured PVC was unaffected by dilution with solvent and decreased by the binders. To verify the ability of the percent pigment equation to fit these data, coefficients a_0, a_1, and a_2 were calculated by a regression fit. The correlation of the PVC then calculated from eq. (11) and the coefficients with the measured PVC is shown in Fig. 4. Overall, this provides a good prediction of the coating PVC, independent of the solvent to binder ratio. Therefore, the experimentally determined coefficients for a given binder system can then be used together with measured ink ϕ and ρ to predict the final PVC or PMC. Finally, we would like to emphasize that the current apparatus is at least ten times faster than any current method available, such as TGA, and is applicable for on-line measurement.

ACKNOWLEDGEMENTS

The author would like to thank IBM Corporation for permission to publish this study. Thanks are also due to Mr. Ali Afrassiabi for technical assistance during his co-operative education assignment, and to Dr. James Lyerla at the IBM Research Division Almaden Research Center, and Dr. Myung S. Jhon at Carnegie Mellon University.

REFERENCES

1. T.E. Karis, IEEE Trans. Magn., MAG-22 , 665(1986).
2. N.L. Ferrari, M.R. Lorenz, and H. Sussner, IBM Technical Disclosure Bulletin, 25 , 3260(1982).
3. R. Boylestad, and L. Nashelsky, "Electricity, Electronics, and Electromagnetics", p. 64, Prentice-Hall, New Jersey, 1983.

MELT RHEOLOGY OF PARTIALLY CROSSLINKED

CARBON BLACK COMPOSITES

C. Mark Seymour and Thomas E. Karis

IBM Research Division
Almaden Research Center
650 Harry Road
San Jose, California 95120

and

George G. Marshall

IBM Information Products Division
P.O. Box 1900
Boulder, Colorado 80301

Styrene/acrylate copolymers containing 7 volume % carbon black over a range of gel content from 0 to 70%, typical of those employed as toners for hot roll fusing in electrophotography, were studied with the Rheometrics Mechanical Spectrometer using the cone-plate geometry at temperatures between 100°C and 180°C. Three tests were used to measure the viscoelastic response at the fusing time scale and temperature: dynamic, step shear, and step shear rate. These tests show a pronounced effect of gel content and shear history on the melt rheology. It is also found that shear must be considered in successive tests on the same sample. A double relaxation model is employed to characterize the results from the step shear and step shear rate tests. This provides the equilibrium recoverable compliance, zero shear viscosity, and characteristic time of the composites.

INTRODUCTION

Laser printers based on the dry electrophotographic toner process have now become an essential element of modern business. It is important that the printout is crisp, well-defined, and adequately fused to the paper. Poorly fused toner, unwanted lines, dark smudges, or ghosts on the paper are common forms of toner-dependent, print quality degradation.

Toner is made up of a blend of thermoplastic resin and pigment, usually carbon black. Much effort has been spent to understand the relation between the mechanical properties of the toner composite during the fusing process, in which the particles are melted and calendered onto the paper by compression in the nip between hot rollers. Often a print defect is caused by some toner adhering to the roll instead of the paper when it is melted. Depending on process speed, the melting and pressing take place in $\simeq 0.6s$. Thus, the transient viscous and elastic properties of the melt could play an important role in determining distribution of the melt between the roll and the paper. This transfer of toner to the fuser roll is referred to as hot offset, and is a problem in toner fusing.

A model for the toner fusing process based on the deformation of cylindrical particles was first derived by Kuo[1,2]. This model incorporates a generalized Newtonian fluid constitutive equation which cannot adequately describe the viscoelastic toner melt rheology. Although some degree of correlation between theory and experiment is reported, the design of toner materials for specific applications demands a more accurate constitutive equation for the melt response to stress or deformation.

In a previous toner rheology study by Prime[3], compressive flow in a parallel plate rheometer was used to measure the deformation of molten toner in response to an applied stress. The data were interpreted with linear viscoelasticity theory, and time temperature superposition was used to obtain the creep compliance $J(t)$ over a wide range of time.

Lakdawala and Salovey[4-6] measured the steady shearing viscosity η as a function of the shear rate $\dot{\gamma}$ to examine the reduction of melt viscosity with increasing shear rate and yield stress properties of styrene and butylmethacrylate copolymers and homopolymers over a range of carbon black (C-B) pigment concentration and particle size, and temperature. These studies illustrated a yield stress attributed to reversible flocculation of C-B. This flocculation was decreased by increasing the polarity of a homopolymer matrix [5], or by increasing the C-B particle size[6]. Gandhi and Salovey[7] employed dynamic mechanical property measurement of the storage modulus G' and loss modulus G'' to examine the toner melt rheology over a limited range of frequency.

Our study introduces the variable of gel content in the resin. We examine the changes in rheology due to increased partial crosslinking, or gel content, with 7 volume % C-B. We use three complementary tests. The dynamic mechanical property measurements are done over a range of temperature and frequency to obtain G' and G'' over eight orders of magnitude in reduced frequency by time-temperature superposition[8]. The other two are transient tests involving rapid changes in shear rate. The results of these tests are consistent with the disruption of the C-B network. A step shear test is done on each C-B composite to measure the shear relaxation modulus $G(t)$. A step shear rate test is then done, and the stress relaxation is observed after the shearing has stopped.

A double relaxation model is adopted to interpret the stress relaxation data. The parameters in the double relaxation model are used to calculate the zero shear viscosity, equilibrium recoverable compliance, and a characteristic time for each sample from the stress relaxation data.

EXPERIMENTAL

Materials

The polymer resins used in this work are random styrene/n-butylacrylate copolymers. The average molecular weight M_w and M_n, glass transition temperature T_g, and weight percent gel in the polymer before processing with the C-B are listed in Table I. Gel content was determined gravimetrically after extraction with tetrahydrofuran. The C-B used in the composites, Raven 1020 from the Columbian Chemicals Co., has N_2 surface area of 95 m^2/g and a mean diameter of 25 nm. The composites with 0, 8, and 70% gel were melt blended with 7 volume % C-B after the dry powders were intimately mixed.

The resins used to prepare the corresponding toners are also listed in Table I. The 8% gel composite was made from the RESIN B, initially containing 30% gel. The gel content was reduced from 30 to 8% during melt mixing in a high shear batch mixer thermostated to 166°C. This reduction was effected by shear degradation of the gel containing copolymer. The 30% gel sample was prepared by solution dispersion to provide a material with the original gel still intact. It was prepared by dissolving the polymer in

Table I. The glass transition temperature, number average, M_n and weight average, M_w molecular weight and gel content of the resin from which the C-B composites were made. The molecular weight listed are for the soluble portion of the polymer.

Resin Code	T_g (°C)	M_w × 10⁻³	M_n × 10⁻³	% gel	Toner Code
RESIN A	66	101	5	0	0% gel
RESIN B	74	266	15	30	8% gel
RESIN B	74	266	15	30	30% gel
RESIN C	48	†	†	70	70% gel ‡

† Insoluble.
‡ Estimated from the initial gel content.

chloroform and then dispersing the carbon black in a ball mill. Solvent was removed at 80°C under vacuum. Complete removal of solvent was confirmed by thermogravimetric analysis, which showed no weight loss prior to polymer decomposition.

Rheological Measurements

For each test, 0.5g of the powdered C-B filled composite was compression molded for 20 minutes at 41.4 MPa and room temperature. This provided a 2.54 cm diameter disk with close to the theoretical density and the volume necessary to fill the gap in the rheometer fixture. The measurements were done on a Rheometrics RMS-705 Mechanical Spectrometer equipped with a 0.196 N-m torque transducer. The fixture geometry was a cone and plate with a diameter and cone angle of 2.54 cm and 0.0985 rad, respectively. The sample was melted into the fixtures at 200°C and then cooled to 100°C. Care was taken to adjust the gap for thermal expansion of the fixtures. The rheological test conditions of time, temperature, and shear rate were selected to approximate those in the hot roll fusing process. Since shear history can alter the sample morphology, the order in which the tests were done was selected in increasing severity of shear. In a series of tests on a given sample, the frequency-temperature sweep for time-temperature superposition was done first. At 20°C increments between 100 and 180°C, the frequency was changed from 0.1 to 100 rad/s (0.016 to 16 s^{-1}) with 5 intervals per decade. Dynamic mechanical properties were measured at each temperature, allowing 5 minutes for temperature equilibration. The amplitude of oscillation was increased from 2 to 15% as the temperature was increased from 100 to 180°C to maintain the torque within the measurement range of the transducer. In addition, strain amplitude sweep tests indicated that the dynamic mechanical properties showed little dependence on strain amplitude within this range. That is, the strain amplitude in the dynamic tests was in the linear region.

Next, a step shear test was carried out to measure the shear relaxation modulus at 160°C. In this test a 100% strain was applied, and the shear relaxation modulus was recorded, for 8 s following the step shear. The last test on a given sample was the transient shear rate test to examine the initial stress overshoot, steady state stress, and relaxation following steady shear. In the transient shear rate test the sample was sheared at 50 s^{-1} for 8 s, and the stress output was recorded during this 8 s and for 8 s immediately following the cessation of the steady shearing.

RESULTS

The storage and loss modulus were measured in oscillatory shear as a function of oscillation frequency and temperature. Time-temperature superposition of these data was done by shifting the log frequency data for each temperature to provide an overlapping function of the log modulus vs log frequency. The data was reduced to 160°C to generate the master curve for each material. This provides the dynamic mechanical properties over

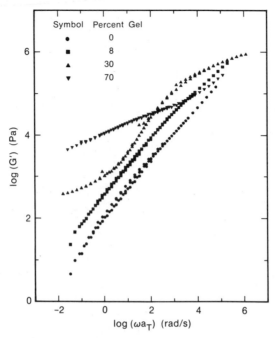

Figure 1. The storage shear modulus G' master curve, reduced to 160 °C, from the dynamic mechanical analysis of the C-B composites over a range of gel content.

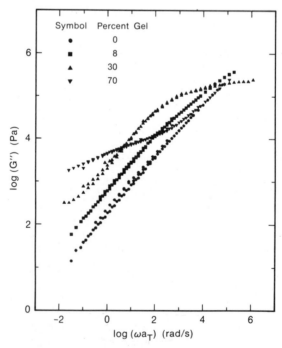

Figure 2. The loss shear modulus G'' master curve, reduced to 160 °C, from the dynamic mechanical analysis of the C-B composites over a range of gel content.

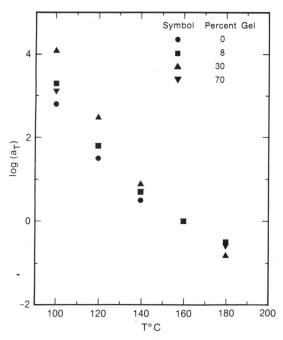

Figure 3. The logarithmic shift factor from time-temperature superposition of the dynamic data, relative to 160 °C.

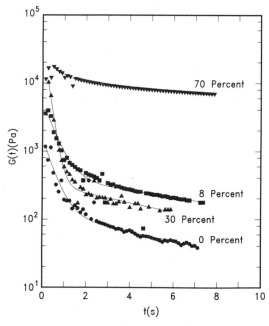

Figure 4. The shear relaxation modulus from the 100% step shear test at 160 °C. The smooth curves are from a regression fit to the double relaxation model.

eight orders of magnitude in frequency. All the dynamic property data were shifted relative to 160 °C. Over the temperature range examined, no vertical shift was required. The master curves are shown in Figures 1 and 2, and the shift factor in Fig. 3.

The shear relaxation modulus following the 100% step strain is shown in Fig. 4. This is plotted with semi-log coordinates because there is such a large range of modulus over the 8 s during which it was measured. The smooth curves in Fig. 4 are from a regression fit to the double relaxation model and will be discussed later.

The shear stress during the step shear rate test at 160 °C is shown in Fig. 5. In this test the shear rate was 50 s^{-1} for 8 s, and then zero for the next 8 s. The curve for the 30% gel composite could not be obtained in this test due to melt fracture. It is apparent that the equilibrium steady shear viscosity is not reached in only 8 s, because there is a distinct downward slope of the stress as a function of time while at steady shear rate. The shear viscosity reached after 8 s, defined as $\eta(8\ s) = \sigma(8\ s)/\dot{\gamma}$, where $\sigma(8\ s)$ is the stress after 8 s of steady shear and $\dot{\gamma}$ is the steady shear rate. Numerical values of $\eta(8\ s)$ are listed in Table II for the 0, 8, and 70% gel.

Interpretation of the Transient Data

Three tests were done to characterize the melt rheology of each sample. First, is a low strain dynamic measurement, which should provide minimal disturbance to any gel or C-B network structure. Second, the 100% step strain is somewhat more severe and may, to some extent, alter the C-B dispersion and polymer entanglement network. Third, the step shear rate for 8 s and 50 s^{-1} can be expected to dramatically alter the initial microstructure of the melt.

The tests can be examined within the context of linear viscoelasticity theory, although the linear theory does not account for the stress overshoot or time dependent viscosity, it can provide some grounds for studying the effect of gel and shear history on the shear modulus and relaxation times. The general linear viscoelastic shear relation[8] for these

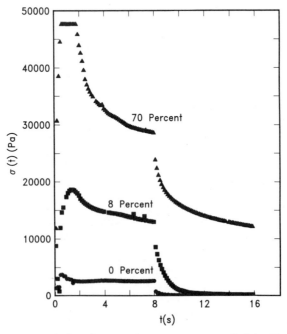

Figure 5. The shear stress during the step shear rate test at 160 °C. The shear rate is 50/s for 8 s and then 0 for the remaining 8 s.

tests gives the stress at a given time t, $\sigma(t)$, in terms of the relaxation modulus $G(t)$ and the time dependent shear rate $\dot{\gamma}(t)$ as:

$$\sigma(t) = \int_{-\infty}^{t} G(t - s)\dot{\gamma}(s)ds. \tag{1}$$

Equation (1) is first applied to the step shear test. The step strain is mathematically expressed as $\gamma(t) = \gamma_0\Theta(t)$, with γ_0 being the amplitude of the step strain and $\Theta(t)$ is the Heaviside step function. Substituting $\gamma(t)$ into eq. (1), we get the relation used to obtain $G(t)$ from the stress relaxation following the step shear:

$$G(t) = \frac{\sigma(t)}{\gamma_0}. \tag{2}$$

Two regions of the stress during the step shear rate test can be interpreted with linear viscoelasticity theory, the initial stress growth before the overshoot, and the stress relaxation after the cessation of the steady shearing. At the onset of steady shear at $\dot{\gamma}$, the shear rate passes through the integral in eq. (1), and we differentiate with respect to t in the limit of $t \to 0$ to get the initial, or instantaneous shear modulus:

$$G(0) = \frac{d}{dt}\left(\frac{\sigma(t)}{\dot{\gamma}}\right)\Big|_{t\to 0}. \tag{3}$$

In addition, at the instant before the steady shearing is stopped, the ratio of the shear stress to the shear rate provides a zero shear viscosity:

$$\eta_0 = \frac{\sigma(0)}{\dot{\gamma}}. \tag{4}$$

At the cessation of steady shear, eq. (1) becomes:

$$\frac{\sigma(t)}{\dot{\gamma}} = \int_{t}^{\infty} G(s)ds. \tag{5}$$

Thus, with some functional form for the dependence of $G(t)$ on t, we can compare the results from the stress relaxation in the step shear test with the stress relaxation after steady shearing.

To interpret the data, we employ an approximate model for $G(t)$. As shown in Fig. 4, it appears that the stress relaxation is dominated by two relaxation times, hence we adopt the double relaxation time model given by:

$$G(t) = G_0 e^{-t/\tau_0} + G_1 e^{-t/\tau_1}, \tag{6}$$

where G_0 and G_1 are shear modulus contributions of the model elements, and τ_0 and τ_1 are the relaxation times ($\tau_0 > \tau_1$).

The instantaneous shear modulus $G(0)$ is then:

$$G(0) = G_0 + G_1, \tag{7}$$

and the stress relaxation after steady shear is obtained from eqs. (5) and (6) as:

$$\frac{\sigma(t)}{\dot{\gamma}} = G_0\tau_0 e^{-t/\tau_0} + G_1\tau_1 e^{-t/\tau_1}. \tag{8}$$

We then employ eqs. (2) and (6) to study $G(t)$ from the step shear test result shown in Fig. 4, and eqs. (3), (5), (7), and (8) to study the result from the transient shear rate test shown in Fig. 5. A regression fit of eq. (6) with the data was done to obtain the model moduli contributions and relaxation times. Eq. (3) was fit to the initial stress growth at the start of steady shearing to find $G(0)$ for the composites with 0, 8, and 70% gel. The stress relaxation after steady shear, shown in Fig. 6, was used to find the parameters in eq. (8). The numerical values of the double relaxation model parameters are summarized

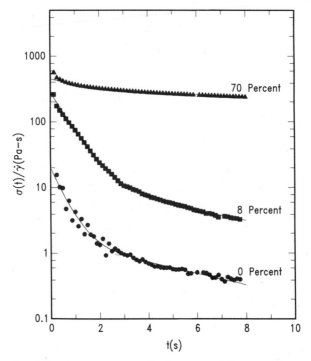

Figure 6. The ratio of the shear stress to the shear rate following the cessation of steady shear at 50/s at 160 °C from the region of Fig. 5 following 8 s. The smooth curves are from a regression fit to the double relaxation model.

in Table II. If the melt structure was unchanged by shearing the moduli and relaxation times calculated from the step shear test should be the same as those from the transient shear test. From Table 2, we see that there were changes induced in the rheological response during the step shear rate test.

Three additional viscoelastic properties can be calculated from the relaxation data in Table II: the zero shear viscosity η_0, equilibrium recoverable compliance J_e^0, and the characteristic time of the fluid τ_c. These are obtained by the moments $G^{(n)}$ obtained from $G(t)$, where:

$$G^{(n)} \equiv <t^n> \equiv \int_0^\infty s^n G(s)ds ,\qquad (9)$$

For the double relaxation model, shown in eq. (6), one can calculate $G^{(n)}$ where:

$$G^{(n)} = G_0\tau_0^{n+1} + G_1\tau_1^{n+1} .\qquad (10)$$

Then, from eq. (10), we obtain η, J_e^0 , and τ_c:

$$\eta_0 = G^{(0)} = G_0\tau_0 + G_1\tau_1 ,\qquad (11)$$

$$J_e^0 = \frac{G^{(1)}}{(G^{(0)})^2} = \frac{(G_0\tau_0^2 + G_1\tau_1^2)}{(G_0\tau_0 + G_1\tau_1)^2} ,\qquad (12)$$

and,

$$\tau_c = \frac{G^{(1)}}{G^{(0)}} = \frac{(G_0\tau_0^2 + G_1\tau_1^2)}{(G_0\tau_0 + G_1\tau_1)} .\qquad (13)$$

These viscoelastic properties are listed in Table III.

Table II. The parameters in the double relaxation model calculated from the transient rheological measurements at 160 °C.

	Step Shear			
Parameter	0%	8%	30%	70%
$G(0)\ (Pa) \times 10^{-3}$	1.55	5.27	36.2	23.5
$G_0\ (Pa) \times 10^{-2}$	1.23	4.19	3.38	10.7
$\tau_0\ (s)$	6.29	8.7	6.06	16.9
$G_1\ (Pa) \times 10^{-3}$	1.43	4.86	35.8	12.8
$\tau_1\ (s)$	0.433	0.488	0.202	0.746

	Initial Transient at the Start of Shear			
Parameter	0%	8%	30%	70%
$G(0)\ (Pa) \times 10^{-3}$	0.21	1.41	–	1.92

	Stress Relaxation After Steady Shear			
Parameter	0%	8%	30%	70%
$G(0)\ (Pa) \times 10^{-3}$	4.77	7.67	–	0.59
$G_0\ (Pa) \times 10^{-2}$	0.14	2.21	–	0.178
$\tau_0\ (s)$	0.909	0.952	–	20.0
$G_1\ (Pa) \times 10^{-3}$	4.76	7.45	–	0.57
$\tau_1\ (s)$	0.0637	0.0662	–	0.455

DISCUSSION

Some qualitative observations can be made about the effects of gel content on the dynamic and transient response of the toner melts. From the master curves of G' and G'' in Figures 1 and 2, the composites are clearly separated by orders of magnitude with respect to gel content; whereas, as the frequency is increased, G' and G'' are less affected by the differences in gel content. This can be understood in terms of the time scale of motions that contribute to the relaxation mechanisms at low and high frequency. At the low frequency range we are seeing large scale motions probably associated with the gel network superimposed on the relaxations of the surrounding uncrosslinked polymer. At the high frequency range we are beginning to look at shorter scale motions that are faster than the gel network response and are nearly independent of the gel content. These are probably the relaxation motions that take place over a length scale that is shorter than the distance between crosslinks in the gel.

On this basis we can look for two distinct relaxation times in the stress relaxation data. From step shear test shown in Fig. 4, it does seem that there are two regions of relaxation: one at short times (high frequency) and the other at long times (low frequency), with a fairly distinct transition between the two.

Table III. The shear viscosity, recoverable compliance and characteristic time calculated from the transient rheological measurements with the double relaxation model at 160 °C.

	Step Shear			
Parameter	0%	8%	30%	70%
$\eta_0 \, (Pa-s) \times 10^{-3}$	1.39	6.02	9.28	27.6
$J_e^0 \, (Pa^{-1}) \times 10^3$	2.66	0.908	0.161	0.411
$\tau_c \, (s)$	3.69	5.47	1.50	11.3

	After Shearing at 50 /s for 8 s			
Parameter	0%	8%	30%	70%
$\eta(8 \, s) \times 10^{-3}$	0.054	0.260	–	0.575

	Stress Relaxation After Steady Shear			
Parameter	0%	8%	30%	70%
$\eta_0 \, (Pa-s) \times 10^{-3}$	0.316	0.703	–	0.615
$J_e^0 \, (Pa^{-1}) \times 10^3$	0.308	0.471	–	19.1
$\tau_c \, (s)$	0.0975	0.331	–	11.7

Two relaxation times should also show up in the stress relaxation after cessation of steady shear from the step shear rate test. The relaxation part of the curve (after 8 s) from Fig. 5 is replotted in Fig. 6, relative to the time at which the shear was stopped. Here, the two relaxation times are most apparent with the 0 and 8% gel composites.

A significant stress overshoot was observed at the inception of steady shear in Fig. 5, with a gradual decay in stress (viscosity) over time for the 8 s observation time. Generally such stress overshoot and decay is due to a particle alignment process or the breakdown of a transient network structure.

The primary results of the rheological tests and data analysis are the viscoelastic properties listed in Table III. Here, the properties from the step shear test should represent those of the undisturbed composite melt, and those after steady shear represent the response of a melt structure altered by the steady shearing. Since the 30% gel sample was solution mixed, while the 0, 8, and 70% were melt mixed, the 30% has different properties and cannot be directly compared with the melt mixed composites. Considering the melt mixed composites before steady shearing, η_0 and τ_c increased, while J_e^0 decreased, with increasing gel content. Comparing the properties before and after shearing, the shearing decreased η_0 and had a complicated effect on J_e^0 and τ_c. The lowest η values were those measured after 8 s while shearing at 50 /s.

SUMMARY AND CONCLUSIONS

A series of three tests was developed to measure the melt rheology of toners (C-B composites) on the time scale relevant to that encountered in the fusing process. These tests are (1) the dynamic measurement of G' and G'', (2) step shear measurement of $G(t)$, and (3) step shear rate to measure the stress overshoot and stress relaxation after steady shear.

We believe that these three rheological tests can provide guidelines in development of new C-B composites with optimum viscoelastic response. Not only are these systems of practical interest, but they highlight the need for a deeper understanding of the effect of structure on melt rheology. In this direction, we studied the effect of gel on the melt rheology of some thermoplastic carbon black composites. We can compare our results from dynamic mechanical analysis with those of previous studies of linear polymers. Qualitatively, our results with increasing gel content appear similar to the those obtained[7] for increasing C-B content, with relaxation mechanisms strongly influenced by the C-B. We are engaged in further analysis using a continuous relaxation spectrum model. Improved understanding of the influence of C-B and gel content on the toner rheology is a step toward the development of the constitutive equation needed for quantitative modeling of the toner fusing process.

ACKNOWLEDGEMENTS

The authors would like to thank the IBM Corporation for permission to publish this study. We are grateful to Dr. Myung S. Jhon of Carnegie Mellon University for providing valuable insight into the physical interpretation of the results.

REFERENCES

1. Y. Kuo, Polym. Eng. Sci., 24 , 652(1984).
2. Y. Kuo, Polym. Eng. Sci., 24 , 662(1984).
3. R. B. Prime, J. Thermal Analysis, 30 , 1001(1985).
4. K. Lakdawala and R. Salovey, Polym. Eng. Sci., 25 , 797(1985).
5. K. Lakdawala and R. Salovey, Polym. Eng. Sci., 27 , 1035(1987).
6. K. Lakdawala and R. Salovey, Polym. Eng. Sci., 27 , 1043(1987).
7. K. Gandhi and R. Salovey, Polym. Eng. Sci., 28 , 877(1988).
8. J. D. Ferry, "Viscoelastic Properties of Polymers," John Wiley & Sons, 1980.

ABOUT THE CONTRIBUTORS

EDWARD M. BARRALL II is currently employed at IBM General Products Division in San Jose as a Senior Technical Staff Member and Manager of the Materials Laboratory. He received his Ph.D. in Analytical Chemistry from MIT in 1961. He has served as a past President and is currently a Fellow of the North American Thermal Analysis Society. He was winner of the Mettler Award in Thermal Analysis in 1973. He is author of over 130 publications and book chapters in the fields of thermal analysis, chromatography, rheology and polymer fractionation. He is currently an external professor at the University of Connecticut and a member of the external Advisory Board at the University of Southern Mississippi, Department of Chemistry.

FRANZ-JOSEF BORMUTH received Ph.D. with thesis in physical chemistry of molecular dynamics of liquid crystalline side chain polymers by dielectric spectroscopy and is the author of 8 publications.

TREVOR N. BOWMER is a member of the Polymer Chemistry and Engineering Research Group in Bellcore, Red Bank, N.J. He received his Ph.D. in Chemistry from the University of Queensland, Australia. Joining Bell laboratories in 1980, he investigated radiation cured systems and lithographic materials. In 1984, he came to his present position where his interests include degradation mechanisms and characterization of polymeric materials used in telecommunications applications.

RICHARD L. BRADSHAW is currently a Senior Engineer/Scientist with IBM Corp., Tucson, AZ which he joined in 1978. He received his Ph.D. in Organic/Polymer Chemistry from Arizona State University, Tempe in 1978.

MICHAEL-JOACHIM BREKNER has been since Jan. 1987 scientist of the R+E Informationstechnik Division of Hoechst AG. He received his Ph.D. in the field of macromolecular chemistry and physics from the Albert Ludwigs University of Freiburg, W. Germany. During 1985-1986 he was postdoc at the IBM T.J. Watson Research Center. His research interests include chemical synthesis and kinetics in the field of polyimides, thermo-mechanical properties of polymers, and polymer blends, and he is the author of 18 publications.

ANDREW BROAD is currently a Research Fellow at the University of Sussex working on routes to titanate ceramics in conjunction with the MIT Ceramics Processing Research Laboratory. He has M.Sc. in Chemistry from the University of Sussex.

JOHN M. BURNS is currently an Advisory Chemist in IBM General Products Division in San Jose which he joined in 1979. He received his Ph.D. in Organic Chemistry in 1979 from the University of Colorado in Boulder. His current areas of interest include fluorocarbon chemistry, thermal analysis, and nuclear magnetic spectroscopy.

SERGIO CALIXTO is currently a researcher at the Centro de Investigaciones en Optica, Mexico. He has a Ph.D. degree in Optics, and his current research interest is in holographic materials.

PAUL CALVERT is with the Department of Materials Science and Engineering at the University of Arizona in Tucson which he joined in October 1988. Before that he was lecturer in polymer science at the University of Sussex in England. He received his Ph.D. in Materials Science from MIT. His current main research interest is the exploitation of bio-mimetic routes to the formation of synthetic composites and ceramics for a variety of applications.

ADOLPHE CHAPIRO is presently "Directeur de Recherches" (Research Professor) within C.N.R.S. which he joined in 1947. He studied chemistry at the University of Paris 1940-1942 and 1945-1947 and received his Ph.D. from the same university in 1950. He is the author 350 scientific and technical articles and of a book Radiation Chemistry of Polymeric Systems published in 1962.

HEIDI L. DICKSTEIN is an advisory scientist at the IBM Product Development Laboratory in San Jose, California. She received her Ph.D. in analytical chemistry/ polymer science and engineering from the University of Massachusetts, Amherst in 1987. Since joining IBM she has been involved in applied research in the area of magnetic polymer composites and photoresist chemistry.

WILLIAM H. DICKSTEIN became a research staff member at the IBM Almaden Research Center at San Jose in 1986. He conducts basic and applied research in the area of polymer synthesis and characterization aimed at providing strategic materials used in IBM varied applications. In addition, he is pursuing independent liquid crystalline research. He received his Ph.D. in Polymer Science and Engineering from the University of Massachusetts, Amherst in 1986.

S.H. DILLMAN is currently a research engineer with Shell Development Company, Houston, TX where he works in styrenic block copolymer research and development. He received his Ph.D. in Chemical Engineering from the University of Washington in 1988. His graduate work focused on the dynamic mechanical behavior of reacting epoxy resins and composites.

SAMUEL J. FALCONE is currently an Advisory Engineer/Scientist with IBM Corp., San Jose, CA and had joined IBM in Tucson, AZ in 1981. Before that he was with ARCO Chemical Company. He received his Ph.D. in Organic Chemistry from Arizona State University, Tempe, AZ in 1978 and did postdoctoral research at the University of Pennsylvania.

CLAUDIUS FEGER has been a Research Staff Member at the IBM T.J. Watson Research Center since 1984 and his research focuses on polymeric materials for packaging applications, in particular polyimides. He received his Dr. rer. nat. degree in Chemistry at the Institute for Macromolecular Chemistry, Freiburg, FRG in 1981. After a short teaching engagement at the Universidade Federal do Rio Grande do Sul, Brazil, he worked from 1981 to 1984 as a post-doc in the Polymer Sci. and Eng. Dept., University of Massachusetts, Amherst. In 1988 he was technical chairman of the 3rd International Conference on Polyimides in Ellenville, N.Y.

HILMAR FRANKE is since 1980 Assistant Professor in the Physics Department of the University of Osnabrueck, FRG. He received his Ph.D. in

444

Experimental Physics in 1979 from the University of Bochum, FRG. Since 1984 he has spent several months at research centers in the U.S. and Canada. his current research interests encompass polymers in electro-optics.

NOBUHIRO FUNAKOSHI is Senior Research Engineer in Storage Systems Laboratory, NTT Applied Electronics Laboratories, Nippon Telegraph and Telephone Corporation and is a group leader for research on optical disk media. He joined NTT Laboratories in 1973. He graduated from Tokyo University in 1968.

AKIRA FURUSAWA is Research Associate of Nikon Corporation. Finished the Graduate School of University of Tokyo, Department of Applied Physics in 1986 and has Master of Engineering degree.

ROBERT P. GIORDANO is a Senior Lab Specialist at the IBM Product Development Laboratory in San Jose, California. He joined IBM in 1974 in the magnetic disk manufacturing area and moved to the disk development laboratory in 1976. His present assignment entails the detailed studies of magnetic defects.

JAN W. GOOCH is a senior research scientist at the Georgia Tech Research Institute, Energy & Materials Sciences Laboratory, Polymer Science and Engineering Group. He received the Ph.D. in Polymer Science from the University of Southern Mississippi in 1979. He is the author of 62 publications and contributor to 2 books. He has been awarded 6 patents. He directs research in magnetic storage devices and other areas of polymer and coatings applications in electronics.

HANS J. GRIESER is currently Senior Research Scientist with the Australian Government's Research Organization CSIRO. He received his Dr. Sc. Nat. (Physical Chemistry) from the ETH, Zurich and did postdoctoral research at the Australian National University, Canberra, followed by industrial R&D at the Research Laboratory, Kodak (Australasia), Melbourne. His research interests include plasma polymerization, plasma surface modification, diffusion in polymers, biopolymers and membranes.

WOLFGANG HAASE has since 1974 been professor of physical chemistry at Technical University Darmstadt, Darmstadt, FRG. He received his Ph.D. (Dr. rer. nat.) in crystal structure analysis. He is author or coauthor of more than 160 publications and book contributions. His research fields are magnetic exchange coupling and ferromagnetism; and low mass liquid crystals and polymeric liquid crystals using dielectric and structural methods.

J. MICHAEL HALTER joined Optical Data, Inc. in May 1985. His work includes media development, data analysis, and finite element analysis modeling. Before coming to ODI, he was Manager of Mechanical Technology at Flow Industries, Automation and Controls Division. He has also worked at Battelle Northwest Laboratories as a Technical Leader for Mechanical design and as a Senior Research Engineer, Electro-Optics Systems Section. He holds a Ph.D. in Mechanical Engineering from the University of Missouri.

EIICHI HANAMURA has been Professor of Applied Physics at the University of Tokyo since 1977 where he received his Ph.D. degree in Applied Physics in 1965. From 1973 to 1974 he was a Visiting Professor at the Institut fur Theoretische Physik, Stuttgart Universitat, W. Germany. In 1975 he was awarded the Nishina Memorial Prize. His current research interests include nonlinear optical properties of semiconductors and organic materials, general theory of photo-induced structural changes and coherent nonlinear optics.

ROY B. HANNAN is currently the Manager of the Advanced Disk Coatings Department at IBM in San Jose. He received a B.A. in chemistry from the College of the Holy Cross in 1973.

KAZUYUKI HORIE is Associate Professor of Research Center for Advanced Science and Technology, University of Tokyo. Finished graduate school, Department of Chemistry, University of Tokyo in 1966, and has Doctor of Science degree. Received the Award of Society of Polymer Science, Japan (1985).

GEORGE HULL is a member of the Solid State Chemistry Research group at Bellcore. He has more than 40 years experience at Bell Laboratories and Bellcore where he investigated inorganic and metallic materials for use in semiconductor and electronic devices. His recent research has been directed towards measuring the physical properties of high-temperature superconductors.

M. IRIE has been Associate Professor at Osaka University since 1978. Received Doctor of Engineering in 1974 and was recipient of the Award of Polymer Science, Japan (1988). Has authored about 100 publications.

N.E. IWAMOTO joined Optical Data, Inc. in July 1985, and is involved in media development and polymer and IR dye design and synthesis. Previously, she held a post-doctoral position with IBM General Technology Division where she was involved in the design and synthesis of high temperature photoresists. She holds a Ph.D. in Organic Chemistry from the University of Washington.

AKIRA IWASAWA is Senior Research Engineer in Storage Systems Laboratory, NTT Applied Electronics Laboratories, Nippon Telegraph and Telephone Corporation and is presently engaged in research on the substrate materials and processes for optical disk. He joined NTT Laboratories in 1973. He received his Master's Degree in applied chemistry in 1973 from Waseda University.

MYUNG S. JHON is professor in the Department of Chemical Engineering at Carnegie Mellon University. He received his Ph.D. in Physics from the University of Chicago. He was a Visiting Scientist at the IBM Research Division, Almaden Research Center, IBM General Products Division, and the Naval Research Laboratories, and he has served as a consultant for the United Nations. Polymer processing, magnetic information technology, and membrane science are among his current research interests. He is the author of approximately fifty scientific papers. Dr. Jhon is known as one of the most dedicated educators at Carnegie Mellon University, and has been recognized for several years as the finest educator at the university. He has received official recognition for his teaching efforts on several occasions, including the Ladd, Ryan and Teare awards.

R. SIDNEY JONES is currently Program Manager, Media Development in the Hoechst Celanese Optical Disk project in Summit, N.J. He received his Ph.D. in organic stereochemistry from Purdue University. After graduation he joined the Summit Technical Center of Celanese Corporation, now Hoechst Celanese. His research interests have included high temperature polymers, lyotropic and thermotropic liquid crystalline polymers, and since 1983, materials for optical recording. He has received 14 U.S. and numerous foreign patents, and is the author or coauthor of many technical publications. He is currently a principal member of X3B11, the American Committee for optical disk standards.

GUENTHER KAEMPF has been since 1983 head of the Section "Physical Development" in the Department "Central Research" of Bayer AG at Leverkusen, FRG, but he has been at Bayer since 1958. He received his Ph.D. degree in Physics from the University of Darmstadt. He is primarily interested in the physics of inorganic pigments, weather resistance of pigmented polymers, multiphase high polymers and polymeric data storage systems. In 1987 he habilitated at the Technical University of Aachen and received professorship in 1981. He is Associate Professor for "Structural Analysis of Polymers" at Aachen. To date he has published 90 papers, is author of the textbook <u>Characterisation of Plastics by Physical Methods</u> (Hanser Publishers 1986) and holds 32 patents.

KIMIHIKO KANENO joined the Tec. Res. Lab., Hitachi Maxell Ltd. in 1982 and has been engaged in research on binder systems for the magnetic recording media. He received M.Sci. degree in Chemistry from Osaka University in 1982.

THOMAS E. KARIS is a Research Staff Member at the IBM Research Division, Almaden Research Center in San Jose. He received his Ph.D. degree in Chemical Engineering from Carnegie Mellon University. He is the author of 16 scientific papers, and his research interests are presently in suspension and melt rheology, polymer powder compression molding, polymer physics and tribology.

EDMOND G. KOLYCHECK is currently R&D Manager of the Estane Group in the BFGoodrich Company and he has been associated with thermoplastic polyurethane technology in BFGoodrich Comapny for 27 years. He received his Bachelor of Chemical Engineering from Northeastern University in 1959 and M.S. in Statistics from Case Western Reserve University in 1969. He holds a number of U.S. and foreign patents.

RAMANATHAN KRISHNAN has been since 1964 a staff member of the Laboratoire de Magnetisme, Meudon, France. At present he holds the rank of Directeur de Recherche. Since 1977 he is heading a group which is involved in the research on amorphous ribbons and thin films. Since 1983 he is also working on magnetic multilayers. He came to Paris in 1960 and worked on ferrimagnetic garnet and spinel single crystals under Prof. C. Guillaud and obtained his D.Sc. degree from the University of Paris in 1964. From 1952 to 1960 he worked under Sir K.S. Krishnan in the National Physical Laboratory, New Delhi, in the group which pioneered the research on ferrites.

MARK H. KRYDER is Professor of Electrical and Computer Engineering and Director of the Magnetics Technology Center which he founded at Carnegie Mellon University. Before joining Carnegie Mellon in 1978 he was at the IBM T.J. Watson Research Center. He received his Ph.D. degree in Electrical Engineering and Physics from Caltech in 1969. He pioneered the use of high-speed flash laser photography of dynamic magnetic domains in magnetic thin films. His current research includes work on thin film and magnetoresistive recording heads, magneto-optical recording media, advanced bubble memory devices and materials for integrated magneto-optical signal processing devices. He has over 120 publications and nine patents in the field of magnetics. In 1985 he was selected as the Distinguished Lecturer for the IEEE Magnetics Society in the field of Magneto-Optical Recording. He was General Chairman of the 1987 Intermag Conference in Tokyo, Japan.

JAMES E. KUDER joined Hoechst Celanese as a Research Associate and has conducted exploratory studies in laser-induced chemistry, electro-organic synthesis, and chemical modification of membranes. He initiated the Celanese program in nonlinear optics and has been involved in the

development of organic materials for optical recording since 1982. Before
that he was with the research laboratories of Xerox Corp. where he
carried out quantum chemical, spectroscopic and electrochemical studies
aiding in the design of organic imaging materials. He received his Ph.D.
from Ohio University for studies in heterocyclic chemistry. He has served
as an alternate member of the American National Standards Committee on
optical disk standards.

KAZUO KUROKI is a student of the Graduate School of Nihon University.

NELSON R. LAZEAR is with Mobay Corporation in Pittsburgh which he joined
in 1979 in the research department, and is currently manager of technical
marketing in the plastics and rubber division. He has held the position
of manager research and development with responsibilities for
polycarbonate blends, polyesters, nylons and polyphenyline sulfide. He
received his Ph.D. in organic chemistry in 1974 from Villanova
University.

PETER E.J. LEGIERSE has been working since 1985 in the Philips Optical
Disc Mastering department, now part of Philips and DuPont Optical (PDO).
He joined the Philips Gloeilampenfabrieken, Eindhoven in 1966 where he
worked on photochemical milling. From 1976 to 1984 he worked on optical
discs, and from 1984 to 1985 he was project leader on decorative
techniques. He is the holder of more than 10 patents and coauthor of some
10 papers on optical disc mastering, electroforming and replication.

HARTMUT LOEWER has since 1985 been assigned to the Technical Marketing
Department of Bayer AG's daughter Company, Mobay Corporation, Pittsburgh,
PA. He finished his studies at the Universities of Bochum and Freiburg
(West Germany) with a Ph.D. degree in Organic Chemistry in 1982. From
1982 to 1985 he was working in the Technical Application and Development
Department of Bayer AG's Plastics Division.

GEORGE G. MARSHALL has been with the IBM Boulder Supplies Development
Laboratory since 1978. He attended the University of Arizona where he
received a Ph.D. in natural products chemistry in 1978. His primary
research interest has been in the development of toners for electrophotographic devices, with particular emphasis on toner charging and
fusing.

MILAP C.A. MATHUR is an Advisory Engineer/Scientist at IBM-GPD Tucson as
well as an Adjunct Professor of Materials Science and Engineering at the
University of Arizona in Tucson. During the course of this research work,
he was a Visiting Researcher at the Magnetics Technology Center at
Carnegie Mellon University in Pittsburgh, PA. He received his Ph.D. in
Chemistry in 1970 from the University of Florida, Gainesville. From 1970
to 1982 he served the faculties at the Universities of Florida,
Mississippi, and S.E. Missouri State University, and in 1983 joined IBM
in Tucson. He has published in areas of inorganic particles, magnetics
and magneto-optics.

TSUNETAKA MATSUMOTO is an emeritus professor at Kobe University which he
joined in 1951 as an associated professor. He received Dr. Eng. degree
from Kyoto University in 1962. He has been engaged in research on polymer
colloids, polymer composites and magnetic composites. He received the
award of the Chemical Society of Japan in 1979.

WILLIAM D. McINTYRE is project leader, Flexible Packaging Product
Development, James River Corp., San Leandro, CA which he joined in 1988.
From 1980 to 1985 he was with Exxon Chemical Co., Baytown, TX. He
received his M.S. in Chemical Engineering from University of California
at Berkeley in 1988.

448

ITARU MITA is Professor of Research Center for Advanced Science and Technology, University of Tokyo. Finished the Graduate School of University of Tokyo, Department of Chemistry in 1956 and has Doctor of Science degree.

KASHMIRI LAL MITTAL * is presently employed at the IBM US Technical Education in Thornwood, N.Y. He received his M.Sc. (First Class First) in 1966 from Indian Institute of Technology, New Delhi, and Ph.D. in Colloid Chemistry in 1970 from the University of Southern California. In the last 16 years, he has organized and chaired a number of very successful international symposia and in addition to this volume, he has edited 26 more books as follows: Adsorption at Interfaces, and Colloidal Dispersions and Micellar Behavior (1975); Micellization, Solubilization, and Microemulsions, Volumes 1 & 2 (1977); Adhesion Measurement of Thin Films, Thick Films and Bulk Coatings (1978); Surface Contamination: Genesis, Detection, and Control, Volumes 1 & 2 (1979); Solution Chemistry of Surfactants, Volumes 1 & 2 (1979); Solution Behavior of Surfactants: Theoretical and Applied Aspects, Volumes 1 & 2 (1982); Adhesion Aspects of Polymeric Coatings, (1983); Physicochemical Aspects of Polymer Surfaces, Volumes 1 & 2 (1983); Surfactants in Solution, Volumes 1, 2 & 3 (1984); Adhesive Joints: Formation, Characteristics, and Testing (1984); Polyimides: Synthesis, Characterization and Applications, Volumes 1 & 2 (1984); Surfactants in Solution, Volumes 4, 5 & 6 (1986); Surface and Colloid Science in Computer Technology (1987); Particles on Surfaces 1: Detection, Adhesion and Removal, (1988); and Particles in Gases and Liquids 1: Detection, Characterization and Control (1989). Also he is Editor of the Series, Treatise on Clean Surface Technology, the premier volume appeared in 1987. In addition to these books he has published about 60 papers in the areas of surface and colloid chemistry, adhesion, polymers, etc. He has given many invited talks on the multifarious facets of surface science, particularly adhesion, on the invitation of various societies and organizations in many countries all over the world, and is always a sought-after speaker. He is a Fellow of the American Institute of Chemists and Indian Chemical Society, is listed in American Men and Women of Science, Who's Who in the East, Men of Achievement and many other reference works. He is or has been a member of the Editorial Boards of a number of scientific and technical journals, and is the Editor of the Journal of Adhesion Science and Technology, which made its debut in 1987.

NAOTO NAGAOSA has been Assistant Professor in the Department of Applied Physics, University of Tokyo since 1987, and currently he is postdoctoral staff in the Department of Physics, MIT. He received his Ph.D. degree in Physics from the University of Tokyo in 1986.

KATSUHIKO NAKAMAE has been in the Faculty of Engineering, Kobe University since 1966 and since 1985 he has been a professor. He received Dr. Eng. degree (Polymer Chemistry) from Kyoto University in 1967. He has been engaged in research on polymer properties, surface properties of polymers and magnetic composites. He received the award of the Society of Fiber Science and Technology, Japan in 1988.

---

* As the editor of this volume.

MICHAEL E. OXSEN is currently an Advisory Engineer in Manufacturing Engineering, IBM General Products Division, San Jose which he joined in 1981. Before that he was with IBM Research which he joined in 1977. He received his B.S. degree in Organic Chemistry from San Jose State University in 1977. His technical interests include cure chemistry of magnetic coatings, synthetic organic and polymer chemistry, thermally stable polymers, and synthesis of electrochromic materials.

JAN H.T. PASMAN joined the Philips Optical Disc Mastering department in 1984 where he became responsible for Optical Engineering and the development of the mastering process for preformatted recordable optical media. After his graduation in physics at the Twente Technical University in 1979 he started to work for Philips in their Central Research Laboratory in Eindhoven where he worked mainly on optical disc mastering and optical diffraction theory. He holds 4 patents in the field of optical recording, and is one of the authors of the book <u>Principles of Optical Disc Systems</u> edited by G. Bouwhuis.

IRENE M. PLITZ is a member of the Polymer Chemistry and Engineering Research Group at Bellcore. She received her B.S. degree in Chemistry from Morgan State University in 1970 and then joined Bell Laboratories. Since transferring to Bellcore in 1984 her interests have centered on the chemical and structural analysis of degraded organic materials.

M. PORTE is Research Engineer and is involved in the investigations of anisotropy in magnetic materials. In the past has been working on the induced and cubic anisotropies in garnets and spinels. Has Engineering degree form CNAM (Paris).

R. BRUCE PRIME is currently a senior scientist at the IBM Laboratory in San Jose and affiliate professor of Chemical Engineering at the University of Washington. He received his Ph.D. in Chemistry in 1968 from Rensselaer Polytechnic Institute and joined IBM in 1968 in Endicott, N.Y. He is the author of a book chapter entitled "Thermosets" and has been a guest lecturer in the course on "Thermal Analysis in Research and Production." He is a Fellow of the North American Thermal Analysis Society (NATAS). During 1987-1988 he served as Chairman of the Plastics Analysis Division of the Society of Plastics Engineers (SPE).

RAINER REUTER is currently a Ph.D. student in the Physics Department of the University of Osnabrueck, FRG. He received his diploma in physics from the University of Bonn, FRG, in 1984. His research interests include optical waveguiding, photoconducting organic systems, and integrated detectors.

GEOFFREY RUSSELL is currently an Advisory Scientist in the Materials Technology Center at Optical Data, Inc. Prior to joining ODI, he was a scientist in the Kodak Research Laboratories. He received his Ph.D. in Materials Science and Engineering from the University of Utah. His interests include polymer physical chemistry, materials characterization and study of polymer structure/property relationships.

C. MARK SEYMOUR is a Research Staff Member at the IBM Research Division, Almaden Research Center in San Jose. He received his Ph.D. in Chemistry from the University of Texas. His research area includes polymer processing and the study of triboelectrical and rheological properties of novel polymer and composite systems.

WOLFGANG J. SIEBOURG joined in 1978 the Application Development Department of the Bayer AG in Leverkusen, W. Germany. Having spent several years with the Bayer daughter company Mobay Corporation in

Pittsburgh, PA, he took in 1985 the position as manager for the project "Resins for Optical Disk Substrates" at Bayer AG. He studied chemistry at the University of Freiburg and wrote his thesis (Dr. rer. nat.) in Macromolecular Chemistry.

W. EUGENE SKIENS is currently Manager of Materials Research at Optical Data, Inc. Beaverton, Oregon where he is participating in the development of erasable optical recording media for data storage and other uses. Previously he spent 11 years at Battelle Memorial Institute as Senior Research Scientist working in the biomedical field, and Dow Chemical Company for 15 years carrying out research in permselective membranes. He received his Ph.D. in Physical Chemistry from the University of Washington and is the author of many articles in scientific journals and chapters in 13 books.

DAVID SOANE has been Professor, Department of Chemical Engineering, University of California at Berkeley since 1978, and Vice Chairman since 1986. Also he has been Associate Faculty Scientist, Lawrence Berkeley Laboaotry since 1984. In 1984 he was recipient of Camille & Henry Dreyfus Teacher-Scholar grant. He received his Ph.D. in Chemical Engineering from University of California at Berkeley in 1978.

ATSUHIKO SUDA joined the Tec. Res. Lab., Hitachi Maxell Ltd. in 1984 and has been engaged in research on tribology of magnetic recording media. He received M. Eng. degree (Precision Engineering) from Kanazawa University in 1984.

KENJI SUMIYA is a Senior Staff at the Tec. Res. Lab. of Hitachi Maxell Ltd. and has been engaged in research on magnetic recording materials and magnetic composite both in the Tec. Res. Lab. and at Kobe university. He received Dr. Sci. degree in Physical Chemistry from University of Kwansei Gakuin in 1982.

SATOSHI TANIGAWA received M.Sc. degree (Polymer Chemistry) from Kobe University in 1987. He has been engaged in research on surface chemistry of magnetic composites.

M. TESSIER is technical assistant. He has a diploma in engineering and his work involves preparation of thin films.

H. SCOTT TSENG is currently a Senior Scientist in the Estane R&D Group of BFGoodrich which he joined in 1985. He received his Ph.D. in Materials Science and Engineering from the University of Utah in 1984. He had broad industrial experience in several chemical companies prior to his graduate studies. His research interests include the fundamental understanding of the role of thermoplastic polyurethane in magnetic media applications.

ROBERT S. TU is Principal Chemist and Manager of Chemical Technology, Magnetic Tape Division, Ampex Corporation in Redwood City, California. He received his Ph.D. from University of Cincinnati. He was Guest Professor of Chengdu University and Shandong Chinese Academy of Science. He is the author of a number of publications.

J.P. VITTON is with Kodak Pathe in Chalon-sur-Saone, France.

TRACY H. WALLMAN is currently a graduate student at the Magnetics Technology Center of Carnegie Mellon University in the area of magneto-optics. He received his Bachelor's degree from Milwaukee School of Engineering in 1987. He held internships at 3M St. Paul and IBM GPD San Jose in 1986 and 1987 respectively.

STEVEN J. WRIGHT is currently a Staff Engineer/Scientist at IBM General Products Division, San Jose. He received MS degree in Chemistry from San Jose State University in 1981. His technical interests are in the areas of thin film magnetic disk coating defect analysis and process characterization.

YOSHINORI YAMAMOTO joined the Tec. Res. Lab., Hitachi Maxell Ltd. in 1976 and has been engaged in research on binder systems for the magnetic recording media. He received M.Eng. degree (Synthetic Chemistry) from Nagoya University in 1976.